# Lecture Notes in Artificial Intelligence 7352

## Subseries of Lecture Notes in Computer Science

### LNAI Series Editors



### LNAI Founding Series Editor

Philip Cox Beryl Plimmer
Peter Rodgers (Eds.)

# Diagrammatic Representation and Inference

7th International Conference, Diagrams 2012
Canterbury, UK, July 2-6, 2012
Proceedings

Springer

Series Editors

Randy Goebel, University of Alberta, Edmonton, Canada
Jörg Siekmann, University of Saarland, Saarbrücken, Germany
Wolfgang Wahlster, DFKI and University of Saarland, Saarbrücken, Germany

Volume Editors

Philip Cox
Dalhousie University, Faculty of Computer Science
6050 University Avenue, Halifax, NS, B3H 1W5, Canada
E-mail: pcox@cs.dal.ca

Beryl Plimmer
University of Auckland, Department of Computer Science
Private Bag 92019, Auckland, New Zealand
E-mail: beryl@cs.auckland.ac.nz

Peter Rodgers
University of Kent, School of Computing
Canterbury, CT2 7NF, UK
E-mail: p.j.rodgers@kent.ac.uk

ISSN 0302-9743 e-ISSN 1611-3349
ISBN 978-3-642-31222-9 e-ISBN 978-3-642-31223-6
DOI 10.1007/978-3-642-31223-6
Springer Heidelberg Dordrecht London New York

Library of Congress Control Number: 2012939642

CR Subject Classification (1998): H.5.2, H.5, H.4, I.3, K.4, F.4.1, G.2.2, I.2, G.3, J.4, J.3

LNCS Sublibrary: SL 7 – Artificial Intelligence



*Typesetting:* Camera-ready by author, data conversion by Scientific Publishing Services, Chennai, India

Printed on acid-free paper

Springer is part of Springer Science+Business Media (www.springer.com)

# Preface

The 7th International Conference on the Theory and Application of Diagrams—Diagrams 2012—was held in Canterbury, UK, during July 2012.

Diagrams is the only conference series that provides a united forum for all areas that are concerned with the study of diagrams, including architecture, art, artificial intelligence, cartography, cognitive science, computer science, education, graphic design, history of science, human–computer interaction, linguistics, logic, mathematics, philosophy, psychology, and software modelling. The conference attracts a large number of researchers from these fields, positioning Diagrams as a major international event in interdisciplinary research.

Diagram 2012 solicited long papers, short papers and posters. We received 54 long submissions, 17 short submissions and 12 poster submissions. This represents an increase on the previous conference, Diagrams 2010. Many long submissions were accepted as short papers or poster abstracts, similarly, some short papers were accepted as posters. The final totals in these proceedings are 16 long papers, 6 short papers and 21 poster abstracts, giving a long-paper acceptance rate of 30%.

Workshop proposals and tutorial suggestions were solicited. Three workshops and one tutorial were accepted. The workshops were held the day before and day after the main conference, and gave an opportunity for specialists in those research areas to converge and discuss their shared interests. The organizers of the workshops managed their own peer reviewing. A two-hour tutorial added further to the vitality of the conference program. The conference also included a Graduate Symposium which provided research students and recent graduates a forum to present their work and to network.

The conference was fortunate to include two excellent invited keynote speeches. Catherine Plaisant, a Senior Research Scientist at the Human–Computer Interaction Lab of the University of Maryland Institute for Advanced Computer Studies, discussed interactions with temporal event sequence representations. Maxwell Roberts, an experimental psychologist at the University of Essex gave his talk on the usability, aesthetics and evaluation of transport schematics. The conference also coincided with the opening of his travelling exhibition, "Underground Maps Unravelled," at the University of Kent.

We appreciate the contribution to the peer-review process of the Program Committee's 29 members and of the additional reviewers. Each paper was considered by at least three reviewers, followed by a discussion phase. This peer review was organized using the EasyChair system. The quality and substance of the reviewers' contributions allowed the Program Chairs to make decisions about acceptance with confidence.

The University of Kent Hospitality Department made an important contribution to the smooth running of the conference. We would like to thank the conference Organizing Committee for managing their responsibilities effectively. They also provided valuable support in running the conference. Nathaniel Miller applied for and administered a National Science Foundation (NSF) award to support the Graduate Symposium, and Lisa Best was awarded a grant from her institution, the University of New Brunswick, also in support of the Graduate Symposium. The Cognitive Science Society provided the best student paper prize in the main conference.

July 2012

Philip Cox
Beryl Plimmer
Peter Rodgers

# Conference Organization

## Conference Chair

| | |
|---|---|
| Peter Rodgers | University of Kent, UK |

## Program Chairs

| | |
|---|---|
| Philip Cox | Dalhousie University, Canada |
| Beryl Plimmer | University of Auckland, New Zealand |

## Tutorials Chair

| | |
|---|---|
| Gem Stapleton | University of Brighton, UK |

## Workshops Chair

| | |
|---|---|
| Nathaniel Miller | University of Northern Colorado, USA |

## Graduate Symposium Chair

| | |
|---|---|
| Lisa Best | The University of New Brunswick, Canada |

## Publicity Chair

| | |
|---|---|
| Aidan Delaney | University of Brighton, UK |

## Program Committee

| | |
|---|---|
| Gerard Allwein | Naval Research Laboratory, USA |
| Dave Barker-Plummer | Stanford University, USA |
| Alan Blackwell | Cambridge University, UK |
| Rachel Blagojevic | University of Auckland, New Zealand |
| Dorothea Blostein | Queen's University, Canada |
| Paolo Bottoni | University of Rome, Italy |
| B. Chandrasekaran | Ohio State University, USA |
| Richard Cox | University of Edinburgh, UK |
| Frithjof Dau | University of Wollongong, Australia |
| Richard Davis | Singapore Management University, Singapore |
| Jim Davies | Carleton University, Canada |
| Aidan Delaney | University of Brighton, UK |
| Max J. Egenhofer | University of Maine, USA |

| | |
|---|---|
| Stephanie Elzer | Millersville University, USA |
| Jacques Fleuriot | University of Edinburgh, UK |
| Jean Flower | Autodesk, UK |
| Ashok Goel | Georgia Institute of Technology, USA |
| Kirstie Hawkey | Dalhousie University, Canada |
| Mary Hegarty | University of California, Santa Barbara, USA |
| John Howse | University of Brighton, UK |
| Mateja Jamnik | Cambridge University, UK |
| Unmesh Kurup | Rensselaer Polytechnic Institute, USA |
| Richard Lowe | Curtin University of Technology, Australia |
| Kim Marriott | Monash University, Australia |
| Mark Minas | Universität der Bundeswehr, Germany |
| N. Hari Narayanan | Auburn University, USA |
| Luis Pineda | Universidad Nacional Autónoma de México |
| Helen Purchase | Glasgow University, UK |
| Derek Reilly | Dalhousie University |
| Frank Ruskey | University of Victoria, Canada |
| Metin Sezgin | Koç University, Turkey |
| Atsushi Shimojima | Doshisha University, Japan |
| Nik Swoboda | Universidad Politécnica de Madrid, Spain |

## External Reviewers

Jim Burton
Alejandro Erickson
Scott Fleming
Michael Helms
Veronika Irvine
Johann M. Kraus
Ludwig Lausser
David Majerich
Stefan Marks
Keith McGreggor
Simone Paolo Ponzetto
Phil Scott
John Taylor
Sean Wilson
Michael Wybrow
Peter Chapman
Andrew Fish
Melanie Grieb
Christian Hirsch
David Joyner
Maithilee Kunda
Sonja Maier
Khalegh Mamakani
Markus Maucher
Petros Papapanagiotou
Sattiraju Prabhakar
Ryo Takemura
Matej Urbas
Bryan Wiltgen

## Sponsoring Institutions

National Science Foundation, USA
The University of New Brunswick, Canada
The University of Kent, UK
Cognitive Science Society

# Table of Contents

## Keynote



## Tutorial



## Workshops



## Graduate Student Symposium



## Psychological and Cognitive Issues

## Diagram Layout



## Diagrams and Data Analysis



## Venn and Euler Diagrams

## Reasoning with Diagrams



## Investigating Aesthetics



## Applications of Diagrams



## Posters

# Life on the Line: Interacting with Temporal Event Sequence Representations

Catherine Plaisant

Human-Computer Interaction Lab, University of Maryland
plaisant@cs.umd.edu

**Abstract.** Sequences of events are part of people's life, their travel, hospital visits, even web browsing experiences. Analysing collections of event sequences can be challenging even for skilled computer professionals. We will review a series of visualization techniques developed at the Human-Computer Interaction lab to handle temporal data.

**Keywords:** visual analytics, time, interaction, temporal patterns.

## 1 Interactive Environments

Our early work on Lifelines (www.cs.umd.edu/lifelines) has shown that a timeline visualization of personal histories can provide benefits over a tabular view; but many tasks involve temporal comparisons across multiple records relative to important events (e.g. a first heart attack). We are exploring novel strategies that facilitate the exploration of temporal patterns in sequences of events. Lifelines2 [1] (Fig. 1) allows for aligning records on important events, ranking, and filtering combined with grouping of results to find common or rare events, e.g.. Other approaches explore query-by-example. Our current project, LifeFlow (Fig. 2) [2]

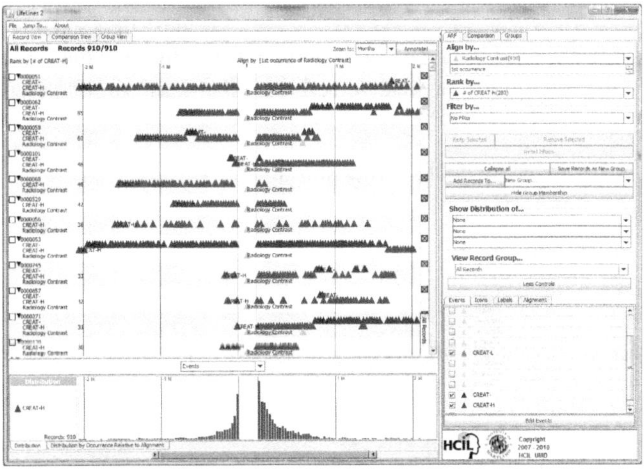


**Fig. 1.** Lifelines2 allows the exploration of multiple records, which can be aligned, ranked and filtered. Temporal summaries allow comparisons of aggregate data across groups of records.

**www.cs.umd.edu/hcil/lifelines2**

P. Cox, B. Plimmer, and P. Rodgers (Eds.): Diagrams 2012, LNAI 7352, pp. 1–2, 2012.

expands on LifeLines2 to provide compact visual summaries of all the sequences found in the data. This breakthrough technique allows users to explore questions such as "what happens before and after hospital patients are admitted to the Intensive Care Unit?

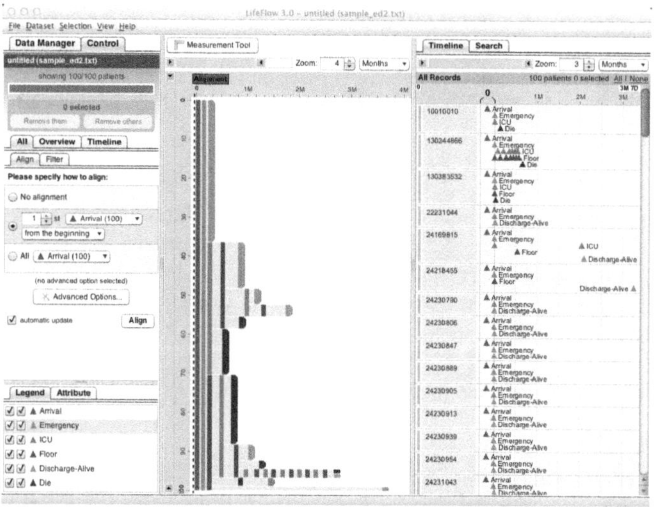


**Fig. 2.** LifeFlow is a novel interactive visual overview of event sequences. It summarizes all possible sequences and the temporal spacing between events.

**www.cs.umd.edu/hcil/lifeflow**

## 2 Evaluation

While controlled studies can quantify the benefits of interaction (e.g. the benefits of providing alignment), we use multi-faceted in-depth longitudinal case studies (MILCs) [3] with clinical researchers. These studies document the discoveries made as participants work on problems over time, and the successes - and struggles- they encounter while using our prototypes with their own data in their own work environment. Case studies seem well adapted to studying the creative activities that users of visual analytics engage in.

**Acknowledgements.** This work is conducted with Ben Shneiderman and many graduate students from HCIL, with support from NIH, Washington Hospital Center and Oracle.

## References

1. Wang, T., Plaisant, C., Shneiderman, B., Spring, N., Roseman, D., Marchand, G., Mukherjee, V., Smith, M.: Temporal Summaries: Supporting Temporal Categorical Searching, Aggregation and Comparison. IEEE Transactions on Visualization and Computer Graphics 15, 1049–1056 (2009)
2. Wongsuphasawat, K., Guerra Gómez, A., Plaisant, C., Wang, T.W., Taieb-Maimon, M., Shneiderman, B.: LifeFlow: Visualizing an Overview of Event Sequences. In: Proc. Conf. on Human Factors in Computing Systems (CHI), pp. 1747–1756 (2011)
3. Shneiderman, B., Plaisant, C.: Strategies for Evaluating Information Visualization Tools: Multidimensional In-depth Long-term Case Studies. In: Proc. of BELIV 2006, BEyond Time and Errors: Novel evaLuation Methods for Information Visualization, a Workshop of the AVI 2006 International Working Conference, pp. 38–43. ACM (2006)

# Learning to Use the Openbox: A Framework for the Implementation of Heterogeneous Reasoning

Dave Barker-Plummer[1], John Etchemendy[1], Michael Murray[1], Emma Pease[1], and Nik Swoboda[2]

[1] CSLI, Stanford University, Stanford, CA, 94305-4101, USA
[2] Universidad Politécnica de Madrid, Boadilla del Monte, Madrid, 28660, Spain

## 1 Description

In this tutorial we will present the Openbox, a framework for constructing *heterogeneous reasoning* systems. Heterogeneous reasoning is reasoning involving multiple representations. A common example is using a map (diagram) together with an address (sentence) to plan a route from one point to another. This kind of reasoning may involve diagrams of multiple types, diagrams and sentences, and/or multiple instances of the same diagram type. Reasoning with sentences, or with a single diagram are special cases of the general heterogeneous setting.

Many research groups within the diagrams community develop software implementations of tools for reasoning with diagrammatic representations. Few of these applications are used outside of the research groups that develop them. We conjecture that part of the reason for this is the difficulty involved in moving software from being an experimental platform usable only by the developers, to a widely available, robust vehicle for research. One major obstacle is in developing the application infrastructure which is not of research interest, and requires effort that is hard to justify within the research/funding community.

The Openbox is a component-based architecture which serves as a container into which different representations can be loaded. A developer interested in building a system to reason with a particular kind of diagram implements the relevant diagram-specific components and loads these into the Openbox to obtain a reasoning tool for those diagrams. Components from other developers can be added to the Openbox to create heterogeneous reasoning systems by mixing and matching the representations.

By providing the Openbox framework, as open source, to the community we hope to encourage the development of more robust, distributable applications for heterogeneous reasoning, encourage sharing of implementations within the community, and reduce redundancy of implementation effort. This will lower the cost of entry to the development of diagrammatic and heterogeneous reasoning applications.

This tutorial is aimed at rseearchers who are considering the use of the Openbox . The tutorial will begin with an overview of the use of the framework, while the core of the tutorial is a program-along exercise to modify or develop a component. The result will be an understanding of the Openbox component life cycle and activities, and preparation for the implementation of original components.

P. Cox, B. Plimmer, and P. Rodgers (Eds.): Diagrams 2012, LNAI 7352, p. 3, 2012.

# 3rd International Workshop on Euler Diagrams

Peter Chapman[1] and Luana Micallef[2]

[1] Visual Modelling Group, School of Computing, Engineering and Mathematics, University of Brighton, UK
p.b.chapman@brighton.ac.uk
[2] School of Computing, University of Kent, Canterbury, UK
lm304@kent.ac.uk

Euler diagrams represent relationships between sets, including intersection, containment, and disjointness. These diagrams have become the foundations of various visual languages and have notably facilitated the modelling of, and logical reasoning about, complex systems. Over the years, they have been extensively used in areas such as biosciences, business, criminology and national security to intuitively visualize relationships and relative cardinalities of sets. This widespread adoption has allowed analysis of complex collections of data.

The workshop will cover all aspects of Euler diagram research, particularly in areas such as:

1. theoretical advances: drawability, layouts, logic and reasoning,
2. software support: diagram generation, automated reasoning, and data exploration,
3. real-world applications: system modelling, information visualization, and education,
4. cognition and perceptual principles: readability, aesthetics, and evaluation including comparison to other representations.

Recently, there have been significant advances in all of the above areas. This workshop of peer-reviewed submissions will afford the growing Euler diagrams community the opportunity to present and discuss new research, and share multi-interdisciplinary expertise. We envisage that this will stimulate collaborations on current and future research needs. This will be the third Euler diagrams workshop (after two successful workshops in 2004 and 2005) and will again bring together researchers with diverse backgrounds, from both academic and industry including: mathematicians, computer scientists, artificial intelligence experts, information designers, visualization experts, human-computer interaction experts and users from various application areas.

P. Cox, B. Plimmer, and P. Rodgers (Eds.): Diagrams 2012, LNAI 7352, p. 4, 2012.

# Technology Enhanced Diagrams Research Workshop

Richard Cox[1] and Jonathan San Diego[2]

[1] School of Informatics, University of Edinburgh
rcox@inf.ed.ac.uk
[2] King's College London
j.p.san_diego@kcl.ac.uk

## Workshop Description

It is an understatement to say that technology has enabled methodological innovations in diagram and spatial reasoning research. New technologies offer opportunities for recording data through video screen recordings; spatial navigation in real and virtual realities, visual attention monitoring, diagram activity on graphics tablets, and recording body position and gestures via position sensors and accelerometers. Whilst the rich data that these techniques yield offer exciting potential for research innovation, researchers face new methodological challenges due to its sheer volume and the challenge of triangulating data from multi-sources.

In the proposed workshop, experienced diagrams researchers will share their knowledge and experiences of diagrams research using innovative technologies including haptic devices, interactive surfaces, eye trackers and virtual reality systems. Interactive sessions on the use of computer-based tools for synchronising and analysing multi-source data will be offered. Workshop participants will have an opportunity to explore and observe the methods in action and discuss their potential for methodological innovation. The workshop will emphasise the need to plan research and design systems that building data collection mechanisms from the outset. The importance of underpinning the analyses with appropriate methodological theory will also be emphasised. Participants will also provide feedback on a proposed platform for locating and sharing useful resources for technology-enhancing their diagrams research.

P. Cox, B. Plimmer, and P. Rodgers (Eds.): Diagrams 2012, LNAI 7352, p. 5, 2012.

# Accessible Graphics: Graphics for Vision Impaired People

Cagatay Goncu and Kim Marriott

Clayton School of IT, Monash University
{cagatay.goncu,kim.marriott}@monash.edu

## 1 Introduction

Graphics are widely used in newspapers, text books, web pages, metro maps, instruction manuals etc. When appropriate they can provide significant cognitive benefits over text. Their use is set to increase as interactive information visualisation applications become more mainstream.

Unfortunately, graphics are not easily accessed by people with severe vision impairment. There have been many different approaches to solve this problem using tactile, tactile-audio, haptic and speech/non-speech audio techniques. However, these approaches have limitations such as the cost of translating into an accessible graphics format, use of expensive tactile graphics or expensive peripheral devices, or lack of congruence with the original visual graphic.

We believe accessible graphics is at a watershed and that recent advances in interface technologies, such as low cost haptic feedback devices, touch screens etc, have the potential to greatly improve access to graphics by the vision impaired in the next decade. How to do this and the potential benefits for people with vision impairment are the subject of this one day workshop. The workshop aims to bring together researchers in the diagrams community, who are interested in the cognitive benefits of graphics and accessible graphics, with researchers and developers of new interface technologies that can support accessible graphics, as well as practitioners from organizations responsible for providing accessible graphics. The workshop will consist of short presentations and demonstrations by workshop attendees intermixed with breakout discussions to answer questions such as:

- What do not we know about cognition of accessible graphics and how could we find out?
- How best can we preserve the cognitive benefits of visual graphics in accessible representations?
- How best can we provide low-cost, portable accessible graphics?
- How can we support interaction and dynamic content in accessible graphics?
- What are the limitations of current technologies and what presentation technologies are feasible for mainstream use in the next 5 years?
- How can we create accessible graphic content quickly and cheaply?
- How useful are accessible graphics in learning?
- How can we use graphics between sighted and blind people in collaborative educational and work places?

P. Cox, B. Plimmer, and P. Rodgers (Eds.): Diagrams 2012, LNAI 7352, p. 6, 2012.

# Graduate Student Symposium of Diagrams 2012

Lisa A. Best

Department of Psychology, University of New Brunswick
100 Tucker Park Road, Saint John, NB E2L 4L5 Canada
Lisa.Best@unb.ca

The Graduate Student Symposium (GSS) was intended to provide student researchers the opportunity to present their research and interact with other students and researchers interested in different aspects of diagrammatic research. We were committed to encouraging participation from a diverse group of students and received submissions from students typically underrepresented in science and engineering, such as members of minority groups (9 submissions), women (7 submissions), and students from institutions not previously represented at the diagrams conference, were encouraged to participate.

The goals of the 2012 GSS was twofold. Firstly, the Symposium was intended to provide senior graduate students and recent graduates with the opportunity to present their research. Feedback from established researchers was provided after each presentation. Secondly, the Symposium provided students with the opportunity to network with each other as future colleagues. Because Diagrams is the only conference series that provides a forum for all areas relevant to the study of diagrams, students were exposed to multidisciplinary research and met other researchers interested in areas closely related to their research.

Three groups of presenters attended the Graduate Symposium: (1) students who presented papers at Diagrams 2012; (2) students who presented posters at the main conference, gave an oral presentation at the graduate symposium; and, (3) students who made submissions directly to the GSS. Submissions were received from students from seven different countries, including UK, USA, Canada, Sweden, Japan, and Sweden. There was a presentation by the GSS chair on how to prepare for a successful academic career.

A grant from the National Science Foundation (totaling $20,000 USD) provided all students attendees with funds to cover their conference fees and partially offset their travel expenses.

To close, the GSS provided students with the opportunity present their research and network with each other as future colleagues. In keeping with the multidisciplinary nature of the Diagrams conference series, students from diverse disciplines, such as psychology, computer science, engineering, education, design, and philosophy, were invited to present their research. The variety of topics presented by students provides researchers with information about the future direction of diagram research.

P. Cox, B. Plimmer, and P. Rodgers (Eds.): Diagrams 2012, LNAI 7352, p. 7, 2012.

# Automatically Recognizing Intended Messages in Grouped Bar Charts

Richard Burns[1], Sandra Carberry[1], Stephanie Elzer[2], and Daniel Chester[1]

[1] Dept of Computer Science, Univ. of Delaware, Newark, DE 19716 USA
{burns,carberry,chester}@cis.udel.edu
[2] Dept of Computer Science, Millersville Univ., Millersville, PA 17551 USA
elzer@cs.millersville.edu

**Abstract.** Information graphics (bar charts, line graphs, grouped bar charts, etc.) often appear in popular media such as newspapers and magazines. In most cases, the information graphic is intended to convey a high-level message; this message plays a role in understanding the document but is seldom repeated in the document's text. This paper presents our methodology for recognizing the intended message of a grouped bar chart. We discuss the types of messages communicated in grouped bar charts, the communicative signals that serve as evidence for the message, and the design and evaluation of our implemented system.

## 1 Introduction

Information graphics are non-pictoral graphics that display information, such as bar charts, line graphs, grouped bar charts, and pie charts. The purpose of an information graphic in popular media—national and local newspapers *(USA Today, Philadelphia Inquirer)* and magazines *(Time, Newsweek)*—is usually to communicate a high-level contextual message to the graph viewer, as opposed to merely displaying data for analysis.

Grouped bar charts are a type of information graphic. They are similar to *simple* bar charts in that they visually display quantifiable relationships of values; however they contain an additional *grouping* dimension. Despite this additional complexity, they still convey intended high-level messages. For example, the grouped bar chart in Figure 1 ostensibly conveys the high-level message that *"China has a greater rate of software piracy than the rest of the world."*

Clark [6] noted that language is more than just words, but rather is any "signal" or lack thereof, where a signal is a deliberate action that is intended to convey a message, such as gestures and facial expressions. We can view information graphics as a form of language. This paper presents a methodology for automatically reasoning about the most likely intended message of a grouped bar chart using their communicative signals as evidence; that is, we predict the graphic designer's high-level intention in designing the graphic.

Carberry [4] studied graphics from popular media and observed that the graphic's message was very often not repeated in the caption or headline, nor

P. Cox, B. Plimmer, and P. Rodgers (Eds.): Diagrams 2012, LNAI 7352, pp. 8–22, 2012.

in any article text. Thus, it is infeasible to perform only natural language processing techniques on the caption and headlines of the graphic and expect to consistently recognize high-level messages.

Some research has already considered the communicative intent of information graphics. Kerpedjiev et al. [14] proposed a methodology for automatically generating graphics that realize desired intentions. Fasciano [11], in the Postgraphe system, generated graphics based on the input of a communicative intention and a data set. Mittal [16], in the SAGE system, implemented a process which automatically generates captions which can be used to explain data in novel or creative graphics. Although the concept of generating good captions bears some similarity to identifying the intended message of a graphic, Mittal is given the data points that will be displayed and the communicative goal of the graphic. In our work, the communicative goal must be inferred by reasoning about the communicative signals in the graphic.

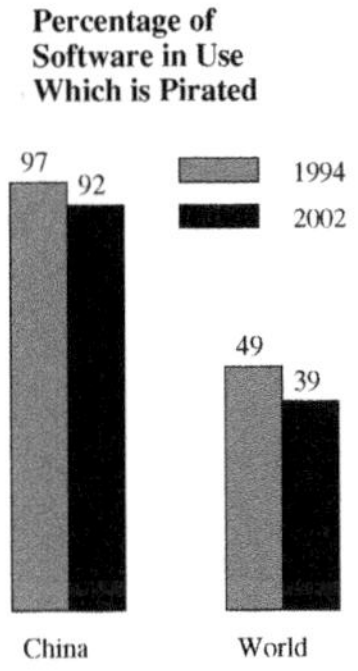


**Fig. 1.** From *NewsWeek*, "Microsoft Cozies Up to China", June 28, 2004

Both Elzer [9] and Wu [21] have implemented systems which automatically recognize the most likely high-level message in simple bar charts and line graphs, respectively. However, grouped bar charts are much more complex than simple bar charts and line graphs; thus they convey a much richer and varied set of messages, the kinds of communicative signals are different, and inferring the intended message requires more complex reasoning.

At least three applications can greatly benefit from this research. The first is a system which provides sight-impaired individuals with alternative access to information graphics in multimodal documents by conveying the high-level content of its intended message via speech. The second is to use a graphic's intended message to index it for retrieval from a digital library. The third is to use the intended message of a graphic as its high-level content and take it into account during the summarization of a multimodal document. Our colleagues are actively investigating all three applications.

Section 2 describes our grouped bar chart collection, the identification of the types of high-level messages that graphic designers overwhelmingly convey in grouped bar charts, and the annotation of our corpus. Section 3 describes the communicative signals that appear in grouped bar charts. Section 4 presents our implemented Bayesian reasoning framework, describes how the extracted communicative signals are used as evidence in inferring the intended message of a grouped bar chart, and discusses the system's evaluation. Section 5 discusses future work motivated by the inherent additional complexity in grouped bar charts.

## 2 Messages

Our corpus is a collection of 222 grouped bar charts from popular media (mainstream newspapers and magazines).[1] We analyzed the corpus to identify the types of high-level messages that graphic designers communicate using grouped bar charts and generalized these into message categories. This section discusses our identified high-level message categories and presents examples from our grouped bar chart corpus; Table 1 lists all of the message categories, and their constraints and instantiated parameters.

### 2.1 Messages

**Trend Messages.** Trend messages convey a general trend (rising, falling, or steady) over a set of ordinal data points. For example, the grouped bar chart in Figure 2 ostensibly conveys the high-level message that *"China increased spending on education, social security, military, and rural support from 2004 to 2006."*, a *Rising-Trends-All* message category. Note that trends can be **within-groups** in which case each group of bars comprises a data series or **across-groups** (as in Figure 2 with the $i^{th}$ bar in each group comprising the $i^{th}$ data series). Table 1 shows that the *Rising-Trends-All* message category requires at least 3 data points for the trend and that data series be over a set of ordinal entities. The trends hold for each *series* and the overall message of Figure 2 can be represented as: *Rising-Trends-All(across-groups: {Education, Social security, Military, Rural support})*.

**Relationship Messages.** Relationship messages capture the consistency of the relative values for a set of entities, or the inconsistency of one set of relative values with respect to the other sets. For example, the grouped bar chart in Figure 8 ostensibly conveys the high-level message that *"The increased funding to Life Sciences is in contrast to the steady or decreased funding to the other research areas."*, a contrasting message that we can represent with the message category *Entity-Relationship-Contrast*. As with trend messages, the set of entities may be **within-groups**, as in Figure 8, or **across-groups** (the $i^{th}$ bar from each group). The parameter <i> as listed in Table 1 is instantiated with the contrasting entity, in this case: *1st group (Life Sciences)*. Thus for Figure 8, the intended message is *Entity-Relationship-Contrast(within-groups:{Life Sciences, Psychology, ..., Other}, 1st group: Life Sciences)*.

**Gap Messages.** Gap messages recognize a high-level message involving either one *gap*, or a trend in the size of multiple *gaps*, where a *gap* is the approximate absolute difference between two values within the same entity. For example, the grouped bar chart in Figure 3 ostensibly conveys that *"There is an increasing gap between the number of patents filed and the number of patents issued, over the period from 1994 to 2003."*, and can be represented as *Gap-Increasing(across-groups:{'94,'95,...,'03})*. Figure 4 shows a *Gap-Crossover* message: *"The gap*

[1] The corpus is available online at http://www.cis.udel.edu/~burns/corpus

**Table 1.** Our grouped bar chart analysis identified numerous messages that graphic designers convey via grouped bar charts. Message categories have constraints (for example, a trend must exist over at least three entities) and one or more parameters which are instantiated. Constraints Key: O (ordinal entities are required), 3+ (at least three entities are required), 2 (two entity limit).

| Message Category <Parameter(s)> | Constraints | Gloss |
|---|---|---|
| Rising-Trends-All <p> | O, 3+ | There is the same trend (rising, falling, steady, or changing) for all entities in the <p> data series where <p> is "within groups" or "across groups". |
| Falling-Trends-All <p> | | |
| Steady-Trends-All <p> | | |
| Changed-Trends-All <p> | | |
| Opposite-Trends <p> | O, 2 | The two entities have opposite trends in the <p> data series where <p> is "within groups" or "across groups". |
| Contrast-Trend <p, i> | O, 3+ | The <i>th trend is contrasting to all of the other trends in the <p> data series where <p> is "within groups" or "across groups". |
| Rising-Trends-Mostly <p> | | There is some trend (rising, falling, steady) for a majority but not all entities in the <p> data series where <p> is "within groups" or "across groups". |
| Falling-Trends-Mostly <p> | | |
| Steady-Trends-Mostly <p> | | |
| Same-Relationship-All <p> | | Each entity in the <p> data series has the same relative ordering of bar values where <p> is "within groups" or "across groups". |
| Opposite-Entity-Relationship <p> | 2 | The two entities in the <p> data series have a different relative ordering of bar values where <p> is "within groups" or "across groups". |
| Entity-Relationship-Contrast <p, i> | 3+ | The <i>th entity has a contrasting relative ordering of bar values compared to all of the other entities in the <p> data series where <p> is "within groups" or "across groups". |
| Same-Relationship-Mostly <p> | | The majority but not all entities in the <p> data series have the same relative ordering of bar values where <p> is "within groups" or "across groups". |
| Gap-Increasing <p> | O, 3+ | The gap between two entities is trending (increasing, decreasing) over the <p> data series where <p> is "within groups" or "across groups". |
| Gap-Decreasing <p> | | |
| Gap-Crossover <p> | | The gap between two entities decreased so much that one entity caught-up-to and then crossed the other where <p> (the sequence of gaps) is "within groups" or "across groups". |
| Gap-Comparison-Single <p, i> | 3+ | The gap in the <i>th entity in the <p> data series is being compared to the gaps of the other entities where <p> is "within groups" or "across groups". |
| Gap-Comparison-Pair <p> | 2 | The two gaps in the <p> data series are being compared with each other where <p> is "within groups" or "across groups". |
| Entity-Comparison <p, i> | | The <i>th entity in the <p> data series is being compared to the other entities where <p> is "within groups" or "across groups". |
| Rising-Entities-All <p> | O, 2 | All entities in the <p> data series have two data points and are (rising, falling, steady) where <p> is groups of series. |
| Falling-Entities-All <p> | | |
| Steady-Entities-All <p> | | |
| Rising-Entities-Mostly <p> | | The majority but not all entities in the <p> data series have two data points and are (rising, falling, steady) where <p> is groups of series. |
| Falling-Entities-Mostly <p> | | |
| Steady-Entities-Mostly <p> | | |

*between the number of Internet users in the US and the number in China has steadily decreased until now China has more Internet users than the US."* As we observe in Table 1, the *Gap-Increasing* and *Gap-Crossover* message categories require similar data constraints to the *Rising-Trend*: namely that there are at least three data points and that the trending is over a set of ordinal entities.

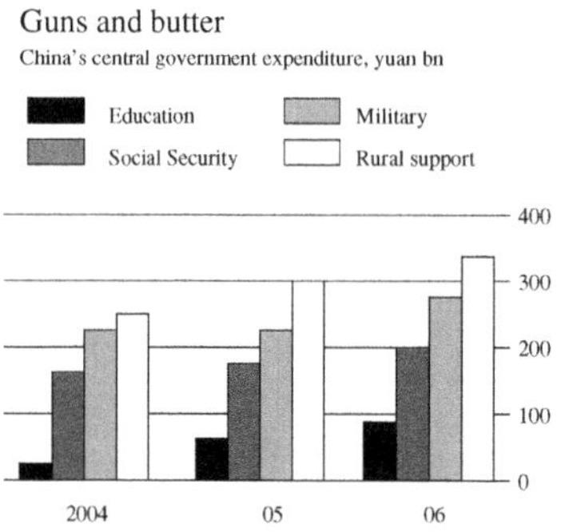


**Fig. 2.** Graphic from *The Economist*, "Planning the new socialist countryside", March 9, 2006

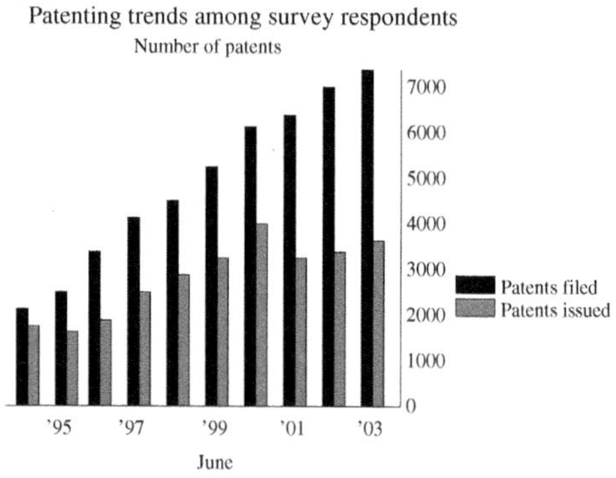


**Fig. 3.** Graphic from *Technology Review*, "A Mixed Bag of U.S. Institutions", July 2005

Some gap messages compare the gaps in one entity with the gaps of the other entities. For example, Figure 5 ostensibly conveys that *"The difference in North American revenue between 2007 and 2008 is much larger than the difference in revenue between 2007 and 2008 for the other areas listed."* This can be represented as *Gap-Comparison-Single(within-groups: {N America, Europe, Latin America, Asia Pacific}, 1st group: <N America>).*

**Entity Comparison Messages.** Entity comparison messages compare one entity against the other entities. For example, the grouped bar chart in Figure 1 ostensibly conveys that *"China has a greater rate of piracy than the rest of the world."*. This is captured by the *Entity-Comparison* message category and is represented as *Entity-Comparison(within-groups:{China, World}, 1st group: China).*

**Additional Messages.** Space limitations preclude us from describing all of our 25 message categories listed in Table 1.

## 2.2 Annotation

Coders individually annotated each graphic in the corpus with the high-level message that it conveyed by determining its message category and the instantiation of its parameter.[2] Where there was disagreement, the coders discussed the graphic until a consensus was reached.

[2] Only a small number of grouped bar charts did not contain an intended message.

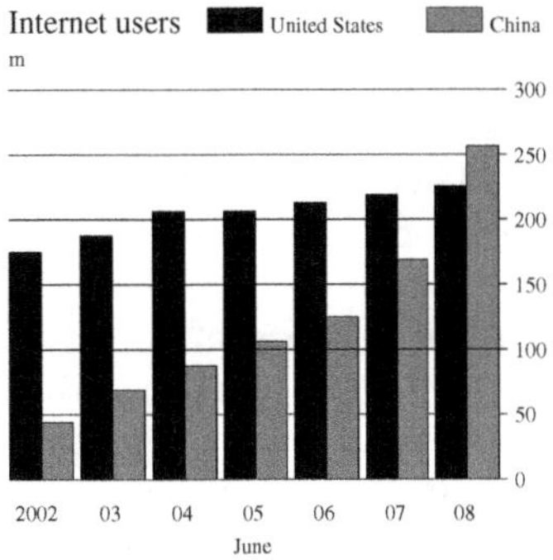


Fig. 4. From *The Economist Daily Chart*, July 31, 2008

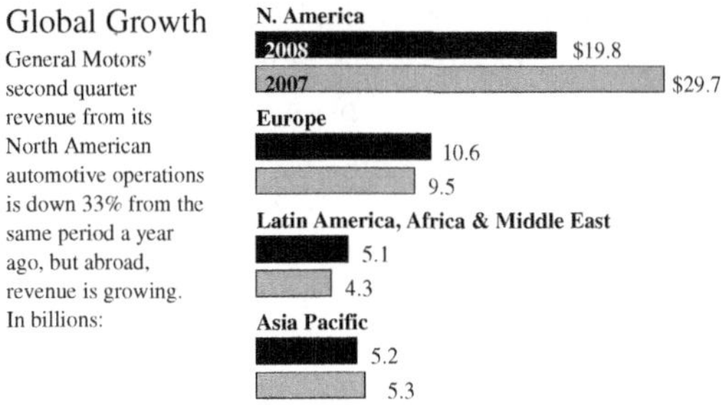


**Fig. 5.** Graphic from *Wall Street Journal*, "GM to Build Diesel Engines in Thailand", August 14, 2008

# 3 Communicative Signals

Graph designers use communicative signals in grouped bar charts in order to help convey their intended message. This section describes the kinds of communicative signals found in grouped bar charts. These signals are exploited as communicative evidence in the intention recognition system presented in Section 4.

## 3.1 Salience via Visual Signals

Visual signals are often used by graphic designers to make an entity (a bar or set of bars) salient. This suggests that the salient entity is an important part of the graphic's intended message—in our terminology, it should be an instantiation of the <i> parameter in the message categories of Table 1.

An entity can be made salient by coloring it differently from the other entities in the graphic. Figure 6 shows a graphic from *Time*, where coloring creates salience. Here the '04 bar in the first group is colored differently from the '04 bars in the other groups, thereby drawing attention to the increased instruction on reading, in contrast with the decreased instruction for other subjects.

Sets of bars can become salient based on their position in the graphic. For example, in Figure 8 the group "Life Sciences" is salient by virtue of its leading position which is not part of a natural (such as alphabetical) ordering of the groups.

A dramatic difference in height between one entity and the other entities can make an entity salient. The "Life Sciences" entity in Figure 8 also jumps out because it is so much taller than the other groups.

## 3.2 Linguistic Signals

Although Elzer [7] observed that captions in popular media very often do not capture a graphic's intended message, captions often contain linguistic signals that help convey the message. We observed two kinds of linguistic signals in grouped bar charts: verb signals and linguistic structure signals.

Certain kinds of verbs can signal one or more high-level message categories. For example, in the caption for Figure 7, "Shrinking Giants", the verb *shrinking* suggests the message categories: *Falling-Trends-All*, *Falling-Trends-Mostly*, *Falling-Entities-All*, or *Falling-Entities-Mostly*.

The linguistic structure of the caption can signal the salience of a *specific entity*. For example, in the caption for Figure 9, "Obama captured first-time voters, but Clinton was strong among older voters", three of the four graphed entities (Obama, Clinton, older) are mentioned. They are each mentioned once in the caption in independent clauses; Obama and Clinton are both in subject position; older is in object position; however, Clinton is in a contrastive clause introduced by "but". This suggests that *Clinton* is a salient entity that is to be compared.

### 3.3 Relative Perceptual Effort as a Communicative Signal

Green et al. [12] hypothesized that graphic designers construct graphics that facilitate as much as possible the tasks that the graph viewer will need to perform to understand the graphic's message. Thus, following Elzer [10] we view relative perceptual task effort as a communicative signal: messages that require more perceptual effort than others are less likely to be the message that the graph designer intended to convey. This correlates with Larkin and Simon [15] who observe that informationally equivalent graphics are not necessarily computationally equivalent, and Peebles and Cheng [17] who note that seemingly minor design changes can greatly affect performance on graph reading tasks.

For example in Figure 10, although both graphics contain the same data, individually they convey two different messages. The high-level message conveyed by the left graphic is ostensibly that male salaries are greater than female salaries in all of the subject areas, while the message conveyed by the right graphic is ostensibly a message of rank: that engineering and the physical sciences have the greatest salaries for both men and women. While this information can be obtained from either graphic, the design of the graphic affects the perceptual effort required and thus the intended message of the graphic.

We have built a cognitive model which produces a relative estimate of the perceptual effort required given a message and graphic, which is also considered as communicative evidence in the intention recognition system. This model is built in the ACT-R [2] cognitive framework, following the ACT-R theory as well as graph comprehension work from the psychological literature. Pilot eye-tracking experiments, in which we asked human subjects to perform specific graph tasks on grouped bar charts and subsequently analyzed their eye scan patterns and fixation and attention locations, also helped us identify the factors in grouped bar charts which affect the required recognition effort.

Pinker [18] identified high-level visual patterns such as linear lines and quadratic curves which are easily identifiable for most graph viewers. In our pilot experiments, we found that subjects fixated less on sets of entities whose

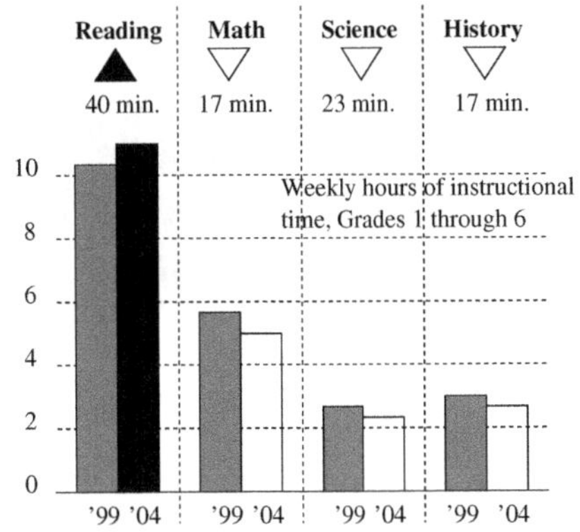


**Fig. 6.** Graphic from *Time*, "How to Fix No Child Left Behind", June 4, 2007

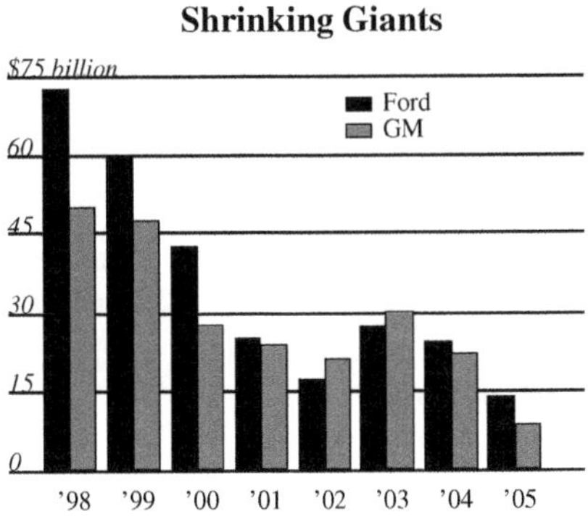


**Fig. 7.** Graphic from Graphic from Wall Street Journal, "Auto Industry, at a Crossroads, Finds Itself Stalled by History", January 2, 2006

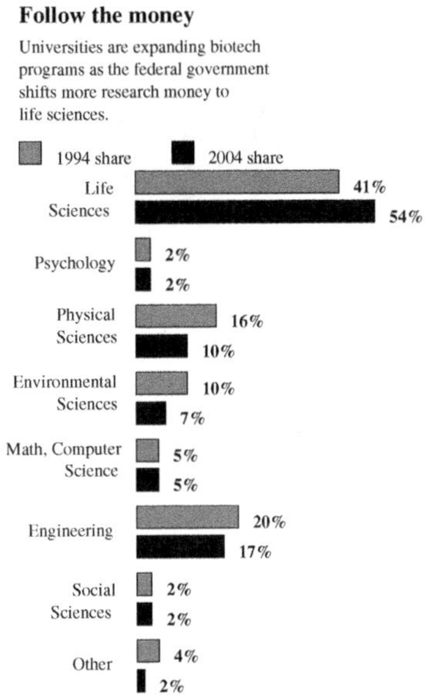


**Fig. 8.** Graphic from *USA Today*, "Universities grid for battle for biosciences supremacy", June 24, 2005

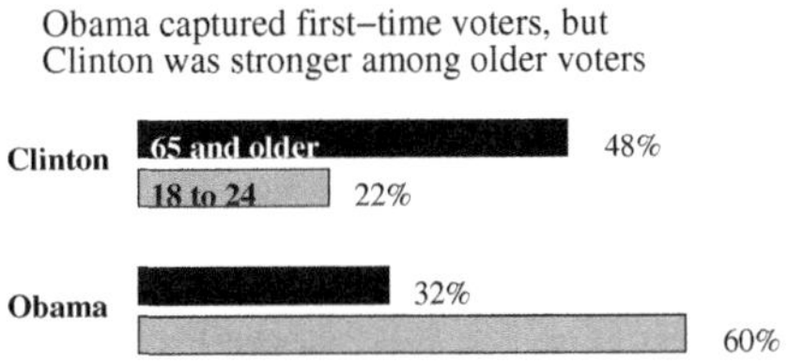


**Fig. 9.** Graphic from Time Magazine, January 21, 2008

bar heights resembled an easily identifiable visual pattern. Shah [19] noted that the grouping of data points will influence the perceived pattern recognition of trends and in our pilot experiments, subjects could still perceive trends despite the presence of *exceptions* (a data point which does not follow the trend). Peripheral vision—the ability for multiple objects to be processed in parallel in a guided search [1]—was present in our experiments: subjects showed the ability to perform some graph tasks without fixating on the first or last groups. Wickens and Carswell [20] defined the *proximity compatibility principle* and showed how close perceptual proximity is advised (the perceptual similarity of two elements) if and only if closeness in processing proximity is intended (the extent to which elements are used as part of the same task). We observed an increase in the time for subjects to perform graph tasks for grouped bar charts with increased noise

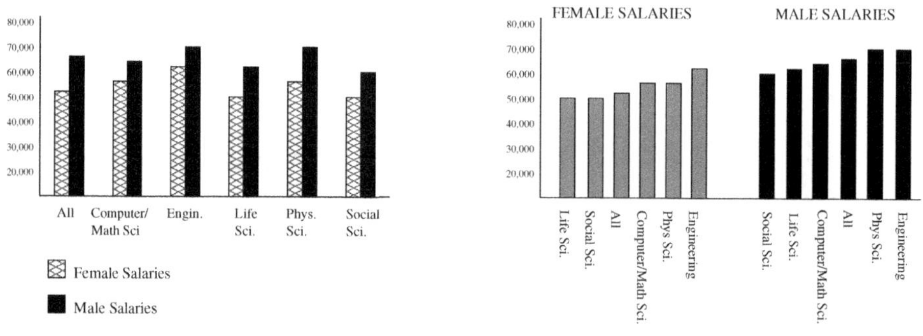


**Fig. 10.** Two bar charts designed from the same data

and visual clutter, where visual clutter is the close spatial proximity of two perceptually or semantically contrasting elements which should not be compared.

The design of our model of relative perceptual effort incorporates these factors so that the presence of high-level visual patterns in a graphic enables the model to process a graph quicker while the presence of visual clutter and exceptions cause an increase in processing time.

Our model was validated for a subset of our message categories in an initial experiment [3] where the relative time required for our model to perform graph tasks on a range of graphics was compared with the relative average for human subjects to perform the same tasks with the same graphics. Our current work includes validating our complete model for all message categories.

# 4 Recognizing the Intended Message

## 4.1 System Architecture

Our system for recognizing the intended message of a grouped bar chart requires an XML representation of a graphic which specifies each bar, each series and group of bars, their heights, colors, annotations, the axes labels, caption, etc. This is the responsibility of a visual extraction module [5]. The generated output is similar to Huang [13] who in addition to representing the graphic also considers vision issues such as identifying an information graphic, locating an information graphic within a noisy pdf document, and performing OCR on the text within the graphic.

Our intention recognition system for grouped bar charts is modeled with a Bayesian reasoning framework which captures the relationship between communicative signals and intended messages. Figure 11 shows the general structure of our Bayesian network. There is an *Intended Message* node at the top whose states are either message categories that still need to be further instantiated or messages that can only have *within-groups* or *across-groups* as their possible instantiation.[3] The five most prevalent states in our corpus are shown along

[3] Thus, we can effectively treat these as message categories in our design.

with their a priori probabilities before any evidence is entered into the network. The communicative signals for a graphic are the evidence for or against possible messages. These are represented in the leaves of the network as evidence nodes.

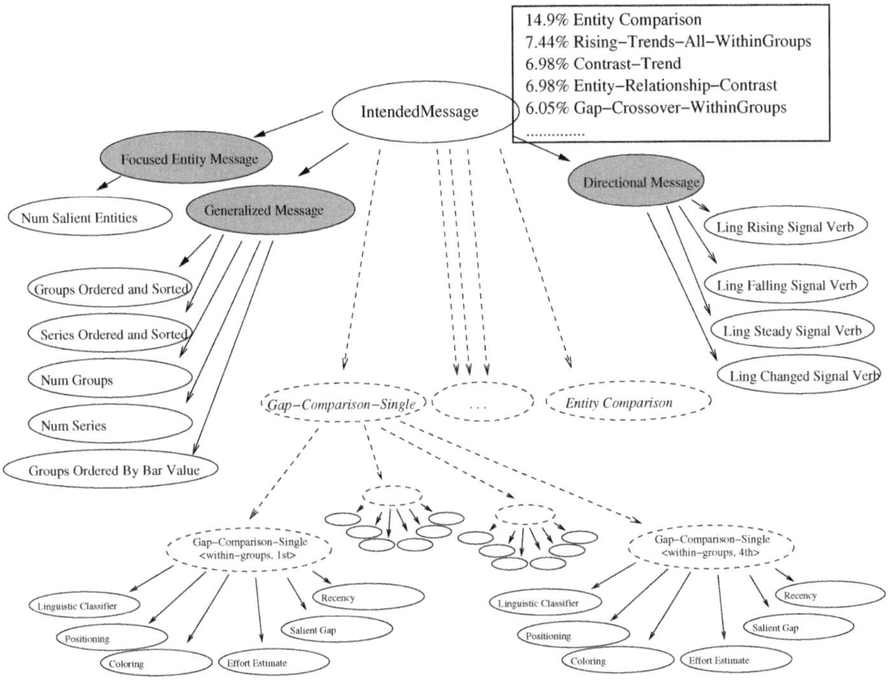


**Fig. 11.** Network structure which captures the probabilistic relationship between intended messages and communicative evidence

The grey nodes in Figure 11 between the top-level and leaves are deterministic, generalizing nodes which group together message categories that have a general feature in common. This alleviates the data sparseness problem that arises due to the limited size of our corpus. For example, the *Directional Message* node groups together message categories which convey the same direction, such as *Rising-Trends-All*, *Rising-Trends-Mostly*, *Rising-Entities-All*, *Rising-Entities-Mostly*, into the *Rising Messages* category. Then, the child evidence nodes (*Ling Rising*, *Ling Falling*, etc.) capture the probabilistic relationships between a *Rising Message* and the "signal verb" communicative signals. Similarly, the *Focused Entity Message* node generalizes those message categories that focus on a *specific entity (<i>)* (such as *Contrast-Trend* and *Entity-Comparison*) and captures the probabilistic relationship between the evidence node *Num Salient Entities* which represents how many entities in a graphic are salient. Finally, the *Generalized Message* node categorizes those message categories together (such as *Rising-*

*Trends-All* and *Falling-Trends-All*) for which we expect naive evidence (such as the number of bars per group) to have the same affect on each.

Message categories at the top level that have a parameter instantiation besides *within-groups* or *across-groups* are instantiated with a *specific entity* lower in the network. In this case, evidence nodes appear as children for each possible instantiation and are able to capture evidence that is only relevant for a specific parameter instantiation. For example, in Figure 11 the *Gap-Comparison-Single* message category node is instantiated for every possible group entity instantiation in Figure 5. The value for the *Salient Gap* evidence node will be positive only for the instantiated node *Gap-Comparison-Single(within-groups, 1st).*

### 4.2 Extracting Evidence

The evidence provided by visual communicative signals (such as whether a group of bars is colored differently from the other bars) is automatically extracted from the XML representation of a graphic and entered into evidence nodes. The text of accompanying captions and headlines is extracted and parsed to identify the presence of any signal verbs.

The extraction of linguistic structure signals is more complex. Headlines and captions are parsed into their clausal structures. A support vector machine was trained on our corpus of captions to produce a learned model that decides which of several mentioned entities is most linguistically salient. It uses features such as the frequency with which the entity is mentioned, the source of the mentioning (main headline, caption, etc.), the ordering of mentions (is one entity preceding), subject position, object position, main/subordinate clausal structure. The decision of the model is entered into the *Linguistic Classifier* evidence nodes.

The relative perceptual effort model takes the XML representation and outputs a relative time for the expected recognition of some message. This relative time is discretized and is entered into the *Effort* evidence nodes.

### 4.3 Training

Associated with each node in the Bayesian network is a conditional probability table that captures the probability of each value for the node given the values for its parent nodes. The conditional probability tables are learned from our corpus of graphics. The Bayesian network applies Bayes' rule to the network constructed for a new graphic to propagate the evidence through the network and compute the posterior probability for each node. Table 2 shows the conditional probability table that captures the probabilistic relationship between the high-level *Gap-Comparison-Single*(within-groups) message category and any visual gap salience of the instantiated <i> entity or the other *within-group* entities.[4]

[4] The *Gap-Comparison-Single*(within-groups) message category generalizes the *Gap-Comparison-Single(within-groups,<i>) messages where <i> is a group entity.*

**Table 2.** Learned conditional probability table for the Gap Salience evidence node under the *Gap-Comparison-Single(within-groups)* message category

| Gap Comparison-Single <WithinGroups, i> | This Entity Only | This Entity Plus Others | Other Entities Only | No Entities |
|---|---|---|---|---|
| Intended | 58.3% | 24.9% | 8.3% | 8.3% |
| Not Intended | 6.9% | 14.8% | 65.3% | 12.9% |

### 4.4 Current Performance and Discussion

We evaluated our system using leave-one-out cross-validation[5], where the XML representation of each of the 222 graphics is in-turn used as a test graphic with the conditional probability tables computed from the other 221 graphs. Results are averaged over all the tests. Currently our system's accuracy rate is 65.6%: that is, the message category and instantiation that the system predicts matches exactly the consensus-based annotation. Table 3 shows our results. As a baseline, we use the message category that appears most often in our corpus; however, note that our system must recognize not only the correct message category but also the instantiated parameters. Our system more than triples the baseline success rate. Although our success rate is lower that that achieved by Elzer or Wu for bar charts and line graphs (78.2% and 73% respectively), grouped bar charts involve more than twice as many message categories and convey far richer messages, making recognition more complex. Note that our success rate improves to 78.6% if we use the top two system hypotheses; this results from grouped bar charts having secondary messages, where occasionally it is difficult to determine which message is primary and which is secondary. (See Future Work).

**Table 3.** Results of our system

| **Grouped Bar Chart System** | | |
|---|---|---|
| Average number of possible messages for a grouped bar chart: 20.2 | | |
| *system* | *criteria* | *accuracy* |
| Grouped Bar Chart System | top message matches annotation | 65.6% |
| Grouped Bar Chart System | either of top 2 messages match annotation | 78.6% |
| Baseline: predict most common possible message | top message matches annotation | 20.2% |
| **Other Systems** | | |
| Simple Bar Chart System (Elzer) [9] | top message matches annotation | 78.2% |
| Line Graph System (Wu) [21] | top message matches annotation | 73.0% |

[5] Leave-one-out cross-validation, as opposed to 10-fold cross-validation, was used to mitigate some sparseness issues in the data set.

**Table 4.** Example showing that removing communicative evidence for Figure 8 affects the network's prediction that *Entity-Relationship-Contrast(within-groups:{Life Sciences, Psychology, ..., Other, 1st group:Life Sciences})* is the intended message

| Likelihood Before | Node | Evidence Before | Evidence After | Likelihood After |
|---|---|---|---|---|
| *only one piece of evidence removed:* | | | | |
| 99.5% | Linguistic Classifier | only entity mentioned | no entities mentioned | 94.1% |
| 99.5% | Salient By Height | only entity that is salient by height | no entities salient by height | 90.1% |
| 99.5% | Positioning | first entity | neither first nor last | 74.3% |
| *evidence removed sequentially, one after another:* | | | | |
| 99.5% | Linguistic Classifier | only entity mentioned | no entities mentioned | 94.1% |
| 94.1% | Salient By Height | only entity that is salient by height | no entities salient by height | 44.2% |
| 44.2% | Positioning | first entity | neither first nor last | 1.25% |

As an example of a graphic processed by our system, consider the grouped bar chart in Figure 8. The graph is processed by the Visual Extraction Module to produce an XML representation. Our system correctly predicts an intended message of *Entity-Relationship-Contrast (within-groups:{Life Sciences, Psychology,...,Other},1st group:Life Sciences)* with an almost certain probability of 99.5%. Three communicative signals are automatically entered in the evidence nodes for the *Entity-Relationship-Contrast* message category node instantiated with *<p>=within-groups, <i>=1st*, namely that it is the only entity mentioned in the caption, that it is the only entity that is visually salient by height, and that it is positioned first in a set of more than two entities. Table 4 shows how the network's prediction of this message decreases if the graphic were altered to eliminate some of the communicative signals and thus alter the evidence in the Bayesian Network. We see that the height salience and leading position of the entity are very important for the system's hypothesis of this message. The effect of removing *Life Sciences* from the text, and thus changing the evidence in the Linguistic Classifier evidence node, follows our intuition that an intended contrasting message may not always be linguistically salient in accompanying text. In each case, the presence of two other kinds of salience compensates when one kind of salience is removed as shown in the top half of Table 4. The bottom half of Table 4 shows the cumulative affect of removing several communicative signals. As we adjust the evidence entered into the network, the system's confidence in this message as the graph's intended message decreases and the likelihood of other possible messages increases.

# 5 Conclusion

## 5.1 Future Work

Sparseness of data and reference resolution issues (such as determining that *USA* refers to *United States*) are two major causes of our system's errors. We are working to address these problems.

Grouped bar charts are much more complex than simple bar charts because of their additional "grouping" dimension. This facilitates the communication of additional high-level messages, such as in Figure 1, *"the rate of Chinese piracy decreased less than the rest of the world"* which supplements our previous observed message that *"China has a greater rate of software piracy than the rest of the world."* Such secondary messages are novel to grouped bar charts and were not observed in the work of simple bar charts and line graphs by Elzer [8] and Wu [21], respectively. In general, secondary messages were not as apparent during the annotation of our corpus and their realization produced more disagreement among the coders. We are currently working on expanding our framework to automatically identify secondary messages in grouped bar charts.

## 5.2 Summary

We have presented an implemented system which automatically hypothesizes the high-level intended message of a grouped bar chart. To our knowledge, no one has previously investigated the communicative signals in grouped bar charts, the wide variety of messages that grouped bar charts can convey, and a methodology for recognizing these messages. Our system automatically extracts communicative evidence from the graphic and incorporates it as evidence in a Bayesian network that hypothesizes the graphic's intended message. This work has several significant applications: (1) a system which provides sight-impaired individuals with alternative access to information graphics in multimodal documents, (2) indexing and retrieving grouped bar charts in digital libraries, (3) and the summarization of multimodal documents.

# References


1. Anderson, J.R., Lebiere, C.: The Atomic Components of Thought. Lawrence Erlbaum Associates, Mahwah (1998)
2. Anderson, J.R., Matessa, M., Lebiere, C.: Act-r: A thoory of higher level cognition and its relation to visual attenion. Human-Computer Interaction 12, 439–462 (1997)
3. Burns, R., Elzer, S., Carberry, S.: Modeling relative task effort for grouped bar charts. In: Taatgen, N., van Rijn, H. (eds.) Proceedings of the 31st Annual Conference of the Cognitive Science Society, pp. 2292–2297. Cognitive Science Society, Austin (2009)
4. Carberry, S., Elzer, S., Demir, S.: Information graphics: An untapped resource of digital libraries. In: Proceedings of 9th International ACM SigIR Conference on Research and Development on Information Retrieval, pp. 581–588. ACM, New York (2006)

5. Chester, D., Elzer, S.: Getting Computers to See Information Graphics So Users Do Not Have to. In: Hacid, M.-S., Murray, N.V., Raś, Z.W., Tsumoto, S. (eds.) ISMIS 2005. LNCS (LNAI), vol. 3488, pp. 660–668. Springer, Heidelberg (2005)
6. Clark, H.: Using Language. Cambridge University Press (1996)
7. Elzer, S., Carberry, S., Chester, D., Demir, S., Green, N., Zukerman, I., Trnka, K.: Exploring and exploiting the limited utility of captions in recognizing intention in information graphics. In: Proceedings of the 43rd Annual Meeting of the Association for Computational Linguistics, pp. 223–230 (2005)
8. Elzer, S., Carberry, S., Demir, S.: Communicative signals as the key to automated understanding of bar charts. In: Proceedings of the International Conference on the Theory and Application of Diagrams (2006)
9. Elzer, S., Carberry, S., Zukerman, I., Chester, D., Green, N., Demir, S.: A probabilistic framework for recognizing intention in information graphics. In: Proceedings of the International Joint Conference on Artificial Intelligence, pp. 223–230. Association for Computational Linguistics, Morristown (2005)
10. Elzer, S., Green, N., Carberry, S., Hoffman, J.: A model of perceptual task effort for bar charts and its role in recognizing intention. International Journal on User Modeling and User-Adapted Interaction 16, 1–30 (2006)
11. Fasciano, M., Lapalme, G.: Intentions in the coordinated generation of graphics and text from tabular data. Knowledge and Information Systems 2(3) (August 2000)
12. Green, N.L., Carenini, G., Kerpedjiev, S., Mattis, J., Moore, J.D., Roth, S.F.: Autobrief: an experimental system for the automatic generation of briefings in integrated text and information graphics. International Journal of Human-Computer Studies 61(1), 32–70 (2004)
13. Huang, W., Tan, C.L.: A system for understanding imaged infographics and its applications. In: Proceedings of the 2007 ACM Symposium on Document Engineering, DocEng 2007, pp. 9–18. ACM, New York (2007)
14. Kerpedjiev, S., Green, N., Moore, J., Roth, S.: Saying it in graphics: from intentions to visualizations. In: Proceedings of the Symposium on Information Visualization (InfoVis 1998). IEEE Computer Society Technical Committee on Computer Graphics, pp. 97–101. IEEE (1998)
15. Larkin, J.H., Simon, H.A.: Why a diagram is (sometimes) worth a thousand words. Cognitive Science 11, 65–99 (1987)
16. Mittal, V.O., Carenini, G., Moore, J.D., Roth, S.: Describing complex charts in natural language: A caption generation system. Computational Linguistics 24(3), 431–467 (1998)
17. Peebles, D., Cheng, P.C.H.: Modeling the effect of task and graphical representation on response latency in a graph reading task. Human Factors 45, 28–45 (2003)
18. Pinker, S.: A theory of graph comprehension. In: Artificial Intelligence and the Future of Testing, pp. 73–126. Lawrence Erlbaum Associates, Hillsdale (1990)
19. Shah, P., Mayer, R.E., Hegarty, M.: Graphs as aids to knowledge construction: Signaling techniques for guiding the process of graph comprehension. Educational Psychology 91, 690–702 (1999)
20. Wickens, C.D., Carswell, C.M.: The proximity compatibility principle: Its psychological foundation and relevance to display design. Human Factors 37, 473–494 (1995)
21. Wu, P., Carberry, S., Elzer, S., Chester, D.: Recognizing the Intended Message of Line Graphs. In: Goel, A.K., Jamnik, M., Narayanan, N.H. (eds.) Diagrams 2010. LNCS, vol. 6170, pp. 220–234. Springer, Heidelberg (2010)

# Representing Category and Continuum: Visualizing Thought

Barbara Tversky[1,2], James E. Corter[1], Lixiu Yu[3], David L. Mason[1], and Jeffrey V. Nickerson[3]

[1] Columbia Teachers College, 525 W. 120th Street
New York, NY 10027 USA
corter@tc.columbia.edu, davidleemason@gmail.com
[2] Stanford University, 450 Serra Mall
Stanford, CA 94305-2130 USA
btversky@stanford.edu
[3] Stevens Institute of Technology, 1 Castle Point on Hudson
Hoboken, NJ 07030, USA
{lisayu1103,jvnickerson}@gmail.com

**Abstract.** Abstract thought has roots in the spatial world. Abstractions are expressed in the ways things are arranged in the world as well as the ways people talk and gesture. Mappings to the page should be better when they are *congruent*, that is, when the abstract concept matches the spatial one. Congruent mappings can be revealed in people's performance and preferences. Congruence is supported here for visual representations of continuum and category. Congruently mapping a continuous concept, frequency, to a continuous visual variable and mapping a categorical concept, class inclusion, to a categorical visual variable were preferred and led to better performance than the reverse mappings.

**Keywords:** diagrams, spatial metaphors, design, networks, information systems, reasoning.

## 1 Introduction

Abstract thought has roots in the spatial world (e. g., Boroditsky, 2002; Lakoff & Johnson, 1980; Shepard, 2001; Talmy, 1983; Tversky, Kugelmass, & Winter, 1991). These abstractions are expressed in the ways people organize space as well as in the ways they speak, gesture, put things on the page, and design the world (Tversky, 2011). External visual expressions of thought, from paintings in caves to bits in computers, go back tens of thousands of years. Expressions of abstract thought have become common only with the widespread use of paper, though apparent in language and gesture far longer. Visualizations of thought are especially apt for conveying information that is intrinsically spatial, like environments, organisms, and objects, where elements and relations in real space can be mapped onto elements and relations on the page. Yet they are also effective for conveying concepts and relations that are metaphorically spatial, including temporal, social, quantitative, and more, in part

P. Cox, B. Plimmer, and P. Rodgers (Eds.): Diagrams 2012, LNAI 7352, pp. 23–34, 2012.

because such concepts have "natural" mappings to space (e. g., Landy & Goldstone, 2007; Tversky et al., 1991). These natural mappings seem to come from the ways that we arrange space to suit our needs as well as the ways that space governs our behavior (Tversky, 2011). They are also evident in language, in common expressions and metaphors (e. g., Cooper & Ross, 1975; Lakoff & Johnson, 1980). For example, people, trees, and more grow stronger as they grow taller; taller piles of money are greater; taller towers, buildings, and bridges must be stronger than smaller ones. In particular, it takes strength to overcome gravity. Such associations provide a worldly foundation for the many metaphors associating *up* with *good*, *more*, *strength*, *health*, and *power*.

By mapping abstract concepts and relations congruently to space and spatial relations, visualizations not only promote comprehension but also promote inference (cf. Bertin, 1981; Norman, 1993; Tversky, 2001; Zahner & Corter, 2010; Zhang, 2000). They allow users to apply highly-practiced skills of spatial reasoning to abstract reasoning (e. g., Tversky, 2001; 2011).

Despite natural mappings, representing abstract relations graphically is not always straightforward. Several alternative means of visual expression are often available, and, typically, each of these has several possible interpretations. Many common and useful devices, like dots, lines, boxes, and arrows, are ambiguous, with multiple meanings, not unlike related spatial terms like *link*, *frame*, *field*, and *relationship*, which also have multiple meanings (e. g., Tversky, Zacks, Lee & Heiser, 2002). Arrows, for example, can indicate order, direction, movement, causality, and more (Heiser & Tversky, 2006). Yet, choosing the right representation is essential to fast and clear communication, and to effective reasoning with diagrams.

Selecting the right representation for visualizing abstractions does not have to be at the whim of a designer. The Production-Preference-Performance program provides empirical methods for design decisions (cf. Kessell & Tversky, 2011; Tversky, Agrawala, et al., 2007). In the 3Ps program, one group of participants *produces* graphic representations for a concept or group of concepts, for example, keeping track of a set of people as they move in time. People's spontaneous graphic productions for representing information reflect their understanding of how that information is structured (e. g., Novick & Hurley, 2001; Zacks & Tversky, 1999). Another group is presented with a set of graphic representations for the same information, for example, a matrix or a graph, and asked which they *prefer*, that is, which is a better or best way to convey the information. In some cases, *interpretation* of the visualization is added or substituted for preference. A third group is asked to make judgments or inferences from one of several graphic representations, allowing comparison of *performance* under each representation. Comparing these measures can help select the right graphic representation and can also provide insight into the cognition underlying the concepts. Ideally, the mappings that are more successful in performance and preference are more successful because they are more congruent.

Designing direct, comprehensible visualizations to express abstract meanings can be more challenging when those meanings are superimposed on a system structure. Structure, especially spatial structure, has priority for the use of space in a diagram over time and abstract relations (e. g., Kessell & Tversky, 2011; Nickerson, Tversky,

Corter, Yu, & Mason, 2010). Some cases are relatively straightforward, for example, superimposing causal relations on the structure of the circulatory system or a bicycle pump or the water cycle by adding arrows indicating the sequence and direction of causality (e. g., Heiser & Tversky, 2006).

Superimposing abstract relations on structural ones is more complicated in other cases. Consider the problem explored here, a network diagram conveying social or computer interrelations. Suppose that we want to show not only the links among the nodes that represent the people or the computers but also how frequently pairs interact or the subgroups that they are part of, issues faced frequently in visualizations, including networks (e. g., Tollis et al., 1998). Effectively diagramming frequency and subgroups are critical in the design of information systems, where balancing efficiency, rooted in frequency of interaction, and security, rooted in subgrouping, are central issues. Representing frequency and grouping are basic to other network problems, and, more generally, to statistical and information graphics. Frequency is a paradigmatic continuous variable and grouping is a paradigmatic categorical variable.

Spatial organization in the world suggests some possibilities for representing frequency and grouping, possibilities that have been produced in practice. All other things equal, individuals who are closer in space interact more frequently; conversely, when high interaction is desired, individuals—and computers—are placed in close proximity. *Distance* is (perhaps arguably) the most common way to use space to represent abstract relations, where distance in space indicates distance on some abstract dimension. Individuals or components that form a subgroup are often put in the same enclosed space; similarly, individuals in the same enclosed space are more likely to form a subgroup. That is, subgroups are often in the same *container*. Finally, thicker pipes carry more water and thicker cables carry more wires. Thickness was used to represent the number of troops in the famous Minard visualization of Napoleon's unsuccessful campaign on Russia. Thus, *thickness* of links connecting components is a natural way to represent the frequency of interactions, especially among components. Each of these real-world and diagrammatic expressions appears in talk as well. We say we've grown apart or *distant*, we talk about band*width*, we say a system or a group *contains* so and so or such and such as members.

Spatial organizations in the world, then, form a basis for abstract thought. They also form a basis for *congruency* of mapping from the conceptual to the spatial world of a diagram (Tversky, Morrison & Betrancourt, 2002). Because frequency is continuous, it is more congruently matched to continuous spatial variables, such as distance or thickness; similarly, because grouping is a categorical variable, it is more congruently matched to a categorical variable such as containment. Some support for congruence in mapping continuous and categorical concepts comes from prior work on line and bar graphs, where participants understood and produced lines for continuous variables and bars for discrete ones (Zacks & Tversky, 1999), but this was for mapping only those variables, not for superimposing that information on a structure, as in the present studies.

Given that several visual expressions of frequency and inclusion have been produced, as in the previous research, we turn to ask whether one or some conceptual mappings to space are more effective in performance or more compelling in preference.

To insure comparability of performance and preference, both frequency and grouping were treated as binary variables: high vs. low frequency and included or not included in a subgroup. Although the more typical ways of regarding these concepts, as categorical or continuous, is expected to affect performance and preference, particularities of the diagrams and the tasks may modulate the predictions derived from congruence.

## 2 Experiment 1: Performance

Will performance with conceptually congruent mappings be superior to performance with less congruent mappings? That is, will people make more inferences about *frequency* when more frequent interactions are represented as closer in proximity or connected by thicker lines than when contained in a common frame? Distance and thickness are continuous, thus more congruent with continuous concepts like frequency, whereas frames are categorical, thus more congruent with categorical concepts like grouping. Will people make more accurate inferences involving *grouping* when groups are contained in the same frame or connected by thicker lines than when they are merely in closer proximity?

### 2.1 Method

**Participants.** 399 volunteers from Amazon's Mechanical Turk website participated, distributed fairly evenly across 6 conditions. The average age was 30, with a range from 18-63. 56% were male, 45% were native English speakers, and 48% had a college degree. Collecting data on a website increases the range of responders, making the data more representative of a general population, but decreases control, which may add variance to the results, decreasing chances of finding significance.

**Design.** There were six groups. Each participant saw one of three visualizations (Figure 1) and answered one of the two questions below (*Frequency* or *Grouping):*

*Frequency.* The diagram below represents computers that can all communicate with each other. S and F communicate with each other six times a second. F and J communicate with each other six times a second. J and B communicate with each other six times a second. E and G communicate with each other six times a second. K and G communicate with each other six times a second. All the other communication links shown indicate that the nodes communicate with each other at the rate of one time a second.

S needs to transmit a message to E. Along which pathway will the message arrive first? (Please list all nodes along the pathway).

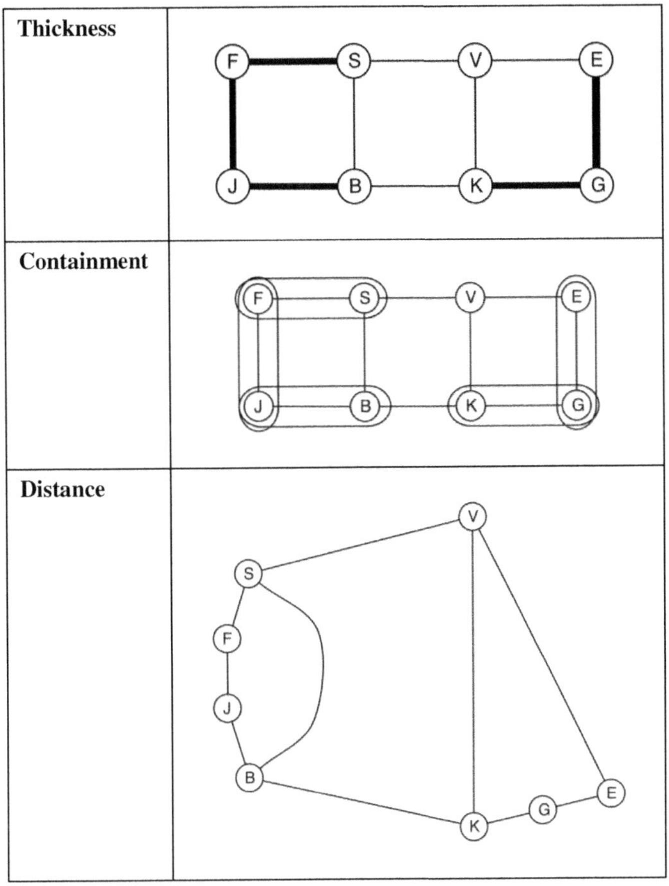


**Fig. 1.** The diagrammatic prompts

*Grouping*. The diagram below represents computers that can all communicate with each other. S and F are part of the same system. F and J are part of the same system. J and B are part of the same system. E and G are part of the same system. K and G are part of the same system. Links within a system are six times as secure as other links.

S needs to transmit a message to E. Which pathway is the most secure? (Please list all nodes along the pathway.)

In pilot experiments using simpler diagrams (such as those in the preference experiment described below), performance was at ceiling. With the more complex diagrams used here, accuracy was about 50%, a level that allowed detection of differences across diagrams and inferences, but makes direct comparison to the preference results more difficult.

## 2.2 Results

Figures 2 and 3 show the proportion of correct responses to the optimal-path inference questions posed in the frequency and grouping problems. In a log-linear analysis, the

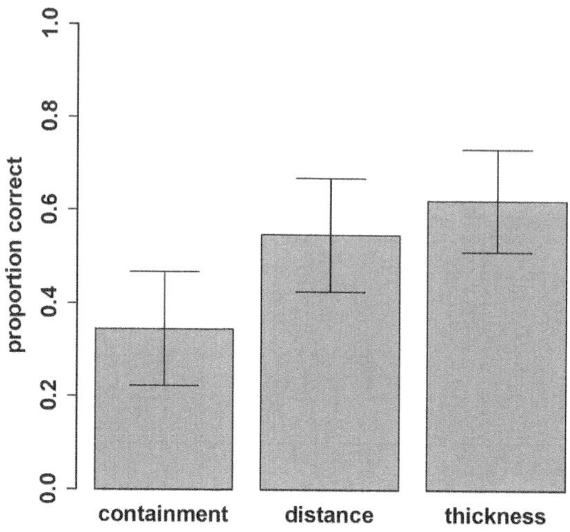


**Fig. 2.** Mean proportions correct for each diagram type for the Frequency problem. Error bars represent the 95% confidence interval for the mean.

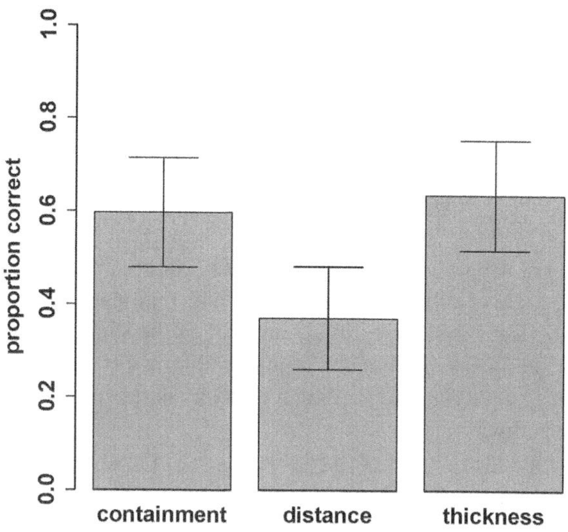


**Fig. 3.** Mean proportions correct for each diagram type for the Grouping problem. Error bars represent the 95% confidence interval for the mean.

three-way association among prompt condition (frequency versus grouping), diagram type (container, distance, line weight), and correctness was significant, $\chi^2(2)=12.16$, $p=.002$, meaning that the effects of the diagram types on correctness differed for frequency and grouping scenarios. Mapping frequency to distance or thickness led to superior performance (Figure 2) compared with mapping frequency to containment, $z = -2.99$, $p=.003$. Distance and thickness mappings did not differ, $z = 0.83$, $p=.407$.

By contrast, mapping grouping to containment or thickness led to superior performance (Figure 3) compared with mapping grouping to distance, $z=3.30$, $p=.001$. Thickness and containment did not differ, $z=1.243$, $p=.214$.

### 2.3 Discussion

Congruence of conceptual mapping can account for the general pattern of results. Participants were more accurate making inferences about frequency when it was mapped to spatial distance or thickness than when it was mapped to containment. Frequency is a continuous conceptual dimension and both distance and thickness are continuous spatial dimensions. Thus the conceptual and visual are congruent.

Inferences about grouping, a categorical relationship, led to a different pattern: participants were more accurate judging groupings of computers when grouping was mapped to containment than to distance. This, too, is a congruent mapping, of a categorical concept, inclusion, to a categorical visual device, a frame. Interestingly, thickness of connection was as good as containment for grouping judgments. In the specific diagrams used in the experiment, thickness had only two levels, so that it could easily be mapped to inclusion, but distance had many levels, hence was more confusing for assessing a categorical concept like inclusion.

## 3 Experiment 2: Preference

Congruence of concept to space accounted for the general pattern of *performance* (i.e., inferences). Here, we examine *preferences* to see if they, too, are congruent. Will people judge the use of boxes to enclose computers belonging to the same system a more natural way to think about grouping and inclusion than putting them close spatially? Will people think putting computer systems that communicate frequently close in space a more natural way to think about frequency than enclosing them?

### 3.1 Method

A total of 377 volunteers from Amazon's Mechanical Turk website participated. 182 participated in the frequency condition (see below), the others in the grouping condition. The average age was 30, with a range from 18-69. 52% were male, 61% were native English speakers, and 70% had a college degree.

In this experiment participants made preference judgments. Participants compared the three diagrams in Figure 4 in one of the two judgment tasks described below, *Frequency* or *Grouping*. *Frequency.* "The diagrams below represent computers that can all communicate with each other. K communicates frequently with Z. H communicates frequently T. Neither K nor Z communicates frequently with either H or T. Choose the diagram that best expresses the description. Choose the second best diagram. Choose the third best diagram." *Grouping*. "These diagrams represent computers that can all communicate with each other. K and Z are part of the same system. H and T are part of the same system. Choose the diagram that best expresses the description. Choose the second best diagram. Choose the third best diagram."

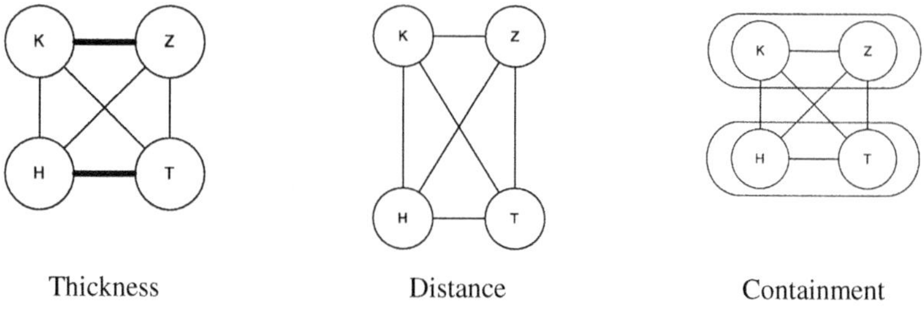


**Fig. 4.** The diagrams presented in the experiment

### 3.2 Results

Figures 5 and 6 show the proportions of participants choosing each diagram type as best in the Frequency and Grouping conditions. In a log-linear analysis, the proportions of "best" choices of the three diagram types differed between the two conditions, $\chi 2(2) = 83.676$, $p<.001$. In the Frequency condition, line thickness was most often chosen as the best representation, by 48% of participants. Containment was chosen as best by 34% of participants, and distance was chosen as best by only 18%. In the Grouping condition, containment was chosen as the best representation by 70% of participants. Line thickness was chosen as best by only 22% of participants, and distance by only 9%.

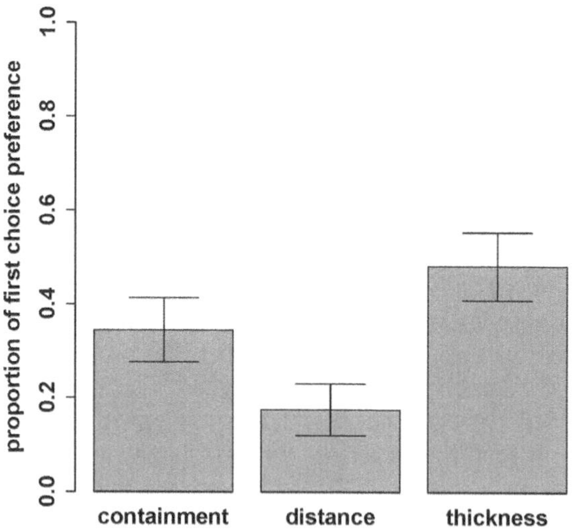


**Fig. 5.** The proportion of participants in the Grouping condition who chose each diagram type as best

### 3.3 Discussion

The importance of the congruence of conceptual content to the visual representation is supported by participants' first-choice preferences. To represent grouping, a categorical concept, most participants chose containment, a categorical visual variable. To represent frequency, a continuous concept, the most common choice was thickness, a continuous visual variable.

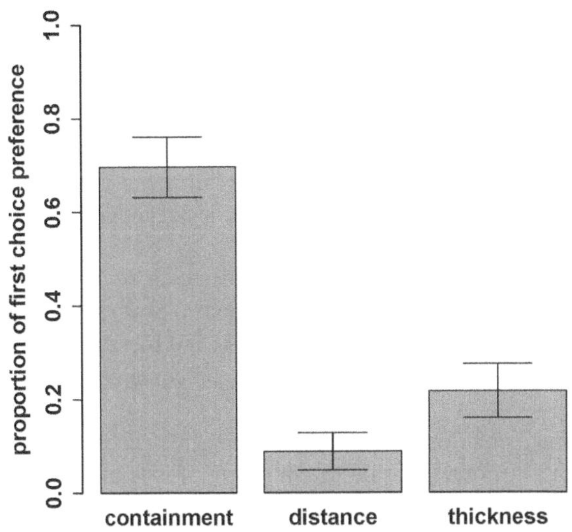


**Fig. 6.** The proportion of participants in the Grouping condition who chose each diagram type as best

## 4 General Discussion

Abstract thought reflected in spatial thinking is all around us, in the world, in talk, in gesture, in diagrams. Communication through diagrams can be fast and efficient, and is increasingly common. Mapping spatial entities and spatial relations in the world to spatial elements and relations on the page is fairly straightforward, as in maps (which is not to say that it is always done well).

But successfully mapping abstract concepts and relations to marks and spatial relations to the page, that is, metaphoric use of diagrams, can be more complicated. The best mappings use space and marks in space in ways that are congruent with the abstract concept. For example, since people think of greater height, health, power, and wealth as going upwards, mapping those concepts upwards on a graph is bound to lead to better and faster comprehension and inference. Sometimes, however, mappings conflict. Think of inflation and unemployment, both undesirable, yet increases in both are typically mapped upwards. Note the *increases*; inflation and unemployment are given in numbers, so that the more general rule that increasing numbers should go upwards overrides the more specific correspondence to content. Congruent mappings can be revealed indirectly in language and gesture, and more directly in experiments eliciting production, performance, and preference (e.g., Kessell & Tversky, 2011, Tversky, 2011; Tversky, et al., 2001). Because of the complexity of thought, there is no guarantee of agreement across people or agreement across tasks, that is, there is no guarantee that production, performance, and preference will align. However, when there is agreement, when the measures converge, the consensus should go far to guarantee successful communication.

Here we explored congruent mappings for a continuous concept, frequency, and a categorical concept, grouping in diagrams of computer networks. These representations were superimposed on a graphic structure, a network, rather than appearing in isolation. Interconnected computer systems are usually visualized as nodes linked in a network, much like social, transportation, and associative networks. Networks are common, perhaps because nodes can represent any idea, and links can represent relations between ideas. A node and a link, then, can be regarded as the minimal representation of thought, a proposition.

It is often desirable to superimpose other information on networks, notably frequency of interaction of nodes and subgroups of nodes. Several ways to superimpose this information on networks have been proposed in the literature, including distance or lengths of link, thickness of link, and frames or containers (e.g., Bertin, 1981; Harel, 1987).

Since frequency is a continuous concept, mapping it to a continuous spatial variable, either distance or thickness, should be congruent. Since inclusion is a categorical concept, mapping it to frames should be congruent. Congruency predictions were borne out both in performance, making inferences from the diagrams, and in preference. Inferences involving frequency were more accurate when frequency was mapped to the continuous representations, distance and thickness, compared with when it was mapped to the categorical aspect of containment. The opposite held for making inferences involving inclusion or group membership, where containment (but also thickness) led to superior performance. Thickness was actually used as a binary variable in the present graphs, encouraging a categorical interpretation.

For preference judgments, containment was preferred to represent grouping; this was predicted because both relationships are categorical. For representing frequency, the continuous aspect of line thickness was most often chosen as the best representation, as predicted. Distance, though, was a distant third choice, perhaps because in the set of diagrams used in this experiment, the distances used in the diagrams were hard to compare.

There are limitations to these experiments, including some inconsistencies in the data for the two studies. In addition, for large networks, these devices, thickness, inclusion, and distance, might not scale well, so that other means of conveying frequency and inclusion might be needed. The present experiments only examined conceptual mappings, and only some of them, between marks in space and discrete and continuous concepts. Perceptual variables, such as the discriminabilty of the marks, are also a factor that is well-known to affect performance. Based on data from psychophysics, Cleveland (1985) has found evidence for a hierarchy of visual features such as length and angle in making inferences from graphs. In addition, experience with using particular kinds of representations or with specific conventions are likely to bias performance, preference, and production.

Networks are familiar and popular because they can represent the structure of relationships among people, objects, or ideas, that is, anything. There are computer networks, social networks, and association networks. Hierarchies, such as corporate or phylogenetic, flow charts, grammatical diagrams, and decision trees can be regarded as special cases of networks. Again, all represent the structure of the relationships among elements. Using spatial relations or marks in space to superimpose other dimensions or variables, such as inclusion relations and frequency of interaction, on a spatial structure can be challenging.

Both preference and performance suggest that congruent mappings of concepts to space are effective. Although the specific mappings to performance and preference differed somewhat between the experiments, for many possible reasons, the overall conclusion holds. Mapping continuous concepts to continuous uses of space and mapping categorical concepts to categorical uses of space were preferred and led to superior performance. The effectiveness of mapping categorical concepts to categorical marks in space and continuous concepts to continuous marks in space is further strengthened by earlier work showing that bars, discrete marks in space, are interpreted as and produced for discrete comparisons, whereas lines, continuous marks in space, are interpreted as and produced for trends (Zacks & Tversky, 1999).

The present findings add support to previous findings of conceptual-spatial congruence in other domains, maps, instructions, graphs, and representations of the actions of agents in time, and space (e. g., Tversky, Agrawala, et al., 2007; Kessell & Tversky, 2011; Zacks & Tversky, 1999). These results also strengthen the case for the general program, of selecting among graphic means to represent abstract concepts and relations by assessing people's productions, preferences, and performance. The program has broad implications for designing diagrams, for information systems as well as for statistical, scientific, and information graphics in the popular and technical media. The reasoning and the techniques provide a model for empirical methods to reveal design principles for many other domains. That is, design can be rooted in research. In turn, the findings have implications for the many arenas of life where understanding diagrams is crucial, including navigation in the world, research in science and engineering, and learning in and out of classrooms, Spatial thinking is pervasive and powerful; visualizations can successfully express a range of abstract concepts, as long as the mappings are congruent with thought.

**Acknowledgments.** Portions of this research were supported by grants from National Science Foundation IIS-0725223, IIS-0855995, and REC-0440103, the Stanford Regional Visualization and Analysis Center, and Office of Naval Research NOOO14-PP-1-O649, N000140110717, and N000140210534.

## References


1. Bertin, J.: Graphics and Graphic Information Processing. Walter de Gruyter, New York (1981)
2. Boroditsky, L.: Metaphoric Structuring: Understanding Time Through Spatial Metaphors. Cognition 75, 1–28 (2000)
3. Cleveland, W.S.: The Elements of Graphing Data. Wadsworth, Monterey (1985)
4. Cooper, W.E., Ross, J.R.: World Order. In: Grossman, R.E., San, L.J., Vances, T.J. (eds.) Papers From the Parasession on Functionalism, pp. 63–111. Chicago Linguistic Society, Chicago (1975)
5. Harel, D.: Statecharts: A Visual Formalism for Complex Systems. Science of Computer Programming 8, 231–274 (1987)
6. Kessell, A.M., Tversky, B.: Visualizing Space, Time, and Agents: Production, Performance, and Preference. Cognitive Processing 12, 43–52 (2011)
7. Lakoff, G., Johnson, M.: Metaphors We Live By. University of Chicago Press, Chicago (1980)
8. Landy, D., Goldstone, R.L.: How Abstract is Symbolic Thought? Journal of Experimental Psychology: Learning, Memory, & Cognition 33, 720–733 (2007)
9. Nickerson, J.V., Tversky, B., Corter, J.E., Yu, L., Mason, D.: Thinking with Networks. In: Proceedings of the 32nd Annual Conference of the Cognitive Science Society (2010)
10. Norman, D.A.: Things That Make us Smart. Addison-Wesley, Reading (1993)
11. Novick, L.R., Hurley, S.M.: To Matrix, Network, or Hierarchy: That is the Question. Cognitive Psychology 42, 158–216 (2001)
12. Shepard, R.N.: Perceptual-cognitive Universals as Reflections of the World. Behavioral and Brain Sciences 24, 581–601 (2001)
13. Talmy, L.: How language Structures Space. In: Pick Jr., H.L., Acredolo, L.P. (eds.) Spatial Orientation: Theory, Research and Application, pp. 225–282. Plenum, New York (1983)
14. Tollis, I.G., Di Battista, G., Eades, P., Tamasssia, R.: Graph Drawing: Algorithms for the Visualization of Graphs. Prentice Hall, New York (1998)
15. Tversky, B.: Spatial schemas in depictions. In: Gattis, M. (ed.) Spatial Schemas and Abstract Thought, pp. 79–111. MIT Press, Cambridge (2001)
16. Tversky, B.: Tools for Thought. In: Benedetti, B., Cook, V. (eds.) Language and Bilingual Cognition, pp. 131–139. Psychology Press, New York (2011a)
17. Tversky, B.: Visualizing Thought. Topics in Cognitive Science 3, 499–535 (2011b)
18. Tversky, B., Agrawala, M., Heiser, J., Lee, P.U., Hanrahan, P., Phan, D., Stolte, C., Daniel, M.-P.: Cognitive Design Principles for Generating Visualizations. In: Allen, G. (ed.) Applied Spatial Cognition: From Research to Cognitive Technology, pp. 53–73. Erlbaum, Mahwah (2007)
19. Tversky, B., Kugelmass, S., Winter, A.: Cross-cultural and Developmental Trends in Graphic Productions. Cognitive Psychology 23, 515–557 (1991)
20. Tversky, B., Morrison, J.B., Betrancourt, M.: Animation: Can it Facilitate? International Journal of Human Computer Studies 57, 247–262 (2002)
21. Tversky, B., Zacks, J., Lee, P., Heiser, J.: Lines, Blobs, Crosses and Arrows: Diagrammatic Communication with Schematic Figures. In: Anderson, M., Cheng, P., Haarslev, V. (eds.) Diagrams 2000. LNCS (LNAI), vol. 1889, pp. 221–230. Springer, Heidelberg (2000)
22. Zacks, J., Tversky, B.: Bars and Lines: A Study of Graphic Communication. Memory and Cognition 27, 1073–1079 (1999)
23. Zahner, D., Corter, J.E.: The process of probability problem solving: Use of external visual representations. Mathematical Thinking and Learning 12, 177–204 (2010)
24. Zhang, J.: External representations in complex information processing tasks. In: Kent, A. (ed.) Encyclopedia of Library and Information Science, vol. 68, pp. 164–180. Marcel Dekker, Inc., New York (2000)

# Elucidating the Mechanism of Spontaneous Diagram Use in Explanations: How Cognitive Processing of Text and Diagrammatic Representations Are Influenced by Individual and Task-Related Factors

Emmanuel Manalo[1] and Yuri Uesaka[2]

[1] Faculty of Science and Engineering, Waseda University, Tokyo, Japan
[2] Graduate School of Education, The University of Tokyo, Japan
emmanuel.manalo@gmail.com, y_uesaka@p.u-tokyo.ac.jp

**Abstract.** Although diagrams are considered effective tools for communication, students have been reported as lacking sufficient spontaneity in using diagrams when explaining what they have learned. This study examined the possible mechanism that relates text to diagram production in the process of providing written explanations. It puts forward the hypothesis that the production of text and diagrammatic representations shares the same cognitive processing resources, the allocation of which is influenced by individual factors like language ability and task-related factors like imageability of what needs to be explained. This hypothesis was tested on Japanese university students who were administered a passage (two versions varying in imageability) to read and subsequently explain in English or Japanese. A significant correlation was found between diagram use and English language competence (measured by TOEIC scores) – but only among students asked to explain the passage with lower imageability, and in English, providing support for the hypothesis.

**Keywords:** spontaneous diagram use, text and diagrammatic representations, explanation, cognitive processing resources, communication.

## 1 Introduction

Diagram use is considered efficacious in many educational situations. This efficacy has been described in terms of diagrammatic representations' capacity to group related information together and support many perceptual inferences – thus rendering such representations more computationally efficient compared to sentential representations [1]. The construction of diagrams serve a number of valuable functions in educational contexts including enhancement of engagement in learning, encouragement of use of more effective strategies, and promotion of the development of a number of crucial skills such as those in creative reasoning, communication, and the use of multiple literacies for knowledge construction and representation [2]. Despite this apparent usefulness, however, researchers have also reported that many

P. Cox, B. Plimmer, and P. Rodgers (Eds.): Diagrams 2012, LNAI 7352, pp. 35–50, 2012.

students manifest problems in the actual use of diagrams [3]. These problems include poor choice of diagrams to use [4, 5], failure to draw appropriate inferences when diagrams are used [6–9], and a lack of spontaneity in the use of diagrams [7, 10–13]. The problem with spontaneity is particularly serious because it means that many students fail to benefit from diagram use in tasks they need to complete in school, as well as outside of school.

Although the lack of spontaneity in diagram use has mainly been reported as a problem among primary and secondary students in math problem solving [7, 10–13], recently Manalo, Uesaka, Kriz, Kato, and Fukaya [14] reported that university students lack sufficient spontaneity in using diagrams when explaining what they have learned. More specifically, they found that although many of the students included some form of diagrammatic representation in the notes they took during the process of learning information contained in a text passage, a significantly lower proportion employed diagrams in a subsequent task of constructing an explanation for an imaginary classmate who did not know anything about the topic of the passage.

The appropriate use of diagrams in communication is considered to be an efficacious strategy: for example, the effectiveness of text-based communication is often enhanced when it is supplemented with diagrams [15]. It is therefore important to develop instructional strategies that could improve students' spontaneity and effectiveness in using diagrams in communication tasks like the provision of explanations. To develop such strategies, however, it would be helpful to understand the mechanisms that may underlie diagram use in such situations.

### 1.1 Cognitive Processing of Text and Diagrammatic Representations

A number of research studies have been conducted to explain the mechanism by which simultaneously presented verbal and visual information (multimedia information) is cognitively processed [16–18]. Mayer and Moreno [19] summarized the proposed mechanism as involving two channels in working memory [20, 21] – an auditory/verbal channel that receives and processes auditory input and verbal representations, and a visual/pictorial channel that receives and processes visual input and pictorial/diagrammatic representations.

There is, however, no equivalent model or explanation for the mechanism involved in *producing* verbal (text) and visual (diagrammatic) representations in communication tasks. No previous considerations had been put forward as to whether the production of text and diagrammatic representations in communicating learned information could be explained using similar processes in working memory. If they could be explained in similar terms, it would be useful to find out whether limitations in cognitive load capacity could explain instances of failure to employ diagrams in such communication tasks.

Figure 1 shows the possible mechanism involved if the dual channel model of working memory is used to portray what happens when a person is required to communicate what he or she has learned. In this model, the requirement to provide an explanation instigates access from long-term memory of relevant knowledge about the material to be explained. If external resources (e.g., notes) are available, these could also be accessed. In working memory, processing occurs through the verbal and visual channels: these organize the portions of the accessed knowledge and/or other information – selecting, recombining, reformulating them – into verbal and/or visual representations. The resulting processing output would then be relayed for motor execution (i.e., script writing, diagram drawing), and hence the relevant knowledge would be represented in the form of written words and/or diagrams.

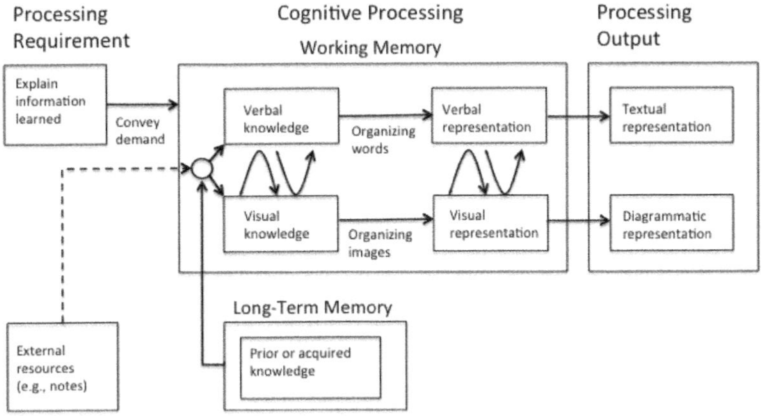


**Fig. 1.** Proposed mechanism involved in the production of text and diagrammatic representations in explaining information that has been learned

In this model, the use of working memory resources for the organization and production of text and diagrammatic representations would depend on two conditions: existence of relevant knowledge and/or skills (i.e., the person possesses both the content knowledge and any requisite skills relating to the material to be explained), and perception of the relevance or value of providing verbal and/or visual representations. For example, if the person lacks semantic knowledge about the material to be explained, or does not have adequate vocabulary to explain it, the person may not be able to use the verbal channel to produce the necessary text representation. This assumption is based on previous research findings that students are unable to use strategies that fall outside the knowledge or skills they posses – even when they are aware that those skills are required [e.g., 13, 22]. Likewise, if the person does not perceive the value or relevance of including a diagram in the explanation, then the visual channel will not be utilized for the production of a diagrammatic representation. This second assumption is based on previous research findings that spontaneous use of a strategy is influenced by perceptions about the efficacy the strategy brings to the task at hand [e.g., 13].

Mayer and Moreno [19] portrayed processing of multimedia stimuli in the verbal and visual channels as more or less parallel, independent processes. Such a portrayal may be appropriate where processing externally presented auditory and visual information (i.e., narration and images), which use different senses, is processed to create internal representations. When students have to provide a written explanation however – processing which essentially needs to proceed in the opposite direction (i.e., from internal to external representation of the information) – greater interactivity and interdependence may be imposed on the verbal and visual channels. To begin with, information accessed from long-term memory would likely contain both verbal and visual components that would need to be relayed to the corresponding verbal and visual channels of working memory. The verbal channel would be responsible for selecting the appropriate words to express the learned information, combining those words into meaningful expressions, and sequentially organizing those expressions into a coherent explanation of what is known. However, encoding that verbal explanation in written form would require interactive processing through both verbal and visual channels (i.e., translation of the words into the written script).

The visual channel, on the other hand, would be responsible for processing of relevant visual information drawn from long-term memory. If modifications are required of the retrieved information, the visual channel would be responsible for its deconstruction, modification, and recombination to produce the required, modified version. Such revisions, however, would likely involve the verbal channel in the articulation, not only of the modifications required, but also the meaning of those modifications. Likewise, the verbal channel is likely to be involved in the production of new diagrammatic representations. It would be involved in the formulation of the propositions that characterize, and instigate the need for, the new representation (e.g., that X is connected to Y which in turn is connected to Z). Those verbal propositions would then require translation for processing via the visual channel (i.e., to construct the corresponding visual representation).

The model proposed here suggests that the construction of text and diagrammatic representations for the provision of explanations entails greater interaction between the verbal and visual channels of working memory. The channels would frequently need to hold information in a shared space in working memory and that information needs to be translated from one channel to the other so that the required forms of representation – script representing verbal information or diagrams representing verbal propositions – could be produced. Determining exactly where this information sharing between the channels occurs is outside the scope of the current paper. It is possible, for example, that the interaction and combination occurs within the episodic buffer [23], a component of working memory responsible for holding integrated episodes or chunks in a multidimensional code, and which acts as a buffer store between the other components of working memory [24]. Possibilities like this would need to be investigated in future research.

It is important to note, however, that there are limited cognitive resources available for such processing. According to both working memory theory [20, 21] and cognitive load theory [25, 26], the human information processing system has a limited capacity. Once processing demand exceeds that limit, the intended or desired outcome will no longer eventuate. In multimedia learning, Mayer and Moreno [19] explained that when cognitive overload occurs (i.e., when the processing requirements of the task exceeds the cognitive resources available for processing), deficient learning outcomes are produced. In the production of written explanations, cognitive overload will likely have similar detrimental effects on performance.

## 1.2 Individual and Task-Related Factors that Influence Resource Allocation

Manalo et al. [14] pointed out that educational socialization in general emphasizes a verbal response when required to provide an explanation of what one has learned. In such situations, therefore, most students would place priority on the production of text/verbal representation of the learned information, with diagrammatic/visual representation likely to be used only as a supplementation or enhancement. This would suggest that initial and possibly greater cognitive processing resources would be utilized for text production compared to diagram construction.

In addition to the socialization issue, however, some individual and task-related factors could affect the allocation and use of cognitive resources. Individual factors include those that are intrinsic to the individual, such as the knowledge, experience, and skills that he or she possesses. Task-related factors are those that are intrinsic to the task itself, such as its length or complexity. In providing a written explanation of information that has been learned, language ability (an individual factor) and imageability (a task-related factor) are two factors that could potentially influence the allocation and use of cognitive resources for text and diagram production.

An individual's language proficiency is likely to have a direct bearing on the amount of cognitive effort required to produce the verbal/text representation of the information that needs to be explained. High proficiency would correspond to less mental effort in producing the necessary text representation. In contrast, lack of proficiency would entail the use of much greater mental effort and resources in producing the same or equivalent text representation.

A similar relationship is likely to exist between imageability of the information that has been learned and the amount of resources required to produce a diagrammatic or visual representation of it. Uesaka and Manalo [27] previously reported evidence that a higher cognitive cost is entailed in constructing an appropriate diagram for some math word problems – more specifically, those that require more abstract representations (e.g., a table or a graph, as opposed to a simple drawing or illustration of key components). The production of a diagrammatic representation for information that is harder to visually imagine (i.e., lower in imageability) is likely to demand a similarly higher cognitive cost, and thus demand the use of more cognitive resources, compared to information that is easier to imagine (i.e., higher in imageability).

If an individual's language ability and the imageability of the target information affect the amount of cognitive resources required to produce a written explanation, situations in which cognitive overload occur could potentially explain poor performance. For example, if a person lacks sufficient proficiency in the language that needs to be used, the production of the verbal/text explanation of the information would place greater demand on available cognitive resources. This would mean that the verbal channel of working memory may end up using most, if not all, of the resources available in working memory for the production of a verbal/text representation. As a consequence, little or no resources may be left for use in the production of a diagrammatic representation.

Likewise, if the target information has low imageability, the requirement to produce an explanation of it could place a greater demand on the cognitive resources available to the verbal and visual channels of working memory. Information that is difficult to imagine could also be more difficult to verbally represent and hence could impose greater processing demands on the verbal channel. But even if the verbal channel did not require greater cognitive resources for the production of the text explanation, the remaining resources available for the visual channel to use for the construction of a diagrammatic representation may prove inadequate for that purpose.

If the above assumptions are correct, they could provide not only a viable means to understand the mechanisms involved in the use of cognitive resources for the production of text and diagrammatic representations in written explanations, but also a possible explanation for why students might manifest insufficient spontaneity in using diagrams when explaining what they have learned. The present investigation sought to experimentally verify these explanations.

### 1.3 Overview and Predictions of the Present Study

The key objective of the present study is to explain how the production of text and diagrammatic representations might be related to each other in the process of providing an explanation of what has been learned. Its primary hypothesis is that the production of these forms of representation shares the same cognitive processing resources in working memory. If correct, this would mean that greater resource demand for the production of one form of representation (text or diagram) could result in inadequate resources for the production of the other form of representation. The secondary, related hypothesis is that some individual and task-related factors influence the allocation and use of cognitive resources for the production of text and diagrammatic representations in explanations. If correct, this would mean that variations in particular kinds of individual and task-related factors would correspond to variations in text and diagrams used in explanations.

To test the above hypotheses, the production of written explanations was examined among Japanese university students. More specifically, the extent to which the students used diagrams in constructing their explanations was examined.

As previously noted, the primary and secondary hypotheses are related. To test the first hypothesis, the resource demand for one form of representation needed to be manipulated so that the resulting effect on the production of the other form of representation could be observed. To manipulate the demand for these resources, individual and task-related factors – the key components of the secondary hypothesis – were selected for manipulation. More specifically, the students' language proficiency in English was chosen as the individual factor, and the imageability of the passages the students had to learn was selected as the task-related factor.

Because English is a second language (L2) for the Japanese students, lower proficiency in English would presumably result in greater cognitive resource demand when the explanation required is in English. In other words, if a student is not so proficient in English, he or she would need to put more effort – and use up more cognitive resources – to construct a written explanation in English of the information he or she has learned. In such cases, cognitive resources remaining for use in diagram construction would be depleted.

However, even if the student is required to explain in L2, the cognitive resources remaining for use in diagram construction may be adequate *if* the task itself is not high in cognitive demand. Hence the relationship between L2 proficiency and diagram use should be more apparent in a task that in itself is more demanding of cognitive resources. In this study, the "more demanding" task was one that required explanation of information that was of lower imageability.

Two main predictions were therefore made. The first prediction was that, overall, the students would use fewer diagrams in producing written explanations of the passage of lower imageability. The second prediction was that an index of the students' level of English language competence would correlate with their diagram use when explaining the passage of lower imageability in English.

Two passages, one about how music is played from a compact disk (CD) (adapted from [28]) and another about the human blood circulation system (adapted from [29]), were used in this study. These passages were deemed as differing in their imageability based on the kinds of diagram considered most appropriate to use in explaining their content. Although both passages concern processes, for the CD passage, a flow chart of the different component parts involved would be appropriate: most of those parts are difficult to imagine (e.g., microscopic indentations on the surface of a CD, photocell detector, binary number decoder, etc.) and the key purpose of the diagram would be to show the sequence involved in generating, coding, and converting data that eventually become music. In contrast, for the circulation passage, a drawing or sketch of the parts of the body involved would be considered appropriate: these (e.g., heart, lungs, blood vessels, etc.) are easier to imagine and the key purpose of the diagram would be to show the directions of blood flow through these parts of the system. The diagram that is appropriate for the CD passage (i.e., the flow chart) is more abstract compared to the one that is appropriate for the circulation passage (i.e., the sketch or drawing of the organs). Uesaka and Manalo [27] had earlier pointed out that constructing a more abstract diagram requires more transformational steps, and therefore involves greater cognitive cost. The first prediction of this study was based on the notion that imageability, as a task-related factor, would affect demand on cognitive processing resources and the consequent use – or otherwise – of diagrams in the provision of explanations. The expectation was that a passage of lower imageability would require more cognitive translational steps to generate a diagram for, resulting in greater cognitive cost. This cost in turn would reduce the spontaneous use of diagrams in written explanations that are produced.

The students' TOEIC test scores (Test of English for International Communication [30]) were used in this study as the index of their English language competence. Students in the faculty of the university where this study was conducted are required to sit the TOEIC test when they first enter the university, and then at regular intervals afterwards as a general assessment of their English language skills development. The second prediction of this study was based on the notion that competence in the language to be used in explaining, as an individual factor, would affect demand on cognitive processing resources and the consequent use – or otherwise – of diagrams in the explanations that are produced. The expectation was that students with lower TOEIC scores would need to use more cognitive resources to produce the verbal/text representation of the passage. If that passage is also of low imageability, the remaining cognitive resources were expected to be inadequate for the additional generation of a diagrammatic representation. In those cases, therefore, a relationship should be observed between TOEIC scores and diagram use.

A possible alternative explanation to the predicted relationship between TOEIC scores and diagram use is that it may not be the students' English language competence but their general language competence – or even their general intellectual ability – that determines the use of diagrams. In other words, students who have higher language or intellectual ability would score higher in the TOEIC test and also perform better in tasks like explaining what they have learned, making effective use of techniques like the inclusion of diagrammatic representations. Verbal/language ability, and intellectual abilities and performance, are generally considered as being related (see, e.g., reviews by Ackerman [31] and Neisser et al. [32]). If this alternative explanation were correct, the predicted relationship between English language competence and diagram use would not be attributable to cognitive resource demand and allocation, but more to general language and/or intellectual ability.

A third prediction was therefore made: that no relationship would be observed between the students' TOEIC scores and their diagram use when the explanation required was in Japanese (the students' native language, or L1). If the relationship is due to general language and/or intellectual ability, its manifestation should not be restricted to when the students have to explain in English: in other words, it should be observable even when they have to explain in Japanese. Thus, if the relationship were found to be absent in the conditions requiring Japanese explanations, it would provide verification for the hypotheses proposed in this study.

It should also be noted here that the predicted relationship between language competence and diagram use was only in terms of the former being a potential limiting factor on the latter. In other words, when language competence is low, it could impose much higher demand on cognitive processing resources to produce the required text (verbal) representation – to the extent that inadequate resources remain for the production of a diagrammatic (visual) representation. However, the converse (i.e., that greater cognitive resources remaining would *necessarily* lead to *more* diagrams being spontaneously produced) was not expected. Thus, for example, it was *not* expected that the students would generally produce more diagrams when explaining in the L1 (Japanese). In such situations, other factors noted earlier (e.g., skills in constructing the required diagrammatic representation, perceptions about the value of including a diagram, imageability) would likely play more significant roles in determining the spontaneous use of diagrams.

## 2 Method

### 2.1 Participants

The participants were 100 undergraduate science and engineering students at a university in Japan (mean age = 19.92 years, SD = .907 years; females = 15). They participated voluntarily in the study and received no monetary compensation for participation, but were allowed to keep a pen supplied for use in the study.

### 2.2 Materials

The participants were provided a booklet containing the experimental materials, and a separate sheet for note taking purposes. There were four versions of the booklet, which were distributed randomly to students in approximately equal numbers. The booklets contained either the passage about how music is played from a CD or the passage about how the human blood circulation system works, both noted earlier. The passages contained only words; no diagrams or illustrations were included. Modifications were made to the original versions of these passages [28, 29] so that the two passages used in the present study were equivalent in length, and contained approximately the same number of discreet information units or segments that convey distinct, meaningful information. To clarify what those units pertained to, the following example sentence was considered as containing four of those units as indicated by the segmentation slashes used here: *The list of numbers representing the music is 'burned' on a CD / using a laser beam / that etches bumps (called "pits") / into the shiny surface of the CD.* In this example, the first unit pertains to what is done, the second indicates what is used, the third states how it is done, and the fourth refers to the location. In each booklet, the passage (CD or circulation) was provided first in the Japanese language, and then in the English language.

Apart from differences in the passage content, the booklets also differed in the language that the participants were asked to use in a subsequent explanation of what they had learned: approximately half of the booklets asked for an explanation in Japanese (the L1) and the other half in English (the L2). Thus, the four versions of the booklet were: (1) CD passage requiring L1 explanation, (2) CD passage requiring L2 explanation, (3) circulation passage requiring L1 explanation, and (4) circulation passage requiring L2 explanation. In all other respects (i.e., the instructions given, the questions asked, etc.), the four versions of the booklet were identical.

## 2.3 Procedure

The factors manipulated in this study (i.e., kind of passage, and language of explanation required) were both between-subject variables. The four resulting groups, with corresponding total number of participants in brackets, were as follows: CD-L1 (25), CD-L2 (27), Circulation-L1 (25), and Circulation-L2 (23). Data collection was carried out at the end of one of the students' regular, scheduled classes. After distribution of the booklets to those who were willing to participate, the students were provided verbal instructions about what to do. Equivalent instructions were provided in the booklets in written form. The students were informed that they would be reading a passage, first in Japanese and then in English, and that later they would be asked questions about it, including explaining its content in Japanese or in English. They were informed that the language they would need to use to explain would be given later.

The students were asked to read the passage they were allotted in Japanese (8 minutes) and then in English (8 minutes). During these 8-minute reading periods, they were informed that they could use the extra sheet of paper they were provided to take notes as they wished, and that they could consult the notes they made during the entire experiment. They were also informed, however, that they were not allowed to return to, or re-read, the original Japanese and English versions of the passage once the time allocated for reading those had expired.

Next, the students were given 2 minutes to answer five questions relating to their perceptions about their understanding of the passage, its level of difficulty, their previous knowledge about it, and how easy or difficult they considered it would be to explain its content using the Japanese language, and the English language. All questions required responses on Likert-type scales.

After this, the students were provided space on the following page to provide an explanation of the passage they had read. For this, they were asked to imagine that their audience was a fellow student who did not know anything about the topic. At this point, the students could see in their booklet whether the language they had to use was Japanese or English. The students were given 10 minutes to complete this task.

Following this, the students were given 4 minutes to answer four questions on the next page of the booklet. The questions were constructed to more objectively assess their understanding of the passage they had read. Finally, on the last page of the booklet, the students were asked to provide some demographic information about themselves (e.g., their age, gender, year at university, etc.), as well as their most recent TOEIC test score.

## 3 Results

Prior to analyses of the data, the explanations that the participants produced were coded for instances of diagram use. For the purposes of this study, a diagram was defined as any representations produced by the participants, other than representations in the form of words, sentences, or numbers on their own. For example, drawings and graphs counted as diagrams, as did arrows and similar symbols when these were used to link three or more concepts or ideas. Examples of diagrams that participants produced for the CD and circulation passages are shown in Figure 2.

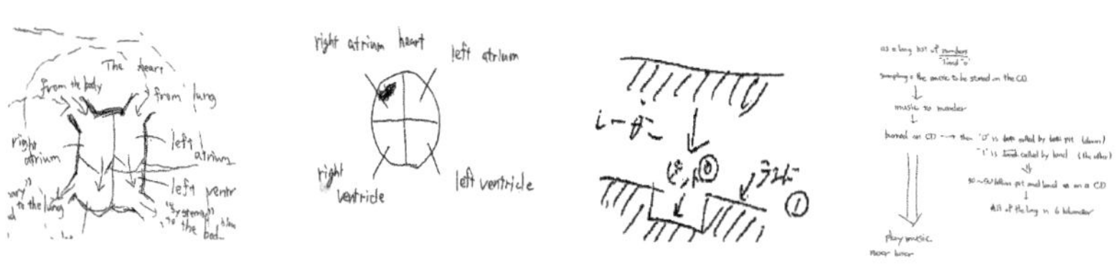

**Fig. 2.** Examples of diagrams that participants produced for the circulation and CD passages

The proportions of participants who spontaneously used a diagram in providing explanations for the CD and the circulation passages in Japanese and in English, as well as in total, are shown in Table 1.

The results of an ANOVA revealed a significant main effect for the kind of passage administered, $F(1, 96) = 20.13$, $p < .001$. However, no effect was found due to the language of explanation required, $F(1, 96) = .43$, n.s.. Likewise, there was no significant interaction between passage administered and language of explanation required, $F(1, 96) = .11$, n.s.. This finding provides support for the first prediction that the participants would use fewer diagrams in the explanations they provide for the passage of lower imageability (the CD passage).

**Table 1.** Proportions of participants who used a diagram in explaining the CD and circulation passages in Japanese, in English, and in both languages

| Passage | Japanese explanation | English explanation | Both languages (combined) |
|---|---|---|---|
| CD | .15 | .12 | .13 |
| Circulation | .57 | .48 | .52 |

The correlations between the participants' TOEIC scores and their diagram use when explaining the CD and circulation passages in Japanese and in English are shown in Table 2.

**Table 2.** Correlation coefficients between participants' TOEIC scores and their diagram use in the CD and circulation passages when explaining them in Japanese and in English

| Passage | Japanese explanation | English explanation |
|---|---|---|
| CD | –.14 | .47* |
| Circulation | –.21 | .28 |

* $p < .05$

These results provide support for the second prediction that the participants' English language competence (as measured by their TOEIC scores) would significantly correlate with their diagram use, but only when explaining the passage of lower imageability (the CD passage) in the English language (the L2). The results also provide support for the third prediction that no similar significant correlation would be found when the explanation required was in the Japanese language (the L1).

Significant *negative* correlations were found between diagram use when explaining the CD passage in the Japanese language and participants' assessment of how much they previously knew about the topic ("prior knowledge", $r = –.54$, $p < .01$), how difficult they considered the passage ("passage difficulty", $r = –.51$, $p < .01$), and how difficult they considered it would be to explain the passage in Japanese ("Japanese explanation difficulty", $r = –.44$, $p < .05$). No other significant correlations were found between the participants' diagram use in the explanations they provided and the other measurements taken through the questionnaire.

## 4 Discussion

In line with the findings of Uesaka and Manalo [27], the results of this study indicate that task-related factors influence the spontaneity with which students use diagrams. In the present study, the imageability of the passage to be explained was found to affect diagram use: significantly fewer diagrams were used in explanations of the passage of lower imageability. More translational steps, and greater cognitive processing resources, would have been required in generating a diagrammatic representation for the passage of lower imageability. Thus, the significantly fewer diagrams that students generated in explaining that passage can be understood both in terms of possible cognitive overload [19, 25, 26] and the view that human performance is based on economic concepts [33, 34]. As explained earlier, for the passage of lower imageability, after the production of text representation has used up a certain portion of the available cognitive processing resources, inadequate resources may remain for use in the generation of a diagrammatic representation. It is also possible that, even if adequate cognitive processing resources remain, the high cognitive cost of constructing the required diagram may prove prohibitive for many students. This is based on the assumption that people are predisposed to avoid or minimize cognitive workload [35], an assumption that has been proven in research studies involving strategy selection [36, 37]. However, prior to the present study, only Uesaka and Manalo [27] had demonstrated the applicability of this assumption in educational contexts. The findings of the present study, therefore, contribute to highlighting the importance in educational contexts of considering the cognitive processing costs that variations in certain task-related factors generate.

The findings of the present study also indicate that individual factors – in this case, language competence – can impact on the likelihood of a diagram being spontaneously used. This impact is again based on the notion of cognitive overload [19, 25, 26]. It is proposed in this paper that, in constructing an explanation of what one has learned, the production of text and diagrammatic representations shares the same cognitive processing resources in working memory. Thus, low language competence could result in an increased demand and use of those resources for the production of text representation: the consequence of this would be a reduction in remaining resources available for the production of a diagrammatic representation. Hence, especially when explaining a passage with low imageability, those remaining resources may prove inadequate for the construction of the required diagram.

The significant negative correlations between students' diagram use when explaining the CD passage in Japanese and measurements of "prior knowledge", "passage difficulty", and "Japanese explanation difficulty" indicate that the more participants felt they previously knew about the CD topic and the less difficult they thought the passage was to understand and to explain in Japanese, the less they used diagrams in their explanations. No such correlations were found in the circulation passage. This finding therefore suggests that, because the CD passage required greater cognitive cost to construct a diagram for, many of those participants opted not to include a diagram in their explanation. They could have felt that explaining the contents of the passage well – using words – was adequate and they did not need to worry about constructing the corresponding hard-to-imagine diagram.

It is important to stress here that a general relationship between language ability and diagram use is *not* being proposed in this paper. The support found for the third prediction, of a lack of relationship between the students' TOEIC scores and their diagram use when the explanation required was in Japanese, indicates that the observed language competence impact on diagram use cannot be attributed to the students' general language ability. It should also be noted here that the authors are by no means suggesting that enhancing students' language competence would directly result in increases in their diagram use. That would only be expected, to a limited extent, if their language competence were so low as to cause a need to use much greater cognitive processing resources for the production of text representations, as explained earlier. In other words, the relationship between language competence and diagram use that is proposed here is only in terms of the former being a potential limiting factor on the latter. For this reason, the authors do not consider the absence of a significant effect due to the language of explanation to be inconsistent with the predictions made. Writing explanations in L1 would not necessarily equate to greater diagram use. A significant difference would only be expected if a much higher proportion of the students who were required to write in L2 had much poorer levels of competence in that language – to the extent that their lower competence became a limiting factor on their diagram use.

The finding that students with lower L2 competence who were required to produce their explanation in that language evidenced lower diagram use may initially sound counter-intuitive. For those students, using diagrams in their explanation could have been easier than verbally explaining using the L2, so the question of why they did not do this arises. The answer may lie with the students' educational socialization which, as noted earlier, generally emphasizes a verbal response when called upon to provide explanations. Thus, even though it may have been more difficult for them, the students with lower L2 competence still strived to verbally explain what they had learned. In the present study, no explicit directions were given about how explanations should be constructed as the authors were interested in the spontaneous responses that the students would make. However, in future research, it may be useful to examine whether a different pattern of results would emerge if the participants were explicitly instructed that they could use words, or diagrams, or a combination of both, in providing their explanations.

Prior to the current research, no previous studies had considered and experimentally investigated the possible mechanism that links text and diagram production in the provision of written explanations. The current paper proposes such a mechanism, and provides experimental evidence to verify predictions derived from that proposed mechanism. The proposed mechanism has two important features. First, unlike the Mayer and Moreno [19] model for multimedia learning, where verbal and visual processing more or less operate independently of each other, the model proposed here suggests greater interaction and interdependence between these modes of processing. To generate text and diagrammatic representations, it suggests greater use of the space that verbal and visual processing share in working memory. As a consequence, processing resource use for the production of one form of representation impacts on the resources that may remain for the production of the other. The second important feature of the proposed mechanism is that individual and task-related factors influence the allocation and use of those resources. Hence, as the findings suggest, individual factors like language competence, and task-related factors like imageability, affected the allocation and use of resources for the production of text and diagrammatic representations.

In its current state, the authors consider the mechanism proposed here as being only at a very basic stage: it simply suggests that cognitive processing resources are shared in the generation of text and diagrammatic representations. Thus, there is an interdependency of sorts between these forms of representation: increased demand on resources for the generation of one form of representation could render remaining resources insufficient for the generation of the other. Potentially, the model can be developed to more accurately represent and promote understanding of the relationships between text/verbal and diagrammatic/visual representations that people generate in teaching and learning situations. Understanding those relationships is important if enhancing students' competence in the use of multi-modal representations to think, learn, reason, and communicate is a desired outcome in science and other educational disciplines [see, e.g., 2, 14].

**Acknowledgments.** This project was supported by a Grant for Scientific Research (Category B) from the Japan Society for the Promotion of Science. The authors would like to thank Tatsushi Fukaya, Fusa Katada, Nilson Kunioshi, Masako Tanaka, and Yoshio Ueno for their help in data collection and/or processing.

## References

1. Larkin, J.H., Simon, H.A.: Why a Diagram is (Sometimes) Worth Ten Thousand Words. Cognitive Science 11, 65–99 (1987)
2. Ainsworth, S., Prain, V., Tytler, R.: Drawing to Learn in Science. Science 333, 1096–1097 (2011)
3. Manalo, E., Uesaka, Y.: Drawing Attention to Diagram Use. Science 334, 761 (2011)
4. Grawemeyer, B., Cox, R.: The Effect of Knowledge-of-External-Representations Upon Performance and Representational Choice in a Database Query Task. In: Blackwell, A., Marriott, K., Shimojima, A. (eds.) Diagrams 2004. LNCS (LNAI), vol. 2980, pp. 351–354. Springer, Heidelberg (2004)
5. Uesaka, Y., Manalo, E.: Active Comparison as a Means of Promoting the Development of Abstract Conditional Knowledge and Appropriate Choice of Diagrams in Math Word Problem Solving. In: Barker-Plummer, D., Cox, R., Swoboda, N. (eds.) Diagrams 2006. LNCS (LNAI), vol. 4045, pp. 181–195. Springer, Heidelberg (2006)
6. Bowen, G.M., Roth, W.-M.: Why Students Not Learn to Interpret Scientific Inscriptions. Research in Science Education 32, 303–327 (2002)
7. Dufour-Janvier, B., Bednarz, N., Belanger, M.: Pedagogical Considerations Concerning the Problem of Representation. In: Janvier, C. (ed.) Problems of Representation in the Teaching and Learning of Mathematics, pp. 110–120. Erlbaum, Hillsdale (1987)
8. Kozhevnikov, M., Motes, M.A., Hegarty, M.: Spatial Visualization in Physics Problem Solving. Cognitive Science 31, 549–579 (2007)
9. Mokros, J.R., Tinker, R.F.: The Impact of Microcomputer-Based Labs on Children's Ability to Interpret Graphs. Journal of Research in Science Teaching 24, 369–383 (1987)
10. Ichikawa, S.: Suugakuteki na Kangaekata wo Megutte no Soudan to Sidou [Case Report of Cognitive Counseling in Mathematical Thinking]. In: Ichikawa, S. (ed.) Gakusyuu wo Sasaeru Nintikaunsering: Shinrigaku to Kyouiku no Aratana Setten [Cognitive Counseling that Supports Learning: A New Approach Bridging Psychology and Education], pp. 36–61. Brain Press, Tokyo (1993)

11. Manalo, E., Uesaka, Y.: Quantity and Quality of Diagrams Used in Math Word Problem Solving: A Comparison Between New Zealand and Japanese Students. Refereed Papers of the NZARE (New Zealand Association for Research in Education) National Conference 2006. NZARE, Wellington. ERIC Document Reproduction Service No. ED518280 (2006)
12. Uesaka, Y., Manalo, E., Ichikawa, S.: What Kinds of Perceptions and Daily Learning Behaviors Promote Students' Use of Diagrams in Mathematics Problem Solving? Learning and Instruction 17, 322–335 (2007)
13. Uesaka, Y., Manalo, E., Ichikawa, S.: The Effects of Perception of Efficacy and Diagram Construction Skills on Students' Spontaneous Use of Diagrams When Solving Math Word Problems. In: Goel, A.K., Jamnik, M., Narayanan, N.H. (eds.) Diagrams 2010. LNCS (LNAI), vol. 6170, pp. 197–211. Springer, Heidelberg (2010)
14. Manalo, E., Uesaka, Y., Pérez-Kriz, S., Kato, M., Fukaya, T.: Science and Engineering Students' Use of Diagrams During Note Taking Versus Explanation. Educational Studies (2012), doi:10.1080/03055698.2012.680577
15. Mayer, R.E.: Multimedia Learning. Cambridge University Press, New York (2001)
16. Mayer, R.E., Anderson, R.B.: Animations Need Narrations: An Experimental Test of a Dual-Coding Hypothesis. Journal of Educational Psychology 83, 484–490 (1991)
17. Mayer, R.E., Moreno, R.: A Split-Attention Effect in Multimedia Learning: Evidence for Dual Processing Systems in Working Memory. Journal of Educational Psychology 90, 312–320 (1998)
18. Mayer, R.E., Sims, V.K.: For Whom is a Picture Worth a Thousand Words? Extensions of a Dual-Coding Theory of Multimedia Learning. Journal of Educational Psychology 84, 389–460 (1994)
19. Mayer, R.E., Moreno, R.: Nine Ways to Reduce Cognitive Load in Multimedia Learning. Educational Psychologist 38, 43–52 (2003)
20. Baddeley, A.D.: Working Memory. Oxford University Press, Oxford (1986)
21. Baddeley, A.D.: Human Memory. Allyn & Bacon, Boston (1998)
22. Cornoldi, C., Gobbo, C., Mazzoni, G.: On Metamemory-Memory Relationship: Strategy Availability and Training. International Journal of Behavioral Development 14, 101–121 (1991)
23. Baddeley, A.D.: Personal communication, February 24 (2012)
24. Baddeley, A.D.: The Episodic Buffer: A New Component of Working Memory? Trends in Cognitive Sciences 4, 417–423 (2000)
25. Chandler, P., Sweller, J.: Cognitive Load Theory and the Format of Instruction. Cognition and Instruction 8, 293–332 (1991)
26. Sweller, J.: Instructional Design in Technical Areas. ACER Press, Camberwell (1999)
27. Uesaka, Y., Manalo, E.: Task-related Factors that Influence the Spontaneous Use of Diagrams in Math Word Problems. Applied Cognitive Psychology (2011), doi:10.1002/acp.1816
28. How CDs and DVDs Work, `http://www.explainthatstuff.com/cdplayers.html`
29. Chi, M.T.H., Siler, S.A., Jeong, H., Yamauchi, T., Hausmann, R.G.: Learning from Human Tutoring. Cognitive Science 25, 471–533 (2001)
30. TOEIC-ETS Home, `http://www.ets.org/toeic`
31. Ackerman, P.L.: Individual Differences in Information Processing: An Investigation of Intellectual Abilities and Task Performance During Practice. Intelligence 10, 101–139 (1986)

32. Neisser, U., Boodoo, G., Bouchard, T.J., Halpern, D.E., Loehlin, J.C., Perloff, R., Sternberg, R.J., Urbina, S.: Intelligence: Knowns and Unknowns. American Psychologist 51, 77–101 (1996)
33. Navon, D., Gopher, D.: On the Economy of the Human-Processing System. Psychological Review 86, 214–255 (1979)
34. Norman, D.A., Bobrow, D.J.: On Data-Limited and Resource-Limited Processes. Cognitive Psychology 7, 44–64 (1975)
35. Allport, G.W.: The Nature of Prejudice. Addison Wesley, New York (1954)
36. Kool, W., McGuire, J.T., Rosen, Z.B., Botvinick, M.M.: Decision Making and the Avoidance of Cognitive Demand. Journal of Experimental Psychology: General 139, 665–682 (2010)
37. Mathews, N., Hunt, E., MacLeod, C.M.: Strategy Choice and Strategy Training in Sentence-Picture Verification. Journal of Verbal Learning and Verbal Behavior 19, 531–548 (1980)

# Orthogonal Hyperedge Routing

Michael Wybrow[1], Kim Marriott[1], and Peter J. Stuckey[2]

[1] National ICT Australia, Victoria Laboratory,
Clayton School of Information Technology,
Monash University, Clayton, Victoria 3800, Australia
{Michael.Wybrow,Kim.Marriott}@monash.edu

[2] National ICT Australia, Victoria Laboratory,
Department of Computing and Information Systems,
University of Melbourne, Victoria 3010, Australia
pstuckey@unimelb.edu.au

**Abstract.** Orthogonal connectors are used in drawings of many network diagrams, especially those representing electrical circuits. Such diagrams frequently include hyperedges—single edges that connect more than two endpoints. While many interactive diagram editors provide some form of automatic connector routing we are unaware of any that provide automatic routing for orthogonal hyperedge connectors. We give three algorithms for hyperedge routing in an interactive diagramming editor. The first supports *semi-automatic routing* in which a route given by the user is improved by local transformations while the other two support *fully-automatic routing* and are heuristics based on an algorithm used for connector routing in circuit layout.

**Keywords:** orthogonal routing, hyperedges, circuit diagrams.

## 1 Introduction

Orthogonal connectors are used in drawings of many network diagrams, especially those representing electrical circuits. Such diagrams frequently include hyperedges—single edges that connect more than two endpoints. While many interactive diagram editors provide some form of automatic connector routing we are unaware of any that support automatic routing for orthogonal hyperedge connectors. This is the problem we address.

In this paper we describe how we have extended the connector routing library `libavoid`[1] to support orthogonal object-avoiding hyperedge routing in a commercial diagramming tool for circuit diagrams and the Dunnart diagram editor.[2]

We give three algorithms for hyperedge routing that support interactive construction and routing of hyperedges in interactive diagramming tools. The first

[1] `http://adaptagrams.sourceforge.net/libavoid/`

[2] Dunnart, including some orthogonal hyperedge routing features, is available for download from `http://www.dunnart.org/`

P. Cox, B. Plimmer, and P. Rodgers (Eds.): Diagrams 2012, LNAI 7352, pp. 51–64, 2012.

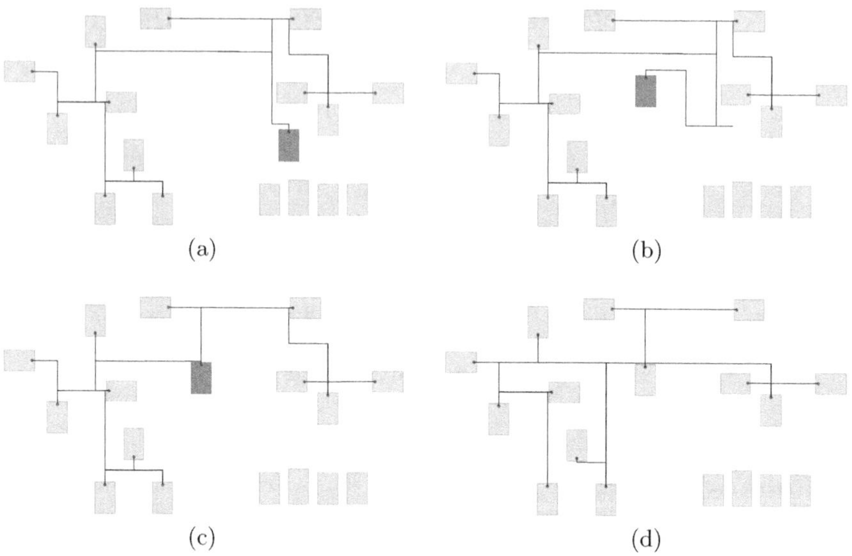


**Fig. 1.** Demonstration of the interaction model. (a) The diagram initially contains a single hyperedge made up of shapes, junction points and the connectors linking them. (b) The user drags a shape to a new location causing the connector between it and the junction point to be rerouted automatically. (c) Semi-automatic routing is performed to improve the hyperedge route. (d) If the user desires, they can perform fully-automatic routing to find a better topology for the hyperedge.

performs *semi-automatic routing* in which a route given by the user is improved by local transformations. The other two algorithms perform *fully-automatic routing* and are heuristic approaches extending an algorithm used for connector routing in circuit layout.

Previous research on connector routing in interactive diagramming tools has focused on poly-line and orthogonal routing for edges, i.e. arcs that connect two nodes [9,10]. Hyperedge routing generalises this by allowing the connector to connect multiple nodes. Here we focus on computing orthogonal routes, i.e. routes composed of horizontal and vertical segments, reflecting the drawing conventions used in circuit design.

Orthogonal hyperedge routing generalizes the problem of finding a minimal length rectilinear Steiner tree (MRST) connecting a set of points in the plane [5]. Computing the MRST is NP-Complete [2] and several heuristics and exact methods are given in [5]. A number of heuristic methods have also been developed for finding obstacle-avoiding rectilinear Steiner minimal trees (OARSMTs) [1,6] for automatic connector routing in VLSI design. Our problem differs from the standard problem studied in the VLSI setting because we are interested in supporting circuit construction in an interactive diagramming tool. This means that the algorithms need to be fast enough to support interaction and that the visual appeal and readability of the routes is important. Thus when computing the routes we penalize the number of bends as well as the total length and the semi-automatic routing step ensures that routes are visually distinct and pleasing in the sense that their paths are not obviously "bad."

Hyperedge routing is loosely related to edge bundling in which edge segments originating at the same node are collapsed together [3,4,8].

The remainder of this paper is organized as follows. In the next section we define our interaction model and formalize the automatic and semi-automatic routing problem. In Section 3 we discuss the semi-automatic routing algorithm which improves a route without changing topology. In Section 4 we give two algorithms for fully-automatic routing. We give experiments showing the effectiveness of the methods in Section 5. Finally in Section 6 we conclude.

## 2 Interaction Model and Problem Statement

Our algorithms are designed to support the following interaction model designed for interactive diagramming tools (Figure 1). It is designed to provide predictable automatic layout but allow the user to guide and override this. The four kinds of interaction are:

**Creation:** The user can create a new hyperedge by defining an initial route through specification of the bends and junctions that comprise the route. This specifies the topology of the route. The route is automatically improved so as to reduce segment lengths or bends but without changing the topology. We call this *semi-automatic* routing.

**Editing:** Whenever the user edits the diagram components, such as moving or deleting a node or diagram object, semi-automatic routing is used to improve the hyperedge routes while preserving the topology. See Figure 1(b) and (c).

**Automatic routing:** If the user is unhappy with a particular hyperedge route they can explicitly request that the tool performs *fully-automatic routing* in which case the system uses heuristic approaches to MRST to find an initial route which is then improved using semi-automatic routing. See Figure 1(d).

**Manual adjustment:** If the user is still unhappy they can manually modify the hyperedge route.

Since explicit support for hyperedges is not common in interactive diagramming tools, users will often work around this by using multiple individual connectors converging at junction points (or small dummy shapes if the software doesn't support junctions). This is actually a reasonably natural representation for hyperedges since it allows for easy incremental construction and alteration by the user, e.g., drawing a connector from a shape to an existing point on the hyperedge path. The difference with our approach is that the junction positions do not need to be tediously managed by the user but can be automatically positioned in response to diagram changes. Without semi-automatic routing its the users responsibility to manually modify Figure 1(b) to become Figure 1(c).

Notice that the interaction model requires that semi-automatic routing is performed very quickly since it must be applied to all hyperedges after most editing

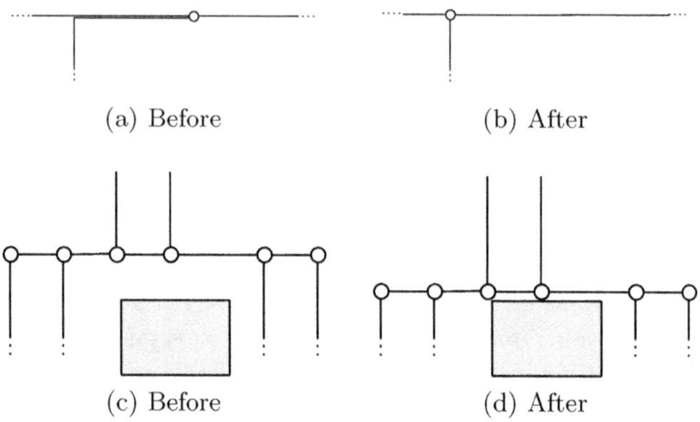

**Fig. 2.** The two local transformations used to improve the initial route. At the top is moving a junction to merge parallel routes, at the bottom moving a segment to reduce overall hyperedge length.

actions. In contrast, automatic routing can be slower since it is only performed for a subset of hyperedges at a time, and only when explicitly requested by the user.

We formalize automatic and semi-automatic hyperedge routing as follows. We have a set of nodes $N$ and a set of hyperedges $H$. Each $i \in N$ has a fixed position, width and height as well as a set of connector ports $P$ on its perimeter. Each $p \in P$ is a connector port with a direction of visibility. Each hyperedge $h \in H$ is a non-empty subset of connector ports. We wish to find a route $R$ for each $h$. This is a set of horizontal and vertical segments $R$ that form a tree whose leaf vertices are the connector ports $h$. The route should not pass through any of the nodes and should minimize a penalty function $\mathsf{p}(R)$ that is a monotonic function $\mathsf{f}$ of the length of $R$, $||R||$, and the number of bends (or equivalently segments) in $R$, $bends(R)$, i.e. $\mathsf{p}(R) = \mathsf{f}(||R||, bends(R))$. We sometimes refer to the leaf vertices of the route as the *terminals* and to the internal nodes as *junctions* and *bendpoints*. In the case of semi-automatic routing we are given an initial route $R'$ for $h$ which we must improve.

## 3 Semi-automatic Routing

Semi-automatic routing has two steps. The first step is to perform *local improvement* on the initial route to improve it by rectifying bad routing that is obvious to the human observer. The local improvement step is novel and is a result of examining many routes and identifying how to improve these manually.

Local improvement is designed to make local changes to the hyperedge which reduce edge length and bends. We first build $R$, a tree representing the routing for the hyperedge, with the root node of the tree being one of the junctions and the leaf nodes being the terminal points. Other nodes within the tree are made up of bendpoints and the remaining junctions from the hyperedge. We use two local transformations on this tree. They are illustrated in Figure 2.

The first transformation is to *remove redundant edges*. This looks at each junction node and if any of the edges in the tree both have another common endpoint (as well as the junction node), then the junction is moved to that node and the redundant edge is removed. This reduces the overall connector length and removes a bend point. The transformation is often not initially necessary but is still important in cases where the user has manually placed junctions at positions that may cause the shortest paths from multiple terminals to converge together before reaching the junction.

The second transformation is first applied in the horizontal dimension, then the vertical. In the case of the horizontal dimension we move a vertical line segment (an edge from our path tree) horizontally within the available space bordered by obstacles, and in the direction with the most divergent paths. This is shown in Figure 2(c–d) for a horizontal line segment in the vertical dimension. A line segment will consist of one or more collinear edges from the path tree, and thus multiple nodes. Using these nodes we maintain for each segment a count of the edges that diverge to each side. We also perform a sweep of the diagram, similar to that described in [10], to give us available space to shift each segment in either direction. We then repeatedly shift unbalanced segments, either to an obstacle boundary, or to the endpoint of the shortest diverging segment in that direction, updating the balance counts and merging segments as we do this. We will also shift balanced segments up to a diverging segment when doing so reduces the overall "length" of the hyperedge taking into consideration the penalty for each bend. Once there are no more segments to shift in that dimension, we remove redundant edges and perform the symmetric vertical process.

For an individual connector the transformations either remove a segment from the route or move the segment against the side of an obstacle (i.e., a shape expanded slightly with some buffer space). Thus, they can be applied no more than $O(n + s)$ times where $s$ is the number of segments in the hyperedge. The sweep takes $O(n \log n)$ time for each dimension. However, in practice local improvement is very fast.

The second step in semi-automatic routing is nudging and centering. This is performed on all hyperedges (and edges) together and is based on that for orthogonal edge routing [10]. We first determine the relative ordering of connectors in shared edges. In order to make the connector route clearer we want to nudge these paths apart to make the paths visually distinct. It is important to do so in a manner which does not introduce unnecessary crossings or bends in segments. Based on this we determine the exact coordinates of the orthogonal connector segments. This nudges connector routes a minimum distance apart to show the relative order of connectors with shared segments and also ensures that connectors pass down the middle of "alleys" in the diagrams when this does not lead to additional cost or additional edge crossings. This is described more fully in [10]. If $n$ is the number of diagram objects and $s$ the total number of connector segments then this step has $O((s + n)^2)$ worst-case complexity.

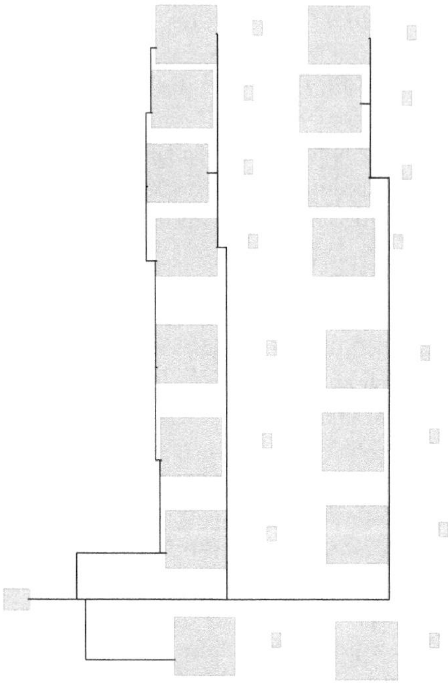

(a) Heuristic 1

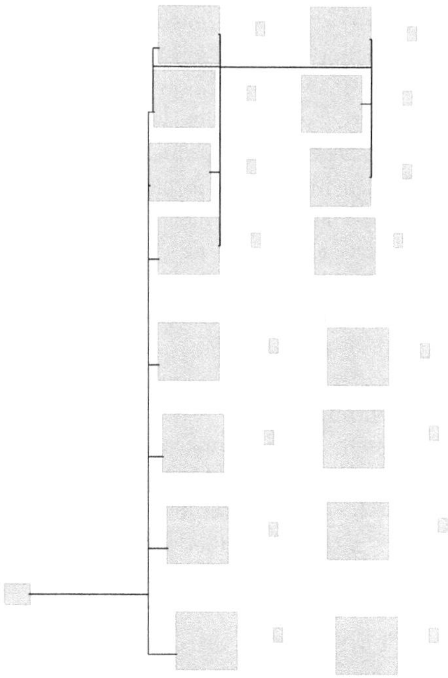

(b) Heuristic 2

**Fig. 3.** Real-world circuit diagram example showing difference between the two fully-automatic routing heuristics. (a) Sequential construction of MTST does not appropriately discount any shared segments on the paths in the MCST. (b) Interleaved construction of SPTF and MTST creates better routes, closer to what a human would draw. These diagrams show the raw output of each heuristic, before the hyperedge improvement step is performed. Note that Heuristic 2 also tends to result in less work needing to be performed in the subsequent improvement stage.

## 4 Fully Automatic Routing

As we have seen, computing an optimal hyperedge route is NP-Hard. In this section we describe two polynomial time heuristics for computing an initial hyperedge route. When combined with the preceding approach for semi-automatic layout they give a method for fully-automatic routing. Computing the initial route is quite an expensive operation. However in our model for user interaction its use is explicitly controlled by the user who must select a set of hyperedges to be rerouted by the tool.

The basis for our heuristics is the observation that when finding routes minimizing the penalty function we need only consider routes in the *orthogonal visibility graph*. This was introduced in [10] and is defined as follows. Let $I$ be the set of *interesting points* $(x, y)$ in the diagram, i.e. the connector points and corners of the bounding box of each object. Let $X_I$ be the set of $x$ coordinates in $I$ and $Y_I$ the set of $y$ coordinates in $I$. The *orthogonal visibility graph* $VG = (V, E)$ is

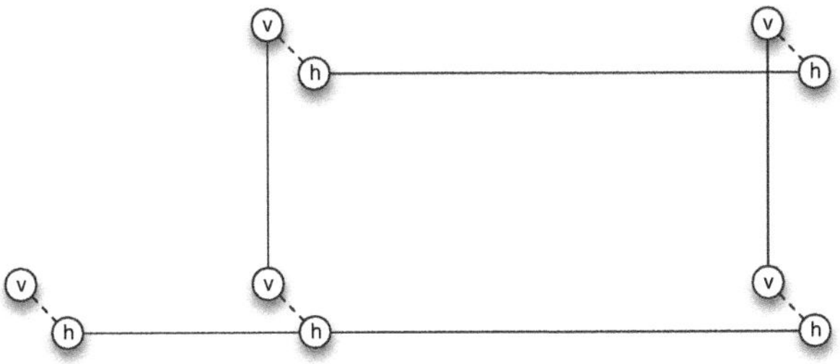


**Fig. 4.** A separated orthogonal visibility graph: $v$ nodes and vertical edges are on a plane above the $h$ nodes and horizontal edges. The dashed edges connect the two planes and correspond to a bend.

made up of nodes $V \subseteq X_I \times Y_I$ s.t. $(x, y) \in V$ iff there exists $y'$ s.t. $(x, y') \in I$ and there is no intervening object between $(x, y)$ and $(x, y')$ and there exists $x'$ s.t. $(x', y) \in I$ and there is no intervening object between $(x, y)$ and $(x', y)$. There is an edge $e \in E$ between each point in $V$ to its nearest neighbour to the north, south, east and west iff there is no intervening object in the original diagram.

We slightly modify the orthogonal visibility graph to produce a *separated orthogonal visibility graph* in which each node is split into two nodes (conceptually on two different planes) corresponding to whether it is connected to horizontally or vertically neighbouring nodes. Additionally, we add a link between these two nodes with a weight representing the bend penalty. This simplifies the algorithms because length and bend penalties are treated uniformly. Figure 4 illustrates the separated graph.

The orthogonal visibility graph can be constructed in $O(n^2)$ time for a diagram with $n$ objects and contains $O(n^2)$ vertices and $O(n^2)$ edges. An example orthogonal visibility graph is shown in Figure 5(a). It is quite different to the standard (non-orthogonal) visibility graph used for poly-line routing. In particular, the standard visibility graph has $O(n)$ nodes if there are $n$ objects in the diagram while the orthogonal visibility graph has $O(n^2)$ nodes. Both have $O(n^2)$ edges.

### 4.1 Heuristic 1: Sequential Construction of MTST

The starting point for our first heuristic is the VLSI routing algorithm of Long et al. [7]. This algorithm uses the standard (non-orthogonal) visibility graph (using the Manhattan distance on visibility edges) and is designed to find the route of minimal length. We have altered it to work with the separated orthogonal visibility graph in order to take the number of bends into account when computing the cost of a route. Figure 5 gives an overview of the main steps in fully-automatic routing with this heuristic for an example layout.

Our heuristic works by first constructing the *shortest path terminal forest (SPTF)* for the orthogonal visibility graph where the terminals are the hyperedge

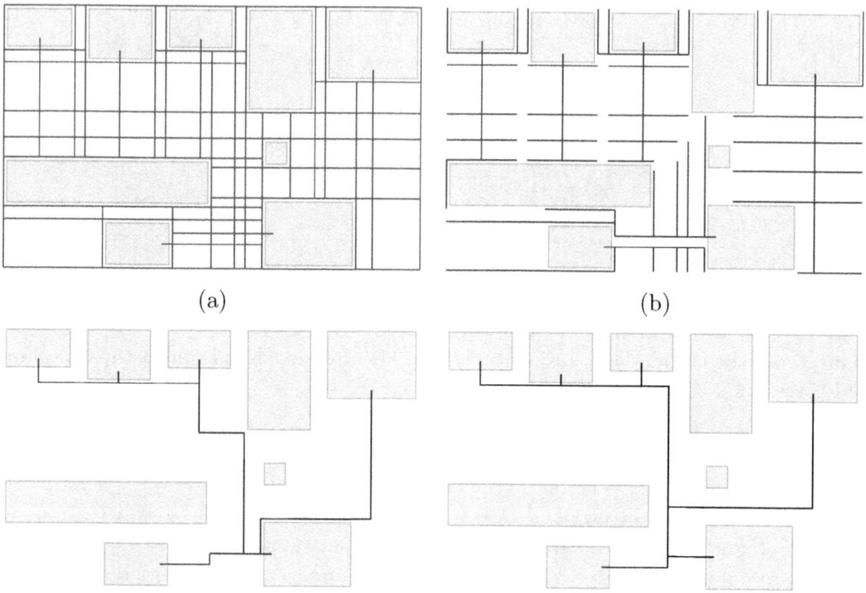

(a) (b) (c) (d)

**Fig. 5.** Heuristic 1: In the sequential fully-automatic hyperedge routing the initial route is found by: (a) computing the separated orthogonal visibility graph, (b) constructing the shortest path terminal forest (SPTF) and using it to compute an (c) initial hyperedge routing from the minimum terminal spanning tree (MTST). Semi-automatic hyperedge routing takes an initial routing and improves it by (d) performing local optimization followed by centering and nudging.

nodes and the edges are weighted by their length. This is computed using an extended Dijkstra shortest path algorithm. This processes edges in order of least distance from a terminal creating a shortest cost tree around each terminal. When processing an edge, if its endpoint is not already in a tree it is added to the tree, otherwise it is marked as a *bridge edge* if its endpoint belongs to a different terminal trees or else ignored if its endpoint belongs to the same tree (*self edge*). An example SPTF can be seen in Figure 5(b).

Importantly, the use of a high bend penalty along with our separated orthogonal visibility graph modification results in shortest cost trees that grow a long way in a straight line before branching. This helps our approach produce ideal two-segment connections between far apart terminals that could otherwise be blocked by the "bushy" growth of traditional SPTFs.

At the end of this step all nodes in the orthogonal visibility graph belong to exactly one terminal's shortest cost tree, and edges between these trees are marked as bridges. Next an extended Kruskal minimum spanning tree algorithm is used to find the *minimum cost spanning tree* (MCST) using bridge edges that connects the terminal trees where the cost of a bridge edge is its weight plus the cost to reach the terminal node from each of its endpoints. The minimum terminal spanning tree *MTST* is then the bridge nodes in the minimum cost spanning tree and the associated path to each terminal. For more details see [7].

As the number of nodes and edges in the visibility graph is $O(n^2)$ the time complexity of computing the SPTF, MCST and MTST is $O(n^2 \log n)$.

### 4.2 Heuristic 2: Interleaved Construction of SPTF and MTST

The main limitation of Heuristic 1 is that for efficiency the SPTF is computed before the MCST. This means that the cost when computing the MCST does not appropriately discount any shared segments on the paths in the MCST, see for example Figure 3. In essence, we can improve the quality of the solution found by interleaving computation of the SPTF with that of the MCST and building the MTST as we go. Our algorithm for this interleaved computation is given in in Figure 6[3] and an example of the algorithm's operation is shown in Figure 7.

The algorithm works by constructing sub-routes connecting disjoint subsets of the terminals in the original hyperedge, repeatedly combining these using a bridge edge from MCST until all of the terminals are connected in which case a MTST has been found.

The algorithm uses two priority queues *MCSTpq* and *SPTFpq* for computing the MCST from the sub-routes and the SPTF. Nodes $n$ in the separated orthogonal visibility graph ($VG$) are annotated to indicate whether they have been reached in the SPTF ($reached[n]$), and if so the cost of the path to them ($cost[n]$), and the sub-route ($route[n]$) from which the path originates (a set of edges). Elements in *SPTFpq* are tuples $(c, n, n', R)$ indicating that node $n$ can be reached from node $n'$ with a path of cost $c$ from sub-route $R$. Elements in *MCSTpq* are tuples $(c, n, w, n')$ indicating that nodes $n$ and $n'$ have been reached in the SPTF from different sub routes, say $R$ and $R'$, and that $R$ and $R'$ can be connected by a path with a total cost of $c$ passing through bridge edge $(n, w, n')$. We assume a function $\mathsf{terms}(R)$ which returns the set of terminal nodes appearing in route (set of edges) $R$.

The algorithm repeatedly does one of two things. It can extend the SPTF around the current set of sub-routes by popping a tuple from the *MTSTpq* and adding non-self edges and non-bridge edges to the *SPTFpq*. Whenever a bridge edge is encountered the appropriate path is added to the *MCSTpq*. Or it can pop a tuple $(c, n, w, n')$ from *MCSTpq* and merge the two sub-routes using the path through $(n, w, n')$ and add nodes on the merged route to the *MTSTpq* with a zero cost. The algorithm stops when it has created a sub-route connecting all of the terminals in the hyperedge.

[3] For the sake of pedagogical clarity the algorithm in Figure 6 omits several details important for implementation. For example, an implementation should use a number greater than any potential path in the diagram (including penalties) in place of $\infty$. Also, the algorithm description assumes that the priority queues will never be empty. This can happen in the case where obstacles prevent all possible paths between sections of the hyperedge. Please see the implementation in `libavoid` for more information.

```
VG := separated orthogonal visibility graph
SPTFpq := {(∞, ⊥, ⊥, ∅)} ; MCSTpq := {(∞, ⊥, 0, ⊥)}
reached[⊥] := true
for each node n in VG do reached[n] := false endfor
for each terminal n in h do
    add (0, n, ⊥, ∅) to SPTFpq
endfor
repeat
    if 2 × cost of top tuple on SPTFpq < cost of top tuple on MCSTpq then
        (c, n, np, R) := pop SPTFpq
        if ¬reached[n] ∧ reached[np] then
            reached[n] := true ; cost[n] := c ; route[n] := R
            for each edge (n, w, n') in VG do
                if reached[n'] then
                    if terms(R) ≠ terms(route[n']) then /* bridge edge */
                        add (c + cost[n'] + w, n, w, n') to MCSTpq
                    endif
                else /* non self + non bridge edge */
                    add (c + w, n', n, R ∪ {(n, w, n')}) to SPTFpq
                endif
            endfor
        endif
    else /* found two sub-routes to connect */
        (c, n, w, n') := pop MCSTpq
        R := route[n] ∪ route[n'] ∪ {(n, w, n')}
        if terms(R) = h then
            return R /* complete hyperedge route */
        for each node n'' where terms(route[n'']) ⊆ terms(R) do
            reached[n''] := false
        endfor
        for each node n'' in R do
            add (0, n'', ⊥, R) to SPTFpq
        endfor
    endif
forever
```

**Fig. 6.** Heuristic for computing a route $R$ for hyperedge $h$ using interleaved construction of the SPTF and MCST

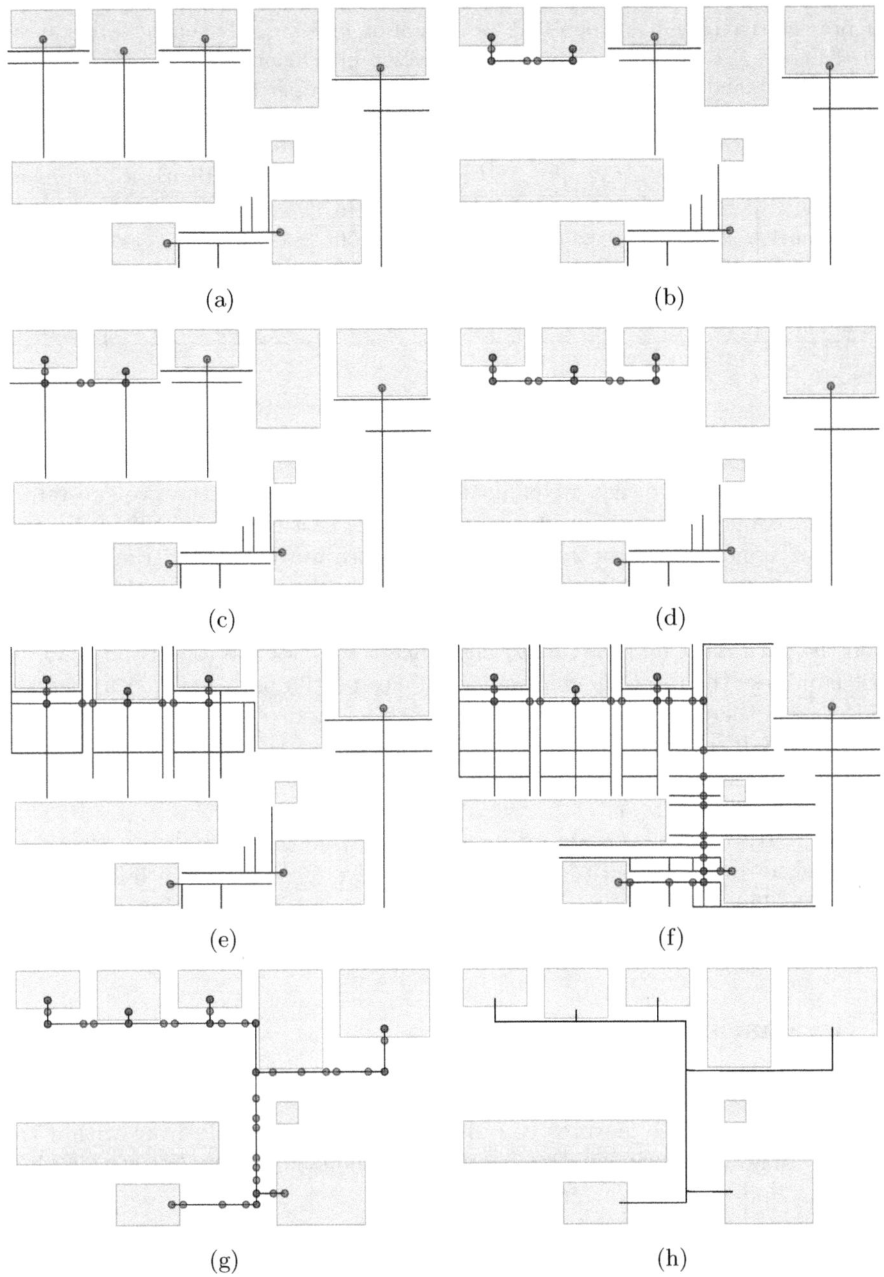


**Fig. 7.** Heuristic 2: In the interleaved approach we (a) incrementally build the SPTF, storing possible bridging edges. (b) We commit to the cheapest bridge when we reach a vertex with a cost more than twice the cost of the bridge. We then remove the bridged terminal's SPTFs and (c) continue, adding new terminals with zero cost for each vertex along the bridged path. (d–f) We repeat this process, until (g) there is just one terminal group remaining. This route is then improved by (h) performing local optimization followed by centering and nudging.

**Table 1.** Average times taken to compute fully-automatic routing for hyperedges in a representative circuit diagram and for several larger randomly generated instances. (Please note, $P$ is the total number of connection pins among all hyperedges $H$, and the VisGraph and Improve times are global rather than per hyperedge).

| | Diagram size | | | VisGraph size | | Times (in msec.) to compute | | | |
|---|---|---|---|---|---|---|---|---|---|
| Diagram | $\|N\|$ | $\|H\|$ | $\|P\|$ | $\|V\|$ | $\|E\|$ | VisGraph | Heuristic1 | Heuristic2 | Improve |
| Circuit | 52 | 5 | 57 | 6,948 | 11,083 | 16 | 155 | 167 | 12 |
| Random-1 | 200 | 1 | 25 | 20,142 | 35,674 | 56 | 47 | 142 | 25 |
| Random-2 | 200 | 1 | 50 | 22,674 | 40,035 | 62 | 61 | 187 | 64 |
| Random-3 | 400 | 1 | 50 | 28,117 | 46,952 | 82 | 70 | 304 | 213 |
| Random-4 | 400 | 10 | 250 | 33,587 | 52,805 | 97 | 74 | 197 | 466 |

The choice of whether to extend the SPTF or to merge two sub-routes depends on the cost of the top tuples of $SPTFpq$ and $MCSTpq$. If the current top tuple on $SPTFpq$ has cost $c$ then we can safely commit to joining the two sub-routes $R$ and $R'$ connected by the bridge edge $(n, w, n')$ in the top tuple of $MTSTpq$ if its cost $c'$ is no more than $2 \times c$. This is because we have found the minimum cost path between the sub-routes connected by $(n, w, n')$. To see this consider any other path $p'$ between $R$ and $R'$. If all nodes on the path have been reached, then the path must be in $MCSTpq$ and since it was not the top of the heap, its cost is no less than $c$. Otherwise not all nodes on the path have been reached. This means there are two nodes on the path $n_R$ and $n_{R'}$ respectively reached from $R$ and $R'$ that are in $SPTFpq$. But this means the cost of getting to $n_R$ from $R$ is at least $c$ and to $n_{R'}$ from $R'$ is also at least $c$ and so the total cost of this path is at least $2 \times c$.

The disadvantage of this algorithm is the increased time complexity. Basically, whenever we process a path from the $MCSTpq$ we recompute the SPTF around that path. This means that in the worst case the algorithm has time complexity $O(kn^2 \log n)$ where $k$ is the number of terminals in the hyperedge.

## 5 Evaluation

We have implemented all algorithms in the open source `libavoid` connector routing library. These features can also be used interactively from within the Dunnart diagram editor.[4] We have used the orthogonal hyperedge routing algorithms to find routes for a variety of diagrams.

To investigate performance of the algorithms we ran the following experiment on a 2008 MacBook Pro with a 2.53 GHz Intel Core 2 Duo processor and 4GB of memory. Our C++ `libavoid` implementation was compiled using `gcc` 4.2.1 with `-O3`. The experiment used a small representative example from our commercial partner as well as some larger, randomly generated examples. We measured the time for each stage of the routing. The results are shown in Table 1. Note the

[4] `http://www.dunnart.org/`

visibility graph construction and the improvement is performed only once for each diagram, whereas the times for the full rerouting are average times per hyperedge.

We found that routing hyperedges of 25–50 terminals in a small diagram of up to 200 hundred nodes can be performed in a fraction of a second. In the largest example of 400 nodes, full rerouting for 10 hyperedges (each with 25 terminals) with heuristic 1 took 1.7 seconds, or just under 3 seconds using heuristic 2. Of course this would be considerably less if the user was requesting rerouting of just a single hyperedge. In general, the interleaved heuristic was approximately 3 times slower than the sequential approach, but resulted in much better hyperedge routes. The interleaved heuristic also lead to less improvement work being necessary, though with insignificant gains. Adding additional hyperedges to any example requires the cost of running the heuristic approach when a user require rerouting be performed, as well as a small time increase for the local improvement stage.

## 6 Conclusion

We have given a practical approach to support hyperedge routing in a diagramming tool designed for electrical circuit design. It produces high-quality routings and is fast enough to be used for interactive diagramming. Our interaction model supports hyperedge creation and semi-automatic routing to improve routes after changes in the diagram, as well as fully-automatic routing, when changes in the diagram suggest the topology of the hyperedge should change. We give two heuristics for fully-automatic routing, one very similar to standard VLSI approaches, and a novel interleaving approach that better captures the cost of resulting hyperedge. The interleaved heuristic creates markedly better routes at about three times the runtime cost of the sequential approach.

**Acknowledgments.** NICTA is funded by the Australian Government as represented by the Department of Broadband, Communications and the Digital Economy and the Australian Research Council. We acknowledge the support of the ARC through Discovery Project Grant DP0987168 and DP110101390.

## References

1. Ajwani, G., Chu, C., Mak, W.K.: FOARS: FLUTE based obstacle-avoiding rectilinear steiner tree construction. IEEE Trans. on Computer-Aided Design of Integrated Circuits and Systems 30(2), 194–204 (2011)
2. Garey, M.R., Johnson, D.S.: Computers and Intractability; A Guide to the Theory of NP-Completeness. W. H. Freeman & Co., New York (1990)
3. Holten, D.: Hierarchical edge bundles: Visualization of adjacency relations in hierarchical data. IEEE Transactions on Visualization and Computer Graphics 12, 741–748 (2006)
4. Holten, D., van Wijk, J.J.: Force-directed edge bundling for graph visualization. Comput. Graph. Forum 28(3), 983–990 (2009)

5. Hwang, F.K., Richards, D.S., Winter, P.: The Steiner Tree Problem. Annals of Discrete Mathematics (1992)
6. Lin, C.W., Chen, S.Y., Li, C.F., Chang, Y.W., Yang, C.L.: Efficient obstacle-avoiding rectilinear steiner tree construction. In: Proc. of the 2007 Int. Symp. on Physical Design, ISPD 2007, pp. 127–134. ACM, New York (2007)
7. Long, J., Zhou, H., Memik, S.O.: EBOARST: An efficient edge-based obstacle-avoiding rectilinear steiner tree construction algorithm. IEEE Trans. on Computer-Aided Design of Integrated Circuits and Systems 27(12), 2169–2182 (2008)
8. Pupyrev, S., Nachmanson, L., Kaufmann, M.: Improving Layered Graph Layouts with Edge Bundling. In: Brandes, U., Cornelsen, S. (eds.) GD 2010. LNCS, vol. 6502, pp. 329–340. Springer, Heidelberg (2011)
9. Wybrow, M., Marriott, K., Stuckey, P.J.: Incremental Connector Routing. In: Healy, P., Nikolov, N.S. (eds.) GD 2005. LNCS, vol. 3843, pp. 446–457. Springer, Heidelberg (2006)
10. Wybrow, M., Marriott, K., Stuckey, P.J.: Orthogonal Connector Routing. In: Eppstein, D., Gansner, E.R. (eds.) GD 2009. LNCS, vol. 5849, pp. 219–231. Springer, Heidelberg (2010)

# Improved Layout for Data Flow Diagrams with Port Constraints

Lars Kristian Klauske[1], Christoph Daniel Schulze[2], Miro Spönemann[2], and Reinhard von Hanxleden[2]

[1] Daimler Center for Automotive Information Technology Innovations, Berlin
lars.klauske@dcaiti.com

[2] Real-Time and Embedded Systems Group, Christian-Albrechts-Universität zu Kiel
{cds,msp,rvh}@informatik.uni-kiel.de

**Abstract.** The automatic generation of graphical views for data flow models and the efficient development of such models require layout algorithms that are able to handle their specific requirements. Examples include constraints on the placement of ports as well as the proper handling of nested models. We present an algorithm for laying out data flow diagrams that improves earlier approaches by reducing the number of edge crossings and bend points. We validate the quality of our algorithm with a range of models drawn from Ptolemy, a popular modeling tool for the design of embedded systems.

## 1 Introduction

With up to ten million lines of code, software-based functions account for 50–70% of the effort in the development of automotive electronic control units [2,24]. To keep up with the growing complexity and tightening time-to-market requirements, embedded software domains such as the automotive, rail or aerospace industry increasingly take advantage of graphical model-based development tools that follow the *actor-oriented* approach of data flow models [10] such as Simulink (The MathWorks, Inc.), SCADE (Esterel Technologies), ASCET (ETAS), or Ptolemy (UC Berkeley). Herein, graphical diagrams are used as input representations for simulators, rapid prototyping systems, and code generators. Fig. 1 shows a typical data flow diagram from Simulink and reveals the basic components of such a diagram, namely *actors* (also called *blocks* or *operators*), connections between the actors, and *ports* specifying the interface of actors and the kind of data that is transported by connections.

While it is generally assumed that graphical diagrams are more readable than textual programs, their readability strongly depends on the diagrams' layout. Therefore, when creating or changing a model, an estimated 30% of a user's time is spent on manual layout adjustments according to Klauske and Dziobek [8]. Additionally, interactive applications employing methods such as automatic model generation and transformation gain importance, requiring diagram layouts to be generated from scratch. Both of these problems imply the need for an adequate automatic view generation using methods of graph layout.

P. Cox, B. Plimmer, and P. Rodgers (Eds.): Diagrams 2012, LNAI 7352, pp. 65–79, 2012.

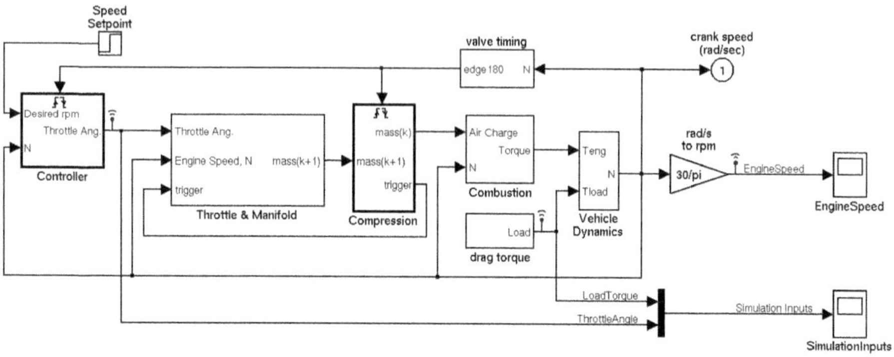


**Fig. 1.** A Simulink model for engine control (example by The MathWorks, Inc.)

*Contributions.* In this paper, we address the problem of automatic layout of data flow diagrams. While the difficulties of port constraints and hyperedges in crossing reduction and edge routing, as well as some basic solutions, have already been introduced [19,9], we show how to further reduce the number of edge crossings and bend points through the creation and handling of additional dummy nodes and extensions of the crossing minimization phase. We also describe an improved method to handle the layout of nested diagrams.

*Outline.* We begin by introducing two example applications in Sect. 2, give an overview of related work in this area in Sect. 3, and continue by defining the necessary mathematical notation in Sect. 4. In Sect. 5 we provide the description of our algorithm, followed by the evaluation and its results in Sect. 6. Finally we conclude in Sect. 7.

## 2 Data Flow Models

For a closer look at the application domains of our graph drawing method, we present two exemplary modeling tools: Simulink and Ptolemy.

### 2.1 Simulink

Simulink is a graphical modeling language based on data flow diagrams with optional Statechart diagrams encapsulated inside data flow diagram nodes. It is of widespread use for embedded software development in the automotive domain. Its models are used for specification, simulation, rapid prototyping, and production code generation. Using real-world Simulink models of automotive body control modules with a total of 40,000 nodes and 50,000 edges as our reference, the average Simulink diagram has about 20 nodes and 30 edges, with 90% of all models counting 60 edges or less. While large diagrams of more than 100 nodes do exist (about 1% of our reference diagrams), they usually follow a very simple structure with few or no potential edge crossings and only a couple of layers.

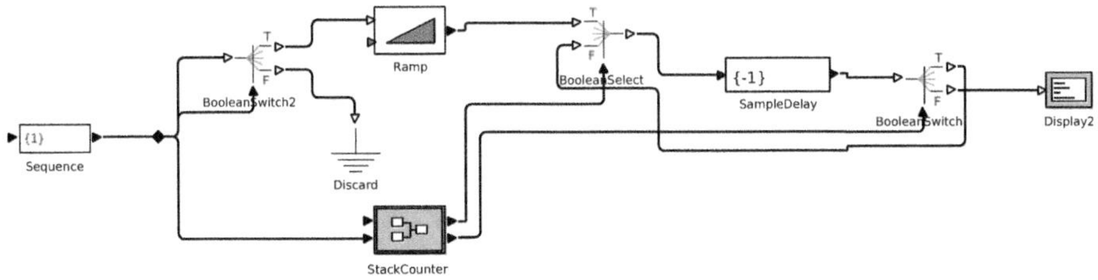


(a) Layout using a previous approach [19] (3 edge crossings, 30 edge bends)

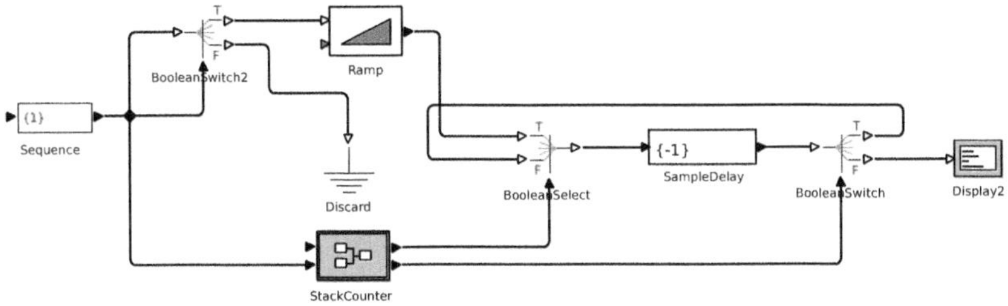


(b) Layout using the method presented here (1 edge crossing, 14 edge bends)

**Fig. 2.** A Ptolemy model representing a stack (example by Edward A. Lee)

The algorithm presented in this paper can directly be applied to Simulink diagrams: Simulink edges can be taken as hyperedges with one source and one or more targets. Ports are arranged on the rectangular node borders, with all output ports on one side, most input ports on the opposite side, and up to three input ports on the remaining sides (see Fig. 1 for a typical example).

The port side and order in Simulink diagrams is always fixed, with port positions that depend roughly linearly on the node size. For the scope of this paper, we assume Simulink node sizes to be fixed, which results in fixed port positions. Simulink diagrams with variable node sizes can be processed using an LP-based edge straightening method [8,9].

## 2.2 Ptolemy

Ptolemy[1] is an open source modeling environment developed at UC Berkeley that targets the modeling and semantics of concurrent real-time systems [4]. Ptolemy models are *actor-oriented* data flow diagrams and can contain nested state machines (*modal models*). A Ptolemy data flow model of the process networks domain is shown in Fig. 2.

The layout algorithm presented in this paper has been integrated in the modeling environment of Ptolemy and is now part of its official distribution. Thereby automatic layout can be used as an aid for the creation of Ptolemy models and for the visualization of generated or transformed models.

[1] `http://ptolemy.eecs.berkeley.edu/`

## 3 Related Work

The foundations for the layout of directed graphs were laid by Sugiyama et al. [21], who introduced the *layered* (a.k.a. *hierarchical*) approach for graph drawing. The basic idea is to organize the nodes in subsequent *layers* such that edges point from layers of lower index to those of higher index. This kind of ordering helps to emphasize the direction of flow, which is quite natural for data flow diagrams. Afterwards the nodes of each layer are reordered so as to minimize the number of edge crossings. This is followed by the calculation of suitable coordinates for node positions, and optionally by an edge routing phase.

The first contributions to the problem of integrating port constraints in the layered approach were motivated by the layout of data structures, where certain fields of a structure may contain pointers to other structures. Gansner et al. showed how node positioning can be extended for including offsets derived from port positions [6]. Sander introduced the idea of handling side ports by adding dummy nodes in order to route the respective edges [14]. The problem of crossing minimization with port constraints was first discussed by Waddle, who adapted the standard node ordering heuristic to consider port positions [22]. These contributions employ FIXEDPOS constraints with spline curve edge routing, but they do not support inverted ports or other port constraints and are not sufficient for the layout of data flow diagrams.

Schreiber proposed different solutions in the context of drawing bio-chemical networks [17]. The crossing minimization phase is adapted by inserting dummy nodes for each port and adding constraints to respect the order of ports. Side ports are handled by routing the incident edges locally for each node, which is done through transformation into a two-layer crossing minimization problem. This suffices for treating FIXEDPOS constraints, but can lead to unpleasant layouts, since the number of resulting bend points is possibly higher than necessary. This can be seen in Fig. 3(a), where the incoming edge at the SOUTH side port of node d has two additional bend points. The approach of Siebenhaller suffers from the same problem, because it also routes edges of side ports locally [18]. However, it supports more flexible port constraints, since constraints are associated with individual edges instead of nodes. The consequence is that a node may have some edges that are constrained to ports, and some that are not. The crossing minimization problem that results from this additional degree of freedom can be solved by reducing it to a network flow problem. This flexibility can be useful for the layout of UML diagrams, where it is possible that only a subset of the edges is connected to fixed points of a node, but data flow diagrams usually do not require such mixed constraints.

A previous approach [19] for layout of data flow diagrams applied local routing not only to the side ports of a node, but also to inverted ports, which leads to layouts such as the one shown in Fig. 3(a). Although it simplifies the crossing minimization phase, the obvious drawback is that it cannot take into account the global structure of the graph, and thus leads to an unnecessarily high number of edge crossings and bend points. In this paper we therefore employ a different approach based on dummy nodes, which is able to route the feedback edge (d, c) in

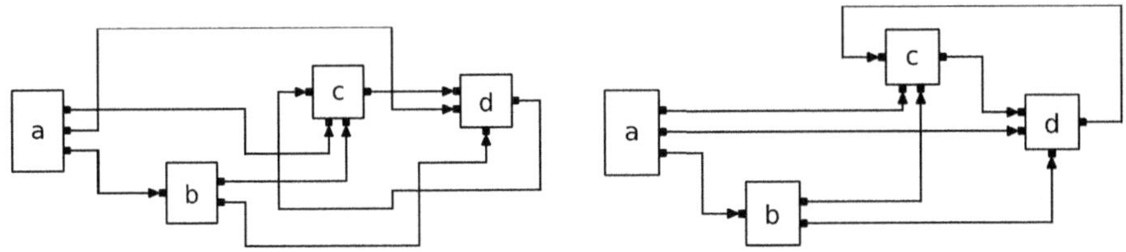


(a) Previous approach [19]: edges at side ports and inverted ports routed locally (5 edge crossings)

(b) Our proposal: edges at side ports and inverted ports routed using dummy nodes (1 edge crossing)

**Fig. 3.** Two alternatives for handling port constraints

Fig. 3(b) with four bend points and without any crossings, as opposed to six bend points and three crossings for the previous approach. Other previous contributions adapt the barycenter heuristic for handling different port constraints [19], and add a specialized node placement for FIXEDRATIO constraints [8,9]. This type of constraint allows changing the size of nodes in order to minimize the number of bend points of incident edges [9].

Orlarey et al. generate data flow diagrams out of textual specifications [12]. Instead of generating a graph and applying a graph layout algorithm to it, they derive the layout directly from an algebraic representation [11]. The compositional nature of this representation implies a geometric node ordering: sequential composition leads to horizontal order, and parallel composition leads to vertical order. Since our contribution is based on graph representations, we will not go into further details on the algebraic approach.

## 4 Definitions

A *directed port-based graph* consists of a finite set of *nodes* $V$ and *ports* $P$, a set of *edges* $E \subseteq P \times P$ connecting the ports, and a function $\mathsf{n} : P \to V$ that maps ports to their nodes. An edge $e = (p_1, p_2) \in E$ is an *outgoing edge* of $p_1$ and $v_1$ and an *incoming edge* of $p_2$ and $v_2$ if $v_1 = \mathsf{n}(p_1)$ and $v_2 = \mathsf{n}(p_2)$; $e$ is said to be *incident* to $p_1$, $p_2$, $v_1$, and $v_2$. We call $p_1$ and $v_1$ the *source* of $e$, while $p_2$ and $v_2$ are called its *target*.

A *layering* of a graph is a partition $\mathcal{L} = (L_1, \ldots, L_k)$ of the nodes into *layers* $L_1, \ldots, L_k$ such that for all edges $e$ with source node $v_1 \in L_i$ and target node $v_2 \in L_j$ we have $i < j$. If we have $i \leq j$, $\mathcal{L}$ is called a *weak layering*, and an edge connecting two nodes in the same layer is called an *in-layer edge*. If $i < j - 1$ then $e$ is called a *long edge* and is split into a sequence of edges that span only consecutive layers by adding *edge dummy nodes*. Each layer has a specific ordering of its nodes which can be altered by the algorithm. The current index of node $v$ in this ordering is written as $\mathsf{idx}(v)$.

Usually ports are drawn on the border of their respective nodes. The function $\mathsf{side} : P \to \{\text{NORTH}, \text{SOUTH}, \text{WEST}, \text{EAST}\}$ assigns ports to one of the node's sides. Ports with only incoming edges are called *input ports* and are usually placed on the West side. Ports with only outgoing edges are called *output ports* and are usually placed on the East side. Input ports that are placed on the East side and output ports that are placed on the West side are called *inverted ports*. Ports on the North or South side are called *side ports*.

Port constraints control how much influence a layout algorithm has over the positioning of the ports of a node. The function $\mathsf{cons} : V \to PC$ maps nodes to their port constraints, with $PC$ containing the available port constraints. They are, in increasing order of strictness:

FREE Ports may be drawn at arbitrary positions on the border of a node.
FIXEDSIDES The side is prescribed for each port, but the order of ports is free on each side.
FIXEDORDER The side is fixed for each port, and the order of ports is fixed for each side.
FIXEDRATIO The side is fixed for each port, and the ratio between the port's position on the side and the side's length is fixed.
FIXEDPOS The exact position is fixed for each port.

Due to constraints from the application domains, we assume the graph to be drawn such that the prevalent direction of edges is from left to right. Hence the nodes of a layer $L_i$ are placed on a vertical line, $L_{i+1}$ is drawn right of $L_i$, and nodes within layers are indexed from top to bottom. Other publications (e. g. Sugiyama et al. [21]) and applications (e. g., UML diagrams) assume a top-down drawing, but our definitions and approaches can be applied symmetrically.

## 5 The KLay Algorithm

The KLay Layered algorithm is part of the *Kiel Integrated Environment for Layout Eclipse RichClient* (KIELER)[2] project, a test bed for layout algorithms and modeling pragmatics. The algorithm expects a set of nodes and edges as its input and computes coordinates and bend points to arrive at a layout. It is structurally based upon the layered approach by Sugiyama et al., being divided into several phases as follows:

1. KLay Layered does not assume the input graph to be acyclic, which requires the first phase to break possible cycles. This is done using the feedback arc set algorithm proposed by Eades et al. [3]. The goal is to have the vast majority of edges point in the same direction, so as to make the flow of data as obvious as possible.
2. As in Sugiyama's approach, a layer assignment phase computes a valid layering for the graph. Long edges are split into segments such that edges only connect nodes in neighboring layers. KLay Layered provides an implementation of the network simplex layering algorithm by Gansner et al., which minimizes the length of edges [6].

[2] `http://www.informatik.uni-kiel.de/rtsys/kieler/`

3. The order of nodes in a layer determines the number of edge crossings. Solving this problem is NP-complete even for two layers [7], making the use of heuristics necessary. A popular heuristic is the barycenter approach, which works with two layers of which one is fixed. Its nodes are assigned rank values reflecting their order in the layer. For the free layer's nodes, rank values are computed based on the ranks of their fixed-layer neighbors. The nodes are then sorted by their computed ranks to arrive at an ordering for the free layer [21].
   KLAY Layered performs forward and backward sweeps through the layers, each time randomizing the order of the sweep's first fixed layer. After a predefined number of sweeps, the result with the least number of edge crossings is chosen.
4. The node placement phase determines the position of nodes inside each layer, making sure not to change the ordering determined by the crossing minimization phase. KLAY Layered uses Sander's method for node placement, which partitions the graph's nodes into *linear segments* whose elements are to be kept on a straight line [15]. Typically, edge dummy nodes inserted to divide long edges form a linear segment to keep long edges free of bend points.
5. While Sugiyama's approach uses straight lines to connect nodes, KLAY Layered routes edges orthogonally, with bend points set in a way that each segment of an edge runs either horizontally or vertically. This requires the addition of a final edge routing phase, which is based on Sander's hyperedge routing algorithm [16].

Having already split the algorith into five distinct phases, it is only a small step to allow the concrete implementations to be exchanged at runtime. This way, the algorithm can also be used for different applications. For instance, our main implementation of the edge routing phase routes edges orthogonally, which is the expected method for data flow diagrams. For other types of diagrams, however, edges may be preferred to simply be straight lines, which is supported by another—in that case rather trivial—implementation of the edge routing phase.

We introduce an additional level of modularity by adding *intermediate processing phases* before, between, and after the five main phases. During these intermediate phases, additional modules can be executed that simplify the main phases by factoring out shared functionality, or by reducing complex layout problems to simpler ones that can be handled by the five main phases. Which modules are executed depends on the graph's features: if there are no inverted ports, no corresponding modules need to be executed.

The remainder of this section describes our methods of handling northern and southern ports, inverted ports, and hierarchical ports.

### 5.1 Side Ports

The usual case for data flow diagrams is for a node to have its input ports on the WEST side and its output ports on the EAST side. The situation becomes

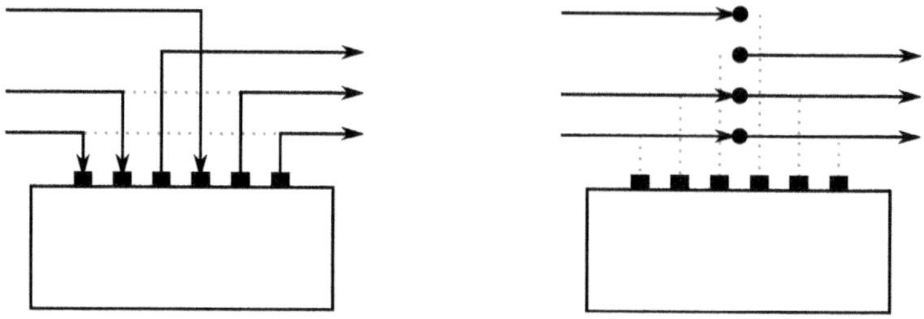

(a) A node with side ports and with edges drawn as KLAY Layered would. Note that the local number of crossings is minimal.

(b) Inserted bend dummy nodes. (solid circles) Edge dummy nodes can be inserted among them without problems, but the relative order of bend dummy nodes must be fixed to avoid ambiguity.

**Fig. 4.** Side ports and how bend dummy nodes are created to handle them

more complicated, however, once nodes are allowed to have ports on the NORTH or SOUTH side, also called *side ports*. Note that this can only happen with port constraints set to FIXEDSIDES or higher. Previous methods transform such nodes to the usual case, either by doing a node-local routing first [17,19], or by adding a dummy node for the northern and for the southern side which encapsulates the necessary edge routing [14]. All of these approaches suffer from the problem that long edges must be routed around edges connected to side ports, introducing unnecessary bend points or crossings, as can be seen in Fig. 3.

Our method resembles the latter approach, but solves its limitations by allowing more than one dummy node to be created for each side: if the side has $x$ ports, we create between $\lceil x/2 \rceil$ and $x$ *bend dummy nodes* for those ports as shown in Fig. 4. These dummy nodes are created just prior to crossing minimization, and are removed after the last phase, inserting bend points at their position.

This method allows edge dummy nodes to be placed between the bend dummy nodes, which was not previously possible. However, it puts two constraints on the result of the crossing minimization phase: First, generated bend dummy nodes must retain their order to avoid ambiguity due to overlapping edges. And second, the bend dummy nodes generated for different nodes must not be interleaved.

To satisfy these constraints, we add appropriate *successor constraints* on the bend dummy nodes and remember which node they were created for. A successor constraint is a tuple $(v_1, v_2) \in V \times V$ which requires $v_1$ to be placed above $v_2$ in a layer. Once an initial order is computed, violated constraints are resolved through a method proposed by Forster [5].

Placing an edge dummy node between two bend dummy nodes causes edge crossings usually not counted by the crossing minimization algorithm, which may lead to inferior results. However, these crossings can be easily counted with a time complexity linear to the number of nodes in a layer.

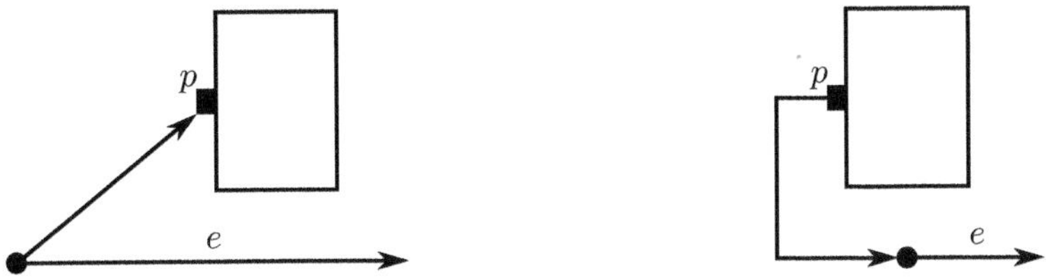


(a) A dummy node placed in the previous layer. The edge $e$ needs to be reconnected and reversed appropriately, and a new edge connects the dummy node with the original target of $e$.

(b) A dummy node placed in the same layer. The edge $e$ is not reversed, but reconnected to the dummy node. A new edge connects the dummy node with its original target.

**Fig. 5.** An inverted port and two approaches for handling it

One limitation of this method is that the way bend dummy nodes are created is designed to minimize edge crossings locally. Future research could go into finding methods to also take surrounding layers into account.

### 5.2 Inverted Ports

With port constraints set to at least FIXEDSIDES, inverted ports may appear in a diagram. Edges connected to inverted ports need to be routed around the port's node to avoid overlapping. There are two basic previous approaches to handle this situation, both based on turning inverted ports into regular ones. The first does so by applying node-local edge routing, as described in Sect. 5.1 [19]. The second approach handles inverted ports through the addition of a dummy node [8,9]. Take $p$ to be an inverted port on the WEST side with an outgoing edge $e$ (Fig. 5(a)). Then a dummy node is added to the preceding layer, and the source of $e$ is changed to the new dummy node. Finally, the dummy node is connected to $p$.

While the problems of the former approach have already been discussed, the latter approach works reasonably well. However, additional work is required to make sure that the dummy node does not take up space in its layer that could well be used by other nodes. In particular, the inserted dummy node needs additional handling when it is later removed, adding complexity.

KLAY Layered therefore uses a different approach, illustrated in Fig. 5(b). After the layer assignment phase, edge dummy nodes are added for edges connected to inverted ports similar to the second approach. The differences are that the dummy node is placed in the same layer, and that the dummy node does not only have outgoing edges. One advantage of this approach is that the inserted dummy node can be treated just like a regular edge dummy node inserted to break long edges.

This of course comes at the cost of turning the layering into a weak layering by the addition of in-layer edges, which has consequences for the crossing minimization phase. For barycenter-based algorithms, it is not immediately clear what to do with in-layer edges. A problem arises when a barycenter value is to be calculated for a node $n_1$ which is connected to another node $n_2$ in the same layer, since $n_2$ does not have a rank value assigned. We solve this by pretending edges incident to $n_2$ to also be incident to $n_1$, and thereby effectively treating $n_2$ as not being there at all. This has the positive effect of making $n_1$ and $n_2$ be closer together, thereby reducing the length and the possibility of crossings due to the in-layer edge connecting them.

This approach also has consequences for cross counting. Usually, cross counting algorithms only count crossings between two layers, not in the same layer. However, a worst-case estimate for crossings caused by in-layer edges can be easily computed in time linear to the number of ports in a layer. First, all ports with incident edges are numbered from top to bottom. Then, for each in-layer edge $e$, we calculate the difference of the numbers of the ports it connects. We get the maximum number of ports between them whose incident edges will cause crossings with $e$.

### 5.3 Hierarchical Ports

In order to control the complexity of large systems, data flow models are broken into hierarchically structured levels using *composite actors*, which are also called *submodules*. Although the nested content of a composite actor is usually displayed in a new window, it is also possible to draw it directly inside the composite actor's bounding box in the containing diagram by enlarging the bounding box accordingly. This leads to a *compound graph* structure [20], where composite actors are represented by *compound nodes*. This kind of visualization allows to directly connect the ports of a composite actor with its content, thus emphasizing the flow of data across hierarchy levels. We call such ports with connections to the inside as well as the outside *hierarchical ports*. Our approach regards each compound node as a separate diagram to be laid out. The hierarchy tree is traversed bottom-up, applying the layout algorithm to the deeper hierarchy levels prior to the containing ones. This method requires the layout algorithm to determine positions for hierarchical ports.

In a previous approach [19], hierarchical ports on the North or South side were handled by routing their incident edges around a diagram's nodes to dummy nodes inserted into the first layer. This produced long edges and a cluttered diagram, two problems that our new approach solves by eliminating the need to route edges around the diagram.

With port constraints set to Free, this is staightforward. Dummy nodes are added to the graph and placed in the first or last layer, depending on how many outgoing and incoming edges they have. In the end, the position of hierarchical ports can be directly inferred from where the algorithm placed their dummy nodes.

With port constraints set to FixedSides or higher, the hierarchical equivalents of side ports and inverted ports can appear. Treating hierarchical ports assigned to the West or East side is similar to the Free case. However, for hierarchical ports assigned to the North or South side, some more work is re-

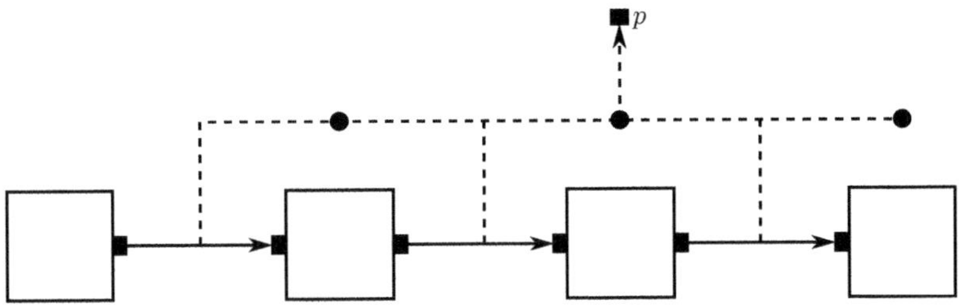


**Fig. 6.** Inserted dummy nodes to handle hierarchical ports on the NORTH side. Solid circles are inserted dummy nodes the regular nodes connect to. The dashed line indicates how our algorithm routes the edges to the hierarchical port $p$.

quired. In these cases, KLAY Layered creates dummy nodes for nodes connected to hierarchical ports to connect to instead, as shown in Fig. 6. These dummy nodes are placed above or below all other nodes inside a layer, depending on whether they belong to a NORTH or SOUTH port. In a separate edge routing phase, these dummy nodes are connected to another dummy node representing the hierarchical port itself, the edges between them routed with the orthogonal edge routing algorithm also used for normal edge routing. The position of the hierarchical ports is derived from the position of their dummy node, calculated using a force-based approach that takes the position of connected nodes into account.

With port constraints set to at least FIXEDORDER, this calculation of dummy node positions can lead to invalid results. In these cases, the positions are corrected to adhere to the given hierarchical port order.

In the FIXEDRATIO and FIXEDPOS cases, the position of the hierarchical ports is explicitly prescribed.

One shortcoming of this approach is that our treatment of hierarchical ports does not take external connections into account. Thus, hierarchical ports can be placed in a way that works well within an actor, but leads to unnecessary crossings in the upper hierarchy levels. Ongoing research within our group aims to solve this problem.

## 6 Evaluation

The quality of layouts is usually measured using a selection of *aesthetics criteria*, of which the number of edge crossings and the number of bend points rank among the most important according to Purchase et al. [13,23]. We evaluated the KLAY Layered algorithm against its predecessor, the KLoDD (*KIELER Layout of Dataflow Diagrams*) algorithm [19], comparing the number of produced crossings and bend points. Since both are meant to be used in interactive applications with users actively waiting for a layout to be generated, we also compared their runtime performance.

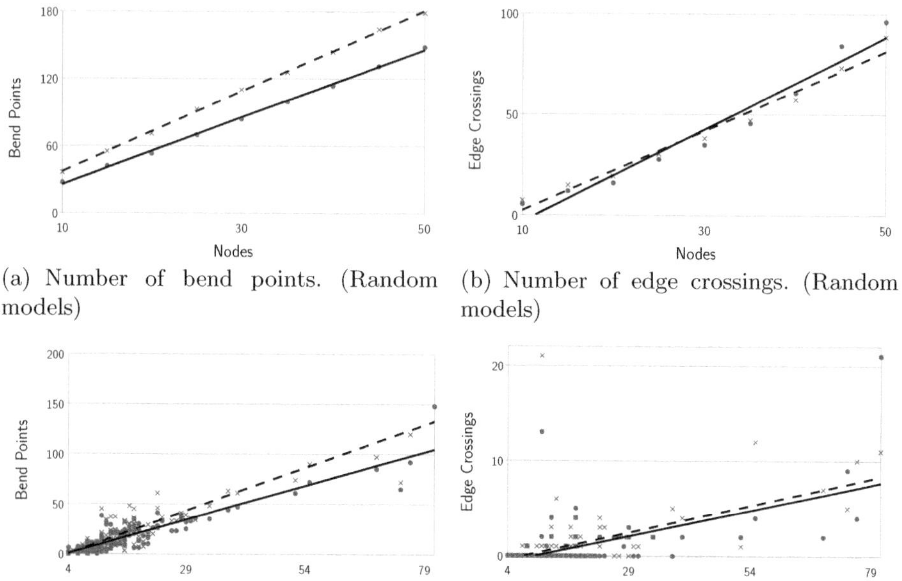


(a) Number of bend points. (Random models)

(b) Number of edge crossings. (Random models)

(c) Number of bend points. (Ptolemy models)

(d) Number of edge crossings. (Ptolemy models)

**Fig. 7.** The number of bend points and the number of crossings produced by the KLAY Layered algorithm presented here (solid lines and circles) and the KLoDD algorithm, which follows a previous approach [19] (dashed lines and crosses), applied to our set of random graphs (a, b) and to our selection of Ptolemy models (c, d)

For a visual impression, Fig. 2(b) shows a drawing created with KLAY, while Fig. 2(a) shows a drawing of the same model created with KLoDD.

We applied the algorithms to two sets of diagrams in order to evaluate the layout quality. The first set consisted of 270 random graphs with 10 to 50 nodes each and an average of 1.2 outgoing edges per node, which is roughly what we find in real-world data flow diagrams. Port sides were chosen randomly: input ports would usually be placed on the WEST side and output ports on the EAST side, with a probability of 0.05 of this being the other way round, and with a probability of 0.2 of a port being placed on the NORTH or SOUTH side. For the second set, we wanted to focus on real-world diagrams. Therefore we used a selection of 141 models taken from the demonstration model repository of the Ptolemy II tool developed at UC Berkeley and imported them into KIELER. Contrary to the set of random graphs, the graph structure of most Ptolemy models was hierarchical, with each compound node averaging 8.98 child nodes, up to a maximum of 43 child nodes.

During the development of KLAY Layered, we placed some emphasis on reducing the number of bend points and thus expected it to be lower compared to KLoDD. Due to improved crossing minimization we also expected the num-

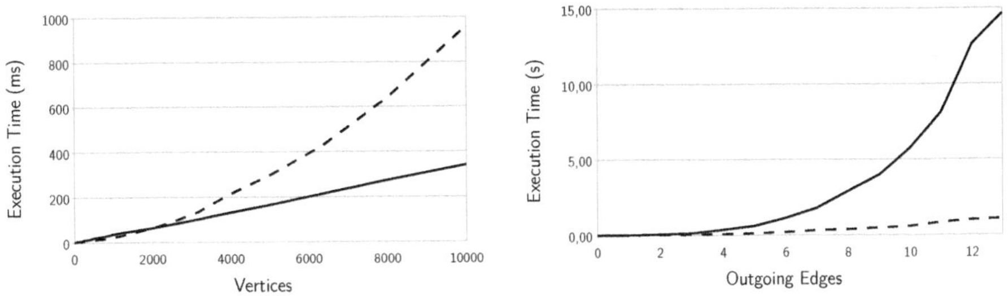


(a) Performance relative to the number of nodes in the graph.

(b) Performance relative to the number of outgoing edges per node.

**Fig. 8.** The runtime performance of KLAY Layered algorithm (solid line) and the KLoDD algorithm (dashed line), plotted against the number of nodes (a) and against the number of outgoing edges per node (b)

ber of crossings to be slightly lower. The results of our quality evaluation are shown in Fig. 7. Indeed they indicate that the number of bend points produced by KLAY Layered is almost consistently lower compared to KLoDD. Regarding the number of crossings, the algorithms average fairly similar results, with KLAY Layered having a slight advantage for smaller diagrams.

For the performance evaluation we used randomly generated diagrams with nearly the same characteristics as the ones already described. Since we wanted to measure the reaction of the algorithms to both changes in the number of nodes and changes in the number of outgoing edges per node, we used two sets of random diagrams. For the first set, we kept the number of outgoing edges per node between 0 and 2, generating graphs with between 10 and 10,000 nodes. The second set was fixed at 100 nodes, with the number of outgoing edges varying between 0 and 15.

As for the results, we expected KLAY Layered to be considerably slower than KLoDD due to its more complex architecture. We were surprised to see that this is not the case, as can be seen in Fig. 8. In fact, for large diagrams, KLAY Layered shows a linear correlation with the number of nodes. It does not react quite as well to the number of outgoing edges per node, however. This is very likely due to its extensive use of dummy nodes, which KLoDD uses more conservatively.

All in all, KLAY Layered performs very well with diagrams from our application domain and is well suited to be used in interactive applications.

## 7 Conclusion

We presented new approaches for handling port constraints as they often appear in data flow diagrams of actor-oriented modeling languages such as Simulink or Ptolemy. These approaches involve the creation and special treatment of dummy nodes. To that end, we introduced enhancements to the crossing minimization phase of the layer-based graph layout method. Compared to previous approaches, our contributions result in significantly lower numbers of bend points and crossings for realistically sized diagrams. However, there is still room for improvements, which we leave for future work:

- The layer-sweep crossing minimization approach requires a method for counting the number of crossings in order to find an appropriate terminating condition. While there exist efficient counting methods for plain graphs [1], these are inaccurate when hyperedges are involved, because their actual number of crossings is determined later in the edge routing phase [16].
- We currently treat hierarchical diagrams by recursively applying the layout algorithm to each hierarchy level, starting with the innermost ones. This procedure is not optimal when the ports of a compound node are rearranged, since the algorithm processing the content of that node does not take into account its external connections.
- *Relation nodes* are hypernodes that are used in Ptolemy to connect an arbitrary number of actors. Treating these nodes in the same manner as data flow actors leads to unsatisfying results, and it is not clear what an optimal solution would look like.

## References

1. Barth, W., Jünger, M., Mutzel, P.: Simple and Efficient Bilayer Cross Counting. In: Goodrich, M.T., Kobourov, S.G. (eds.) GD 2002. LNCS, vol. 2528, pp. 130–141. Springer, Heidelberg (2002), `http://dx.doi.org/10.1007/3-540-36151-0_13`
2. Broy, M.: Challenges in automotive software engineering. In: ICSE 2006: Proceedings of the 28th International Conference on Software Engineering, pp. 33–42 (2006)
3. Eades, P., Lin, X., Smyth, W.F.: A fast and effective heuristic for the feedback arc set problem. Information Processing Letters 47(6), 319–323 (1993)
4. Eker, J., Janneck, J.W., Lee, E.A., Liu, J., Liu, X., Ludvig, J., Neuendorffer, S., Sachs, S., Xiong, Y.: Taming heterogeneity—the Ptolemy approach. Proceedings of the IEEE 91(1), 127–144 (2003)
5. Forster, M.: A Fast and Simple Heuristic for Constrained Two-Level Crossing Reduction. In: Pach, J. (ed.) GD 2004. LNCS, vol. 3383, pp. 206–216. Springer, Heidelberg (2005), `http://dx.doi.org/10.1007/978-3-540-31843-9_22`
6. Gansner, E.R., Koutsofios, E., North, S.C., Vo, K.P.: A technique for drawing directed graphs. Software Engineering 19(3), 214–230 (1993)
7. Garey, M.R., Johnson, D.S.: Crossing number is NP-complete. SIAM Journal on Algebraic and Discrete Methods 4(3), 312–316 (1983), `http://link.aip.org/link/?SML/4/312/1`
8. Klauske, L.K., Dziobek, C.: Improving modeling usability: Automated layout generation for Simulink. In: Proceedings of the MathWorks Automotive Conference, MAC 2010 (2010)
9. Klauske, L.K., Dziobek, C.: Effizientes Erstellen von Simulink Modellen mit Hilfe eines spezifisch angepassten Layoutalgorithmus. In: Tagungsband Dagstuhl-Workshop MBEES: Modellbasierte Entwicklung eingebetteter Systeme VII, pp. 115–126 (2011), `http://www.in.tu-clausthal.de/abteilungen/gi/Forschung/MBEES2011/`

10. Lee, E.A., Neuendorffer, S., Wirthlin, M.J.: Actor-oriented design of embedded hardware and software systems. Journal of Circuits, Systems, and Computers (JCSC) 12(3), 231–260 (2003)
11. Orlarey, Y., Fober, D., Letz, S.: An algebraic approach to block diagram constructions. In: Actes des Journèes d'Informatique Musicale (JIM 2002), pp. 151–158. GMEM, Marseille (2002)
12. Orlarey, Y., Fober, D., Letz, S.: FAUST: an efficient functional approach to DSP programming. In: Assayag, G., Gerzso, A. (eds.) New Computational Paradigms for Computer Music. Editions Delatour, France (2009)
13. Purchase, H.C.: Which Aesthetic has the Greatest Effect on Human Understanding? In: DiBattista, G. (ed.) GD 1997. LNCS, vol. 1353, pp. 248–261. Springer, Heidelberg (1997)
14. Sander, G.: Graph layout through the VCG tool. Tech. Rep. A03/94, Universität des Saarlandes, FB 14 Informatik, 66041 Saarbrücken (October 1994)
15. Sander, G.: A Fast Heuristic for Hierarchical Manhattan Layout. In: Brandenburg, F.J. (ed.) GD 1995. LNCS, vol. 1027, pp. 447–458. Springer, Heidelberg (1996)
16. Sander, G.: Layout of Directed Hypergraphs with Orthogonal Hyperedges. In: Liotta, G. (ed.) GD 2003. LNCS, vol. 2912, pp. 381–386. Springer, Heidelberg (2004)
17. Schreiber, F.: Visualisierung biochemischer Reaktionsnetze. Ph.D. thesis, Universität Passau, Innstrasse 29, 94032 Passau (2001)
18. Siebenhaller, M.: Orthogonal Graph Drawing with Constraints: Algorithms and Applications. Ph.D. thesis, Universität Tübingen, Wilhelmstr. 32, 72074 Tübingen (2009)
19. Spönemann, M., Fuhrmann, H., von Hanxleden, R., Mutzel, P.: Port Constraints in Hierarchical Layout of Data Flow Diagrams. In: Eppstein, D., Gansner, E.R. (eds.) GD 2009. LNCS, vol. 5849, pp. 135–146. Springer, Heidelberg (2010)
20. Sugiyama, K., Misue, K.: Visualization of structural information: automatic drawing of compound digraphs. IEEE Transactions on Systems, Man and Cybernetics 21(4), 876–892 (1991)
21. Sugiyama, K., Tagawa, S., Toda, M.: Methods for visual understanding of hierarchical system structures. IEEE Transactions on Systems, Man and Cybernetics 11(2), 109–125 (1981)
22. Waddle, V.: Graph Layout for Displaying Data Structures. In: Marks, J. (ed.) GD 2000. LNCS, vol. 1984, pp. 241–252. Springer, Heidelberg (2001)
23. Ware, C., Purchase, H., Colpoys, L., McGill, M.: Cognitive measurements of graph aesthetics. Information Visualization 1(2), 103–110 (2002)
24. Wernicke, M.: AUTOSAR auf dem Weg in die Serie. Elektronik Praxis 02 (2008), `http://www.elektronikpraxis.vogel.de/themen/embeddedsoftwareengineering/analyseentwurf/articles/105576/`

# Aesthetic Layout of Wiring Diagrams

Christian Ernstbrunner and Josef Pichler

Software Competence Center Hagenberg,
Softwarepark 21, 4232 Hagenberg, Austria
{christian.ernstbrunner,josef.pichler}@scch.at

**Abstract.** A wiring diagram plays an important role in electrical machine design. The layout of a wiring diagram must facilitate a designer's understanding of the schematic as well as the real electrical machine. Even though wiring diagrams are undirected graphs, standard algorithms and libraries for graph drawing are not sufficient to achieve adequate diagrams that preserve the structure and further characteristics of the real machine. We argue that specialized algorithms are required to achieve adequate and aesthetic diagrams without compromising the characteristics of an electrical machine. In this paper, we describe a new algorithm for positioning diagram elements and a customized algorithm for connector routing for aesthetic wiring diagrams.

**Keywords:** Graph layout, connector routing, wiring diagram.

## 1 Introduction

A wiring diagram is a simplified pictorial representation of an electrical circuit that shows the interconnection of electrical elements such as resistors, inductors, and switches. For an electrical machine, a wiring diagram shows the electrical elements of the machine (e. g. a power transformer) such as windings, switches, terminals, and Y-connections.

A software tool that supports electrical engineers in specifying the wiring of an electrical machine must provide efficient creation and error diagnostics of electrical circuits. Because not all kind of errors may be detected automatically—some require interpretation of an expert—the software tool must display wiring diagrams in a way that electrical engineers identify faulty wirings by brief inspection. This requirement assumes the diagram to be aesthetically and clearly readable. The guidelines for aesthetic diagram drawing as identified by Bennett et al. [2] and readability investigated by Tamassia et al. [14] are mainly applicable to electrical diagrams as well. Even a wiring diagram is a pictorial representation of a real electrical machine, an electrical engineer must be able to close the gap between the real electrical machine and the schematic, pictorial representation. So the wiring diagram must be visualized in a notation that preserves knowledge of the actual electrical machine. This is usually achieved by means of several strategies. To discover symmetric wiring on different phases at a glance, for instance, electrical elements can be assigned to electrical phases. Furthermore, the order of winding elements in the wiring diagram uniquely defines

P. Cox, B. Plimmer, and P. Rodgers (Eds.): Diagrams 2012, LNAI 7352, pp. 80–94, 2012.

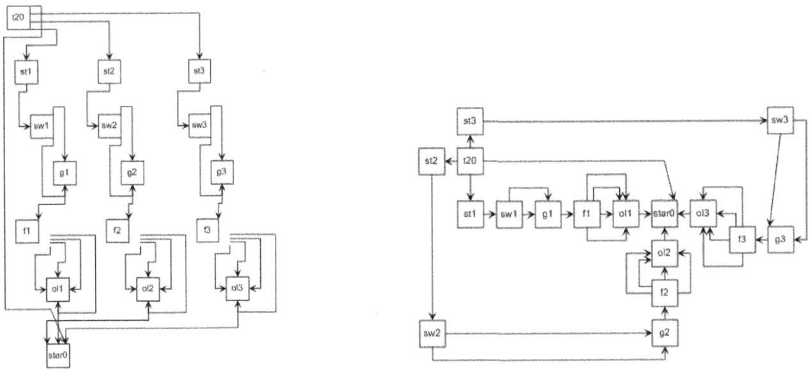


(a) (b)

**Fig. 1.** Wiring diagrams using the yFiles library with (a) swim-lanes and orthogonal routing and (b) with orthogonal layout

the order of windings in the real electrical machine. These knowledge preserving strategies are more important than other graph drawing aesthetics, which either must be adopted or cannot be implemented for electrical wiring diagrams.

Basically, wiring diagrams represent undirected graphs, with electrical elements as graph vertices and connectors as graph edges between vertices. Hence, layout of wiring diagrams can be considered as a graph layout problem. Today, a plethora of algorithms (e.g. [1], [13], [15], [3]) and software libraries (e.g. yFiles[1] and ILOG[2]) exist for this problem. The result quality of standard graph layout algorithms mostly depend on free positioning of vertices. In other words, edge placement heuristics ([2], [11]) such as *minimize the number of edge bends* or *minimize the number of edge crossings* are applied by repositioning graph vertices. Some libraries provide mechanisms to restrict positioning. For instance, the yFiles library provides either swim-lanes which could be used to restrict positioning of phase-specific diagram elements or sketch-driven graph drawing [3]. However, the result is far away from aesthetics expected by domain experts (see Fig. 1). In contrary, *constraint graph layout* used for dynamic graph layout is able to preserve the mental map (also called *layout stability*) of the user [7]. The new layout should not move an existing node unless the current position leads to poor layout [5]. For instance, Dwyer et al. propose an algorithm that preserves the topology of the initial layout based on force-directed style layout [5]. Wybrow et al. [15] propose an algorithm for *orthogonal connector routing* that does not move nodes at all. Hence, it is a potential candidate for routing of wiring diagrams. However, the result is not satisfactory because wiring diagrams requires further restrictions on routing (directions of glue points, variable number of edges between two vertices, etc.) that are not addressed by the algorithm.

[1] http://www.yworks.com
[2] http://www.ilog.com

Due to restrictions of available algorithms and libraries, we have implemented positioning of diagram elements by using simple heuristics to preserve domain-specific aesthetics and extended the *orthogonal connector routing* algorithm [15] in a way that respects most requirements concerning layout of wiring diagrams. We have implemented the algorithms as part of a software tool for the technical design of power transformers (see [4] and [10]). The contribution of this paper is to close the gap between standard algorithms for graph layout and domain-specific requirements (aesthetics) in the electrical engineering domain. The main contribution is a new algorithm for element positioning (Section 3) as well as an extended algorithm for connector routing (Section 4). Both algorithms are defined, elaborated, and evaluated for wiring diagrams. We demonstrate and approve the successful application of domain-specific heuristics for graph layout and connector routing, a field that is less elaborated compared to standard algorithms [2]. The resulting diagrams have been evaluated with potential users of the software system such as electrical engineers (Section 5).

## 2 Problem Statement

In this section, we describe requirements and constraints of wiring diagrams by means of an internal wiring of a power transformer. The software tool that we have developed helps engineers to construct and edit such wiring diagrams in an incremental way. Electrical elements as well as connectors between elements may be added and removed in the diagram directly. Every change triggers an entire layout of diagram. Manual arrangement of elements in the diagram is not yet supported by the tool.

Fig. 2 shows a typical wiring diagram for a three-phase power transformer consisting of two independent electrical cycles that are both applied to a terminal ($T1$ and $T2$). The internal wiring of a power transformer is a network of different types of electrical elements connected to each other by designated entry and exit taps (depicted by circles). A *winding element* ($US$, $ST$, $G$, and $F$) is an electrical coil that is represented by one entry and multiple exit taps. *Terminal elements* ($T1$, $T2$) are ports to connect the power transformer to external elements. *Switch elements* consist of one entry tap and either two (e.g. $S$) or more exit taps (e.g. $O$). Last, a *Y-connection element* ($Y$) is used to join multiple phases to a common point. As depicted in Fig. 2, most elements are symmetric according to the three-phases of the power transformer.

The differences between these element types have influence on the connector routing because of different number of entry/exit taps and preferred directions for connector docking. The preferred direction used to connect to an element on a specific part depends on the element as well as on the entry/exit tap. For instance, taps of a terminal are always connected to the right whereas entry taps of windings (at the top) and a single exit tap at the bottom (e.g. $US$, $ST$) can be connected from arbitrary directions. However, exit taps of a winding with multiple exit taps (e.g. $F$) may be connected from the right side only.

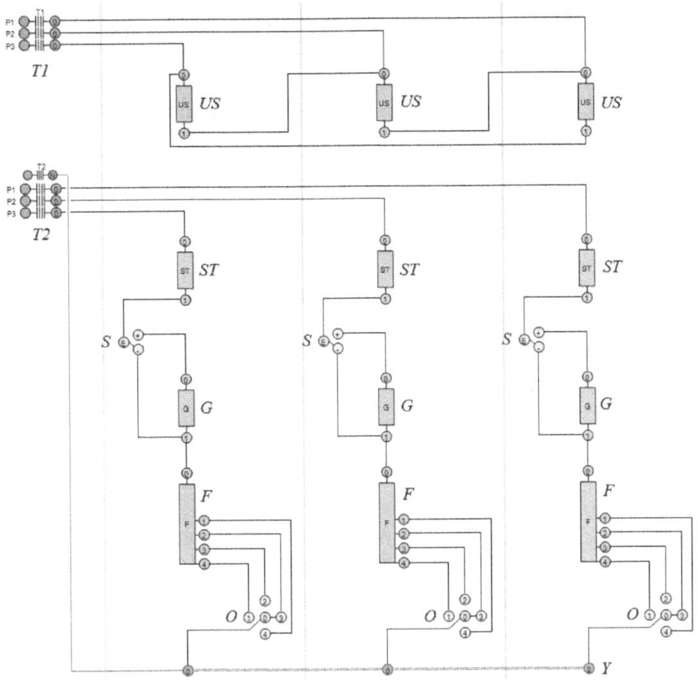


**Fig. 2.** Layout result of a wiring schematic for a three-phase power transformer

The network shown in Fig. 2 presents the positioning of electrical elements expected by electrical engineers. This expectation includes following restrictions concerning horizontal and vertical positioning:

1. Elements that are available on all phases are horizontally aligned.
2. Terminal elements are always placed on the left hand side.
3. A Y-connection element spans all phases of the network.
4. Winding elements are vertically aligned corresponding to their actual geometrical position in the power transformer.
5. The vertical positioning of other element types except windings is free with respect to connectors to other elements, in particular to winding elements.

The enumerated restrictions concerning positioning are used to keep knowledge of elements in the real electrical machine in the schematics diagram as well. Furthermore, positioning of elements must also facilitate and assure aesthetic connector routing as defined in literature [2]. Restrictions concerning reduction of lengths and crossings can only be satisfied by well-coordinated positioning and routing algorithms.

Due to the fact that power transformers commonly consist of less than fifty electrical elements but up to $10^3$ connectors, positioning of elements is not as time critical as routing of connectors. Routing is even more critical because of the requirement that diagrams with same elements and connectors must result in an equal layout which means that modifications of a network (adding, removing, and editing elements or connectors) trigger a new layout of the complete diagram. The next sections present the implementation and adaptation of positioning and routing algorithms to meet all these requirements.

## 3 Positioning of Diagram Elements

We developed an algorithm that respects the requirements and restrictions concerning horizontal and vertical positioning of diagram elements as described in the previous section. The algorithm follows additional rules to assure aesthetic drawing of connectors:

1. Place elements closer to connected elements to keep connectors short. If more than one element is connected to a specific element $E$, the one with more connectors to $E$ will be positioned next to $E$.
2. The vertical position of an element relative to a connected element is determined by the preferred directions of element taps that are connected. For example, if a (right) tap of a terminal element $T$ is connected to the top tap of a winding element $W$, $T$ will be placed above $W$ to keep the connector short.
3. The horizontal position—element is placed left or right from a connected element—can be determined the same way. This helps to reduce bends of connectors. This restriction may not be applied to winding and terminal elements because of the restrictions listed above.

A network (Fig. 3a) is given as undirected graph (Fig. 3b) containing vertices for all diagram elements (e.g. $A1$ and $B1$) and edges representing connectors between diagram elements. For positioning, this graph is transformed to an intermediate undirected graph (Fig. 3c) by collapsing multiple edges between the same elements and collapsing multiple elements with the same text label. Finally, the undirected graph is transformed to a directed graph (Fig. 3d) that contains a single node for symmetric diagram elements. In other words, where the network graph contains three nodes representing symmetric winding elements (e.g.

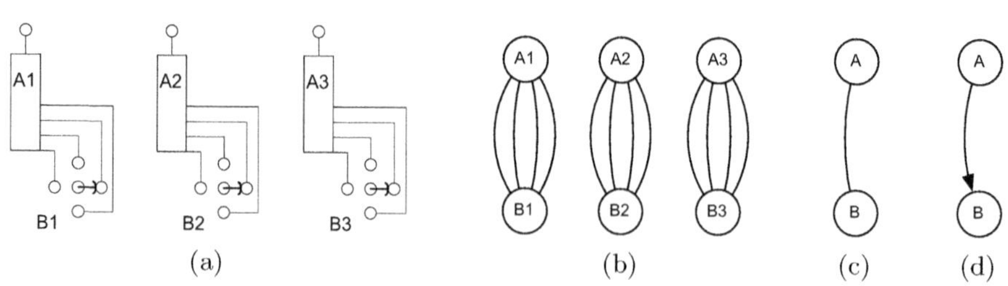


**Fig. 3.** Transformation of a network (a) given as network graph (b) to an undirected graph (c) and, finally, to a directed graph (d)

winding element *US* in Fig. 2) of all phases, the resulting graph contains a single node only, because vertical positioning is equal to all three diagram elements and horizontal positioning is symmetric on all lanes of phases. As consequence, the three diagram elements need not be treated separately for positioning.

The first step of the positioning algorithm is the vertical placement (ranking) of elements. As a connector has no defined start and end point, the resulting edge in the graph is undirected. To simplify ranking of nodes (and hence of diagram elements) and cycle detection later on, edges gets a designated start and end vertex, resulting in a directed graph. The start and end vertex of an edge is determined as follows:

1. If the position of the first tap of the connector is at the bottom, right, or left of element $E_1$ and the position of the second tap is at the top of element $E_2$, the start point of the edge is vertex for $E_1$ and end point is vertex for $E_2$.
2. If position of first tap is bottom of element $E_1$, position of second tap is top, right, or left of element $E_2$, start point is vertex for $E_1$, end point is vertex for $E_2$.
3. If position of the taps is equal (e.g. both are at the bottom), start and end vertices of the edge are determined due to the element types of the vertices by some heuristics as described above.

On the basis of these vertices and edges an adjacency matrix is constructed. Table 1 shows the initial adjacency matrix for the network given in Fig. 2 holding a row and a column entry for every vertex. For each edge, the value in the matrix at column position for start vertex and row position for end vertex is increased by 1. If all values in a column are summed up one gets the number of successors of a specific vertex, summing up a row results in the number of vertex predecessors.

In addition to the adjacency matrix, an ordered set of virtual edges is created for all windings of the transformer. A virtual edge contains a start and an end vertex as well, but start and end are determined by the actual vertical position of the winding and not from a connector given in the wiring diagram. Table 2 shows that, if winding $US$ is directly placed above winding $ST$ a virtual edge is created with start vertex of $US$ and end vertex of $ST$. This strategy ensures that the winding ordering is preserved by the positioning algorithm.

After preparation of these data structures, the actual positioning starts with ranking of unconnected elements (i.e. the sum of column and row values in the adjacency matrix is 0), ignoring unconnected windings. Unconnected elements are simply sorted by the alphabetical order of their names. Note that the order of unconnected elements never affects the resulting, final diagram but intermediate diagrams only where not all elements are connected. Hence, this rather trivial strategy is sufficient for our purpose.

In contrast to unconnected elements, connected elements and virtually connected elements are processed as follows:

1. The elements with the most significant vertex $V_i$ (where $i$ is the 1-based index of the vertex in the adjacency matrix) is determined. The most significant vertex is the vertex with highest number of successors (highest sum of

**Table 1.** Adjacency matrix as graph representation for positioning whereas columns hold successors and rows hold predecessors of graph elements

| | US | ST | G | F | S | O | T1 | T2 | Y | *SUM* |
|---|---|---|---|---|---|---|---|---|---|---|
| US | – | | | | | | 3 | | | 3 |
| ST | | – | | | | | | 3 | | 3 |
| G | | | – | | 3 | | | | | 3 |
| F | | | 3 | – | | | | | | 3 |
| S | | 3 | 3 | | – | | | | | 6 |
| O | | | | 12 | | – | | | | 12 |
| T1 | | | | | | | – | | | 0 |
| T2 | | | | | | | | – | 1 | 1 |
| Y | | | | | | 3 | | | – | 3 |
| *SUM* | 0 | 3 | 6 | 12 | 3 | 3 | 3 | 3 | 1 | – |

**Table 2.** Set of virtual edges between winding elements

| | | | |
|---|---|---|---|
| 1. | US | → | ST |
| 2. | ST | → | G |
| 3. | G | → | F |

columns) and—if there is more than one vertex—with the highest number of predecessors (highest sum of rows).
Example: In Table 1 the vertex with the highest significance is vertex $V_4$, named $F$ (sum of values in column 4).

2. If the most significant vertex is the end vertex of a virtual edge, the algorithm continues with the start vertex (assigning this start vertex to $V_i$) of this virtual edge and repeats this step until no virtual predecessor can be found.
Example: Table 2 shows that vertex $F$ is the end vertex of the virtual edge $G \rightarrow F$ so the newly inspected vertex is $G$. Because $G$ is itself the end vertex of a virtual edge step 2 is repeated until the first start vertex in the set of virtual edges is found – in Table 2 this is vertex $US$.
3. The next step is the identification of predecessors for vertex $V_i$ with the highest local significance, which is the vertex with the highest value in row $i$. This step is repeated until either no predecessor can be found or the most significant predecessor is $V_i$ again. If no predecessor can be found the vertex with no predecessor is assigned to $V_i$. Otherwise if the most significant predecessor is vertex $V_i$ again a cycle in the graph was detected. Breaking this cycle is done by first finding the weakest edge in the cycle. The weakest edge is the vertex tuple with fewest interconnections and, if more than one of these tuples exists, with the least cumulated significance (*in Table 1 the weakest edge in the cycle* $T2 \rightarrow ST \rightarrow S \rightarrow G \rightarrow F \rightarrow O \rightarrow Y \rightarrow T2$ *is the one from* $Y$ *to* $T2$). Then the end vertex of the weakest edge is assigned to $V_i$.
Example: In Table 1 the only predecessor for vertex $US$ is $T1$ (see row 1), which do not have a predecessor on its own (see empty row for $T1$). Hence no cycle in this partial graph is found and the new inspected vertex is $T1$.
4. Vertex $V_i$ is ranked as the bottommost vertex of all already ranked vertices. Because now this vertex is handled, all values in column and row at index $i$ are reset, which primarily results in a loss of significance for its predecessors. Additionally virtual edges with start vertex $V_i$ are removed.

Example: For the graph represented in Table 1, vertex $T1$ is ranked as the topmost vertex.

5. Then the successors of $V_i$, sorted in descending order of local significance, are ranked using a depth first processing [8] to traverse vertex successors. Depth first processing means that successor is assigned to $V_i$ and step 4 and 5 are repeated until either no successor or a successor with another, more significant, predecessor can be found.
   Example: The most significant successor of $T1$ in Table 1 is vertex $US$ (the highest value in column 7) and is therefore ranked after $T1$. Because $US$ do not have any successors and $T1$ does not have any others this partial graph is processed entirely.
6. When depth first processing is stopped, ranking of remaining elements continues with an adjusted adjacency matrix and the reduced set of virtual edges at step 1. Example: The adjusted matrix for Table 1 after processing vertices $T1$ and $US$ is shown in Table 3. Accordingly, Table 4 shows the reduced set of virtual edges.

**Table 3.** Adjusted adjacency matrix after processing $US$ and $T1$

| | ST | G | F | S | O | T2 | Y | $SUM$ |
|---|---|---|---|---|---|---|---|---|
| ST | – | | | | | 3 | | 3 |
| G | | – | | 3 | | | | 3 |
| F | | 3 | – | | | | | 3 |
| S | 3 | 3 | | – | | | | 6 |
| O | | | 12 | | – | | | 12 |
| T2 | | | | | | – | 1 | 1 |
| Y | | | | | 3 | | – | 3 |
| $SUM$ | 3 | 6 | 12 | 3 | 3 | 3 | 1 | – |

**Table 4.** Adjusted set of virtual edges

| | | | |
|---|---|---|---|
| 1. | ST | → | G |
| 2. | G | → | F |

The horizontal arrangement of elements in the phase lanes mainly adjusts the position of windings and terminals due to the alignment constraints for these types of elements. Hence if more taps positioned on the left side of an element are connected to elements of the same phase or more taps are connected to elements of a phase on the right, the element will be positioned on the right side of the lane. On contrary, an element will be positioned on the left side of the lane, if more taps on the left are connected or the element is connected to elements of a phase on the left. If no preferred side could be determined the element is horizontally aligned with windings on the same phase.

## 4 Connector Routing

After arranging elements vertically and horizontally, the connectors can be drawn with respect to a clear and aesthetic visualization of the entire diagram, as proposed by [14]. The principles that apply best to our needs of clear and aesthetic connector routing are following:

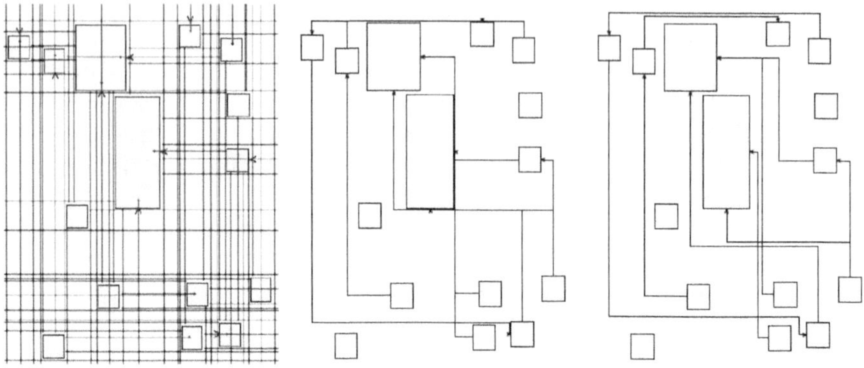

(a) (b) (c)

**Fig. 4.** Orthogonal connector routing by Wybrow et al. [15]

- Route connectors orthogonal
- Reduce length of each connector
- Reduce bends of each connector
- Reduce crossing of multiple connectors
- Reduce overlapping of multiple connectors
- Avoid connectors the overlap elements

These principles are implemented in the *orthogonal connector routing* algorithm [15] and shown in Fig. 4.
The *orthogonal connector routing* algorithm consists of three consecutive stages:

1. Determining the orthogonal visibility graph (Fig. 4a). The orthogonal visibility graph is a set of interesting points including connector points, edge points of the elements, and crossing points of straight lines through these connector and edge points.
2. Calculating routes through the visibility graph (Fig. 4b) using the $A^*$ algorithm [6].
3. Slightly moving of overlapping segments of poly-line routes, called nudging (Fig. 4c).

We mainly adopted this algorithm with few modifications to fit our needs. In contrary to the approach in [15] we can assume that (1) the elements in wiring diagrams are arranged in some kind of a grid (horizontal and vertical alignment of elements), (2) taps (connector points) of the elements may be connected from different directions, (3) connectors are more likely oriented from top-to-bottom than from left-to-right, and (4) symmetric windings, i.e. windings on all phases, are often wired using special connection types like star or delta connection (see wiring of $US$ windings in Fig. 2).

To take these assumptions into account, we perform an additional step before routing a connector. This step follows the idea of *port constraints* (e.g. [12]) and

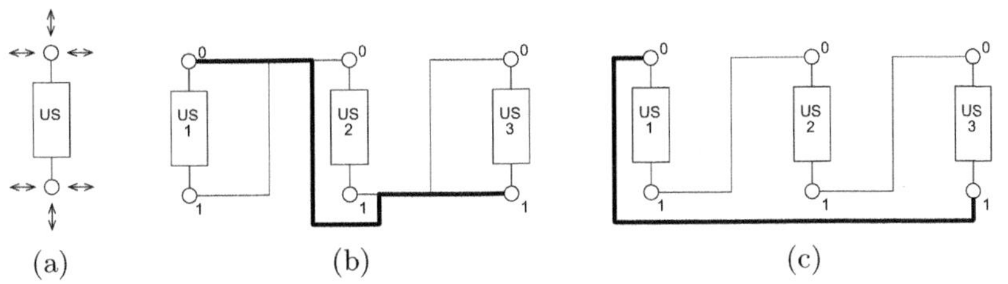


**Fig. 5.** Difference between connector routing by Wybrow [15] (b) and extended connector routing (c) of connector directions (a)

**Fig. 6.** Difference between (a) $A^*$ and (b) $A^*$ with horizontal centering

is used to determine which of the tap directions (see Fig. 5a) is the best one for a given connector. The best direction is the one that allows the route to bypass most elements without additional bends and most likely causes least crossings of connectors. Fig. 5 shows the difference between the results from the standard algorithm by Wybrow (b) and our extension (c). Experiments showed that the most aesthetic results are obtained if the connectors are sorted by ascending distance of their start and end point.

The cost function of the $A^*$ algorithm including length and bend heuristics was adopted and extended as well. The costs for bends were raised so that a longer route will be preferred to a route with more bends [11]. In addition costs for horizontal centering of vertical route segments were applied. Thus visibility graph points with similar horizontal distance to start and end point will be preferred for routing to those with more different ratio. The results are shown in Fig. 6.

Due to performance reasons a penalty is added to the costs if the *Manhattan Route* to the end tap is blocked by a shape in the diagram because it means that the route cannot be finished without adding at least two additional bends. This prevents the algorithm from calculating costs for too many points in the visibility graph that most likely are not part of the best route. When the best route is identified by the $A^*$ algorithm, an entry for every passed point in the orthogonal visibility graph is made holding whether the connector traverses the point horizontally or vertically. Hence this allows to efficiently determine overlapping segments or crossing of route segments after routing of all connectors. If any overlapping segments are detected these segments are moved relatively to a reference segment. This reference segment is either the longest of these segments

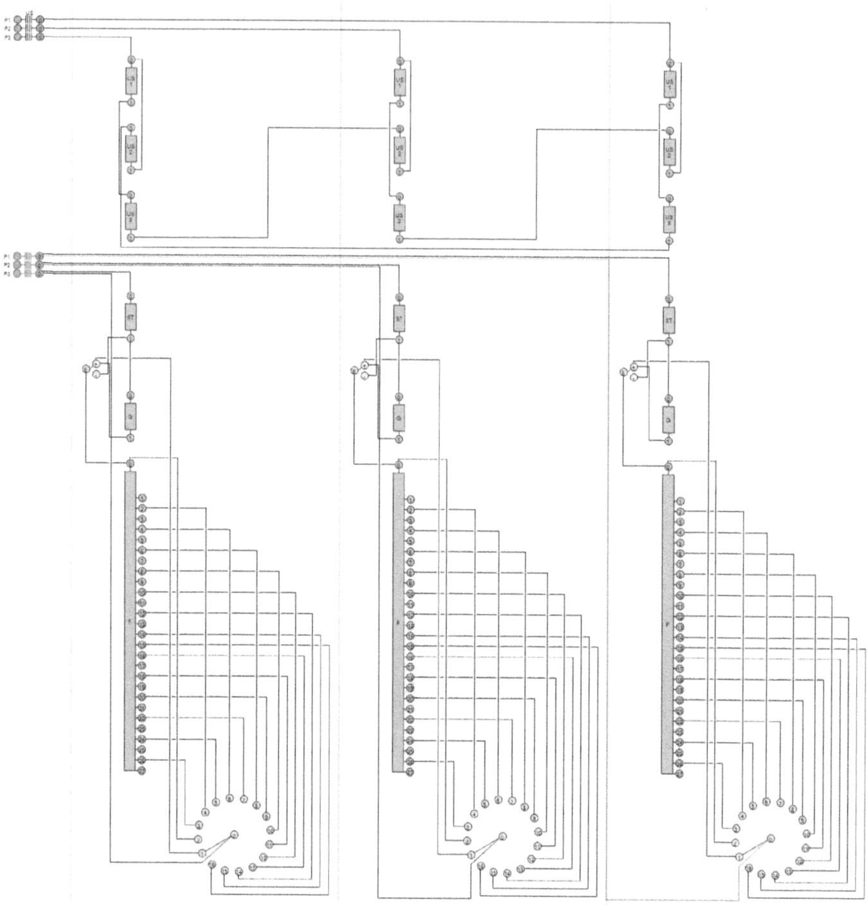

**Fig. 7.** Layouting result for a typical wiring of a three-phase transformer

or—if there is not enough space because shapes or other connectors would be overlapped—another selected segment that allows to move other segments without additional overlappings. Vertical segments are moved either to the left or right from the reference segment depending on whether start point of the dedicated route is located on the left or right of the segment. Horizontal segments are accordingly moved to top or bottom.

## 5 Evaluation

The introduced positioning and routing algorithms were implemented using the .NET platform version 3.5. The results of this implementation as shown in Fig. 2 and Fig. 7 are as defined and expected by domain experts; the only connector that could be routed more aesthetically without crossing other routes is the connector between terminal $T2$ and the $Y$-connection at the bottom of Fig. 2.

Fig. 8 demonstrates nudging and handling of crossings by means of a fictitious wiring. Although multiple connectors lead to the same tap they can be easily distinguished. The complexity of the wiring in Fig. 8 is typical for real wiring diagrams. Electrical engineers have designed wiring diagrams of actual electrical machines with our software tool. The users were able to design all diagrams with minor assistance only. The resulting diagrams were readable and understandable for the engineers as well.

Fig. 9 and Fig. 10 present some open issues concerning positioning of elements and routing of connectors:

– Uncommonly, if a tap is connected with too much other taps there is lack of space for nudging connectors (see tap 6 of element $W$ in Fig. 9) aesthetically. Hence one of these connectors cannot be distinguished from the others.
– Even though all elements and connectors between elements $S$ and $W$ are symmetric the connector of the first phase is routed—in contrary to the other phases—around element $W$ in Fig. 9. This routing behavior makes it difficult to detect the symmetry of the three connectors at a glance.
– The diagram in Fig. 10 contains a cycle with three elements. Each element is linked with exactly three connectors to the other elements, which means that no weakest edge can be determined by means of the criteria presented in section 3. The ordering of elements as is leads to longer routes and more crossings.

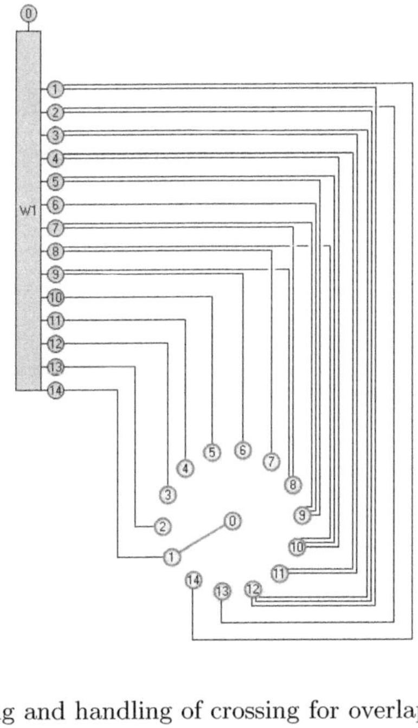


**Fig. 8.** Nudging and handling of crossing for overlapping connectors

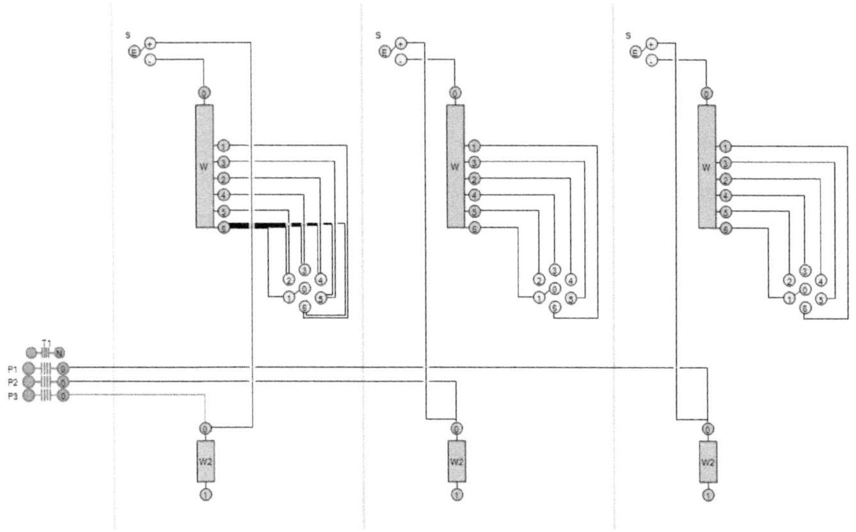


**Fig. 9.** Fictitious wiring example demonstrating routing problems

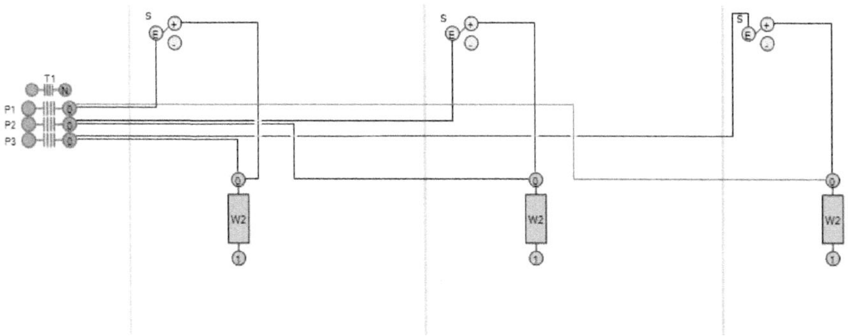


**Fig. 10.** Fictitious wiring demonstrating positioning problem for cycles

**Table 5.** Time measurements for layout of diagrams

| Diagram Size | | VisGraph | Times to compute (in msec.) | | | | |
|---|---|---|---|---|---|---|---|
| Vertices | Edges | Points | Pos | VisGraph | Routing | Nudging | Overall |
| 10 | 40 | 5000 | 20 | 1 | 12 | 1 | 34 |
| 20 | 120 | 17500 | 40 | 2 | 38 | 1 | 81 |
| 30 | 250 | 58000 | 102 | 4 | 260 | 3 | 369 |

The implemented layout algorithm also meets the requirements in terms of performance. Table 5 exposes that layout of diagrams for typical power transformers with less than 20 elements and less than 150 connectors is performed under 100 ms. Indeed, time consumed increases for larger networks, especially if the number of connectors raises. All examples were tested on a PC (dual core with 2.3 GHz and 3 GB of RAM) running Windows XP.

## 6 Conclusion

Wiring diagrams play an important role in electrical engineering. We showed that standard algorithms and libraries for graph layout are not sufficient for aesthetic layouts of wiring diagrams. Hence, we developed a new algorithm for positioning of diagram elements and extended a given algorithm with respect to special requirements concerning layout of wiring. To the best of our knowledge, there is no comparable work for wiring diagrams. However, [9] proposes an algorithm for aesthetic routing for transistor schematics.

The wiring diagrams resulting from our algorithms have been implemented as part of a design tool for power transformers and were evaluated by electrical engineers. Evaluations of the resulting diagrams by electrical engineers acknowledge the aesthetics of the diagrams. Furthermore, time behavior of our implementation is adequate for typical size of wiring diagrams .

Even the current algorithms are sufficient for typical wiring diagrams, we plan to extend our work in following ways. First, typical patterns of wiring shall be recognized and drawn in a more aesthetic way. For instance a delta-connection as shown in Fig. 2 or the connectors from a winding to a multi-tap switch as shown in Fig. 7 are examples of such patterns. Second, the additional step introduced before routing the connectors to determine the best tap direction can be replaced by extending the cost function of the A* algorithm by penalties for crossings. First attempts to implement a non-anticipatory approach resulted in major performance losses because too many points of the visibility graph are processed before the best route was found.

## References

1. Battista, G.D., Eades, P., Tamassia, R., Tollis, I.G.: Algorithms for Drawing Graphs: an Annotated Bibliography. Computational Geometry 4(5), 235–282 (1994)

2. Bennett, C., Ryall, J., Spalteholz, L., Gooch, A.: The Aesthetics of Graph Visualization. In: Computational Aesthetics 2007: Eurographics Workshop on Computational Aesthetics in Graphics, Visualization and Imaging, pp. 57–64 (2007)
3. Brandes, U., Eiglsperger, M., Kaufmann, M., Wagner, D.: Sketch-Driven Orthogonal Graph Drawing. In: Goodrich, M.T., Kobourov, S.G. (eds.) GD 2002. LNCS, vol. 2528, pp. 131–148. Springer, Heidelberg (2002)
4. Czech, G., Ernstbrunner, C., Pichler, J.: A .NET Architecture for Model-Based Domain Expert Programming. In: Fox, R., Golubski, W. (eds.) IASTED International Conference on Software Engineering, IASTED, pp. 25–32. ACTA Press (2010)
5. Dwyer, T., Marriott, K., Wybrow, M.: Topology Preserving Constrained Graph Layout. In: Tollis, I.G., Patrignani, M. (eds.) GD 2008. LNCS, vol. 5417, pp. 230–241. Springer, Heidelberg (2009)
6. Hart, P., Nilsson, N., Raphael, B.: A Formal Basis for the Heuristic Determination of Minimum Cost Paths. IEEE Transactions on Systems Science and Cybernetics 4(2), 100–107 (1968)
7. He, W., Marriot, K.: Constraint Graph Layout. Constraints 3, 289–314 (1998)
8. Hopcroft, J., Tarjan, R.: Algorithm 447: Efficient Algorithms for Graph Manipulation. Communication ACM 16(6), 372–378 (1973)
9. Lee, T.D., McNamee, L.P.: Aesthetic Routing for Transistor Schematics. In: IEEE/ACM International Conference on Computer-Aided Design, pp. 35–38. IEEE Computer Society Press (1992)
10. Pfeiffer, M., Pichler, J.: Trade: A Language and its Tool Support for Programming in Electrical Engineering. In: Fox, R., Golubski, W. (eds.) IASTED International Conference on Software Engineering, IASTED, pp. 33–40. ACTA Press (2010)
11. Purchase, H.C., Carrington, D., Allder, J.: Empirical Evaluation of Aesthetics-base Graph Layout. In: Empirical Software Engineering, vol. 7, pp. 233–255. Kluwer Academic Publishers (2002)
12. Spönemann, M., Fuhrmann, H., von Hanxleden, R., Mutzel, P.: Port Constraints In Hierarchical Layout of Data Flow Diagrams. In: Eppstein, D., Gansner, E.R. (eds.) GD 2009. LNCS, vol. 5849, pp. 135–146. Springer, Heidelberg (2010)
13. Sugiyama, K., Tagawa, S., Toda, M.: Methods for Visual Understanding of Hierarchical System Structures. IEEE Transactions on Systems, Man, and Cybernetics 11(2), 109–125 (1981)
14. Tamassia, R., Battista, G.D., Batini, C.: Automatic Graph Drawing and Readability of Diagrams. IEEE Transactions on Systems, Man, and Cybernetics, 61–79 (1988)
15. Wybrow, M., Marriott, K., Stuckey, P.J.: Orthogonal Connector Routing. In: Eppstein, D., Gansner, E.R. (eds.) GD 2009. LNCS, vol. 5849, pp. 219–231. Springer, Heidelberg (2010)

# Points, Lines and Arrows in Statistical Graphs

Cengiz Acartürk

Middle East Technical University, Informatics Institute,
Cognitive Science Program, 06800 Ankara, Turkey
acarturk@metu.edu.tr

**Abstract.** Widely used statistical graphs (such as line graphs and bar graphs) are usually accompanied by graphical entities other than the graph proper. Those graphical cues, such as point marks and arrows serve for communicative purposes by bringing certain aspects to the foreground over the others. The present study discusses the results of an experimental investigation, in which the participants produced sketches of graphical cues on different types of graphs, given sentential expressions of states and processes. The outcomes of the study have the potential for serving as guidelines for the development of software tools that produce graphical cues.

**Keywords:** statistical graphs, graphical cues, diagrammatic communication.

## 1 Introduction

Statistical graphs (e.g., line graphs, bar graphs) are abundant in communication settings and problem solving settings of daily life. They are used in newspapers and in web blogs, in annual reports of institutions and in management reports, in academic settings such as lecture notes and in scientific articles. The investigation of statistical graphs (henceforth, graphs) from a scientific perspective has attracted interdisciplinary interest. An early study at the intersection between cognitive psychology and usability research is an identification of the 'basic level graphic constituents' and acceptability principles (i.e., design guidelines) for charts and graphs by S. M. Kosslyn [1]. Kosslyn uses the notion of 'basic level' analogous to the conception of the term in categorization hierarchies, thus introducing a general classification of the components of a typical graph (or chart) in addition to keeping a high degree of similarity among different types. All the components of a graph or a chart are presented on the *background*, the first basic component. The background is often blank and it does not play a significant role in communication information. The *framework* represents the domain variables (in graphs), without specifying particular information about the mapping between the values. The *specifier* specifies the mapping between the parts of the framework. The lines in a line graph and the bars in a bar graph, in other words the graph (proper), are the specifier. Finally, the *labels* comprise alphanumerical expressions and depictive labels that contribute to the interpretation of the specifier or the framework. For example, in the time-domain line graph in Figure 1 (a two-year graph for the S&P 500 stock market index) the rectangular grid constitutes

P. Cox, B. Plimmer, and P. Rodgers (Eds.): Diagrams 2012, LNAI 7352, pp. 95–101, 2012.

the background and the framework; the graph line is the specifier (i.e., the graph proper). The alphanumerical expressions on the framework and the depictive labels on the graph proper—in this case a circle and an arrow—constitute the labels. The depictive labels, also called graphical cues, are the major focus of the present study. [1]

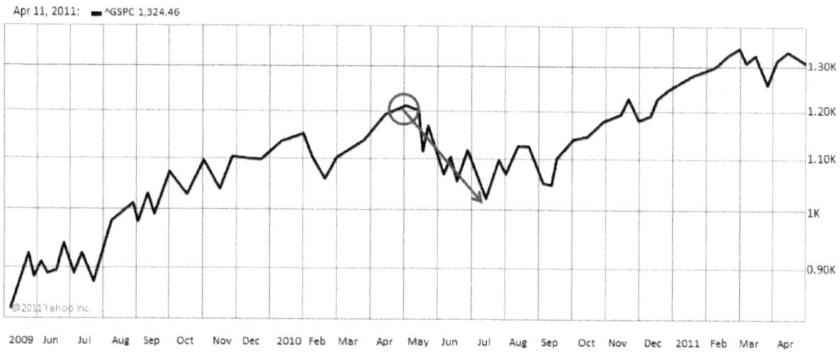


**Fig. 1.** A sample time-domain line graph. Excerpted from and redrawn based on the article "Bull Market Coming to an End?" by Sara Nunnally, Smart Investing Daily, retrieved on Sep. 6, 2011, http://www.smartinvestingdaily.com/articles/smart-investing-042511.html

Graphical cues and alphanumerical labels on the graph proper facilitate comprehension of the reader (or the interlocutor in a communication setting) by highlighting a whole graphical entity, a part of it, and/or domain entities that are referred to by the graphical entities. Human perception in un-cued visual displays is largely directed by perceptual salience of graphical entities. Entities with larger size and brighter colors or unusual shapes tend to be more conspicuous than others [2]. In closed contours, perceptual salience is often determined by a set of critical points (in the terminology of geometry) such as maxima, minima, inflection points [3], discontinuities in curvature and endpoints, among others [4]. These points have the potential for predicting attention and possibly eye movements [5]. Adding graphical cues updates the "natural perceptibility profile" of the display so that perceptibility is better aligned with thematic relevance [6].[2] In time-domain graphs, graphical cues and verbal annotations bring certain aspects of states and processes to the foreground over the others, such as temporal aspect [7] and causal relationships [8]. The use of 'schematic figures' as graphical cues, such as arrows in diagrams, and their influence on comprehension and verbal descriptions have been the subject of research in diagrammatic communication [9], [10], [11], and the relevant domains. In the domain of instructional science, for instance, the role of graphical cues in learning has been investigated in multimedia

---

[1] In the present study, the term 'graphical cue' is used for graphical entities that emphasize comprehension-relevant aspects of other graphical entities in visual displays. In the domain of computer science, the term 'annotation' is used to mean the same type of graphical entities. In the domain of instructional science, the terms 'signaling' and 'scaffolding' are used in addition to 'cueing'.

[2] Thanks one reviewer for emphasizing the relationship between graphical cues, perceptibility and thematic relevance.

learning material that involves picture and text. 'Signaling' is introduced as a technique to foster learning in multimedia material, e.g. [12], [13], as well as in animations without text, e.g. [6]. From the perspective of software development and HCI, although graph-drawing tools in statistical and mathematical software packages are abundant, the methods for designing and producing graphical cues (i.e., annotations) are very limited in the current state of the art, mostly relying on designers' experience and practice rather than research-based design guidelines. The present study reports an experimental investigation of graphical cues by analyzing cues produced by humans in terms of their types and their use in different graph types. The findings are interpreted within the theoretical framework of acceptability principles for charts and graphs proposed in [1], thus leading to a set of basic design guidelines for graphical cues.

## 2 Experiment

### 2.1 Participants, Materials and Design

Sixty participants from the Middle East Technical University, Turkey, participated in the experiment (mean age = 21.6, *SD* = 1.9). The participants were native speakers of Turkish, the language of the experiment. Each participant was presented 21 graph-sentence pairs. Each pair involved a graph (the population of a bird species) and a sentence that emphasized a certain aspect of the information represented by the graph (see Figure 2 for sample stimuli).

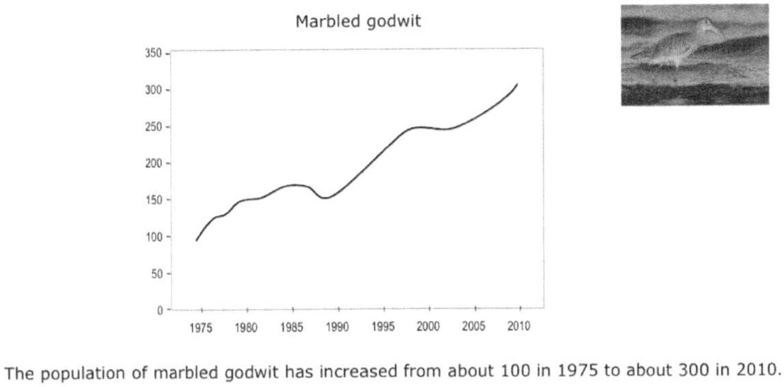


**Fig. 2.** A sample graph-sentence pair, designed based on a bird consensus report by PRBO Conservation Science (http://www.prbo.org). The stimuli were designed and reconstructed where necessary, according to the purposes aimed at the experimental investigation.

The experiment design involved one between-subject parameter (the graph type) and one within-subject parameter (the content of the accompanying sentence, i.e. the sentence type). The graph type parameter involved three experimental conditions (smooth line graph, straight line graph with sharp corners, bar graph). Accordingly, the participants were randomly divided into three groups (20 participants per group) and each group of was presented one of the three graph types. The graphs represented the same statistical data. The within-subject parameter was the type of the accompanying sentence in the presented stimuli. The accompanying sentence represented either a punctual or durative state (‘*For the past 15 years, canvasback population has been about 30 birds*’) or a process such as an increase or decrease with(out) explicit numerical values (‘*Ruddy duck number has decreased to about 50 in 2010.*’), fluctuate (‘*The number of sanderling on the lagoon has fluctuated from about 20 birds to about 120 birds in the past 35 years*’), peak (‘*The number of white-winged scooters peaked at about 90 birds in 1985*’) or cause (the cause sentences were increase / decrease process sentences; however, they involved a causal terminology). The stimuli were presented in printed form and the participants produced graphical cues for graphs on paper. The participants were instructed to improve a seminar presentation by marking and labeling according to the accompanying sentences.

## 2.2 Results

Each participant produced graphical cues for the 21 sentence-graph pairs, thus producing a total of 1260 protocols in the experiment. The graphical cues that were produced by the participants were classified into four groups.

- 0-D: Point-like markings, such as asterisks, stars, dot-circles and dash lines.
- 1-D directional: Straight and curved lines with a single arrowhead.
- 1-D nondirectional/bidirectional: Lines with no- or double-arrowheads.
- 2-D: Region-shapes such as ellipses, circles, rectangles and squares.

In the following sections, the analysis of graphical cues and the projection lines are reported.[3]

**Graphical Cues on the Graph Proper.**
The types of graphical cues on the graph proper were analyzed by a Friedman test, which was conducted to evaluate differences in medians among the cue types. The test was significant, $\chi^2$ (3, $N = 60$) = 70.3, $p < .01$, and the Kendall coefficient of concordance of .39 indicated strong differences among the cue types. The participants produced more 0-D graphical cues than 1-D directional cues, which was followed by 1-D nondirectional/bidirectional cues and then 2-D cues. A comparison between the groups showed that the participants in the bar graph group produced more 1-D nondirectional/bidirectional cues than both line graph groups. In addition, the participants produced different types of cues depending on the sentence type in the stimuli. They produced more 0-D cues when the graph was accompanied by a process sentence in which the domain value was explicitly stated. In the durative states, the participants produced less 0-D cues but more 1-D nondirectional/bidirectional cues and 2-D cues.

[3] The participants also produced alphanumerical annotations, which were numerical expressions of the domain value and time, as well as verbal expressions that were mostly causal statements, where applicable. The investigation of alphanumerical annotations is left for future research and it is not analyzed in the present study.

Finally, in the increase/decrease processes, the participants produced more 1-D directional cues than the other cue types.

**Graphical Cues on the Axes.**
The participants produced less graphical cues on the axes ($M = 9.85$) compared to the graphical cues on the graph proper ($M = 22.6$) per participant. The analysis of the produced graphical cues on the axes revealed significant differences for different types of cues, $\chi^2$ $(7, N = 60) = 99.3$, $p < .01$ (Kendall coefficient .24). The participants produced more 0-D cues (on both the x axis and the y axis) than all the other types of cues. In addition, they produced more nondirectional/bidirectional 1-D cues and more 2-D cues on the y axis than on the x axis. Graphical cues on the axes were more frequently produced by the bar graph group than the two line graph groups. Concerning the relationship between the cue type and the type of the accompanying sentence, the participants produced 0-D cues on the x axis in the durative states (to highlight the duration of the state), whereas in the increase/decrease and in the fluctuate processes, they produced the 0-D cues on the y axis (to highlight the domain values).

**Projection Lines.**
The projection lines serve for facilitating the construction of the mapping between a point on the graph proper (e.g. a point on the line or the tip of a bar) and the relevant axis. In the experiment, the smooth line group produced more vertical projections compared to the straight line group (the participants in the bar graph group produced almost no vertical projection lines because the bars themselves served for vertical mapping to the x axis). Moreover, the participants produced horizontal projection lines if the domain value was explicitly stated in the accompanying sentence. Finally, the participants produced more vertical projection lines in the durative states than they produced in the increase/decrease processes.

## 3 Discussion

The results of the experimental investigation showed that the production of graphical cues by the participants was influenced by several factors, such as the type of the accompanying sentence (state vs. process), the occurrence of an explicit domain value in the sentence and the type of the graph (bar vs. line). In particular, when the graph was accompanied by a durative state sentence, the participants produced more nondirectional/bidirectional lines and region shapes on the graph proper and they used more vertical projections on the framework. On the other hand, when the graph was accompanied by a process sentence, the participants produced more arrows than nondirectional/bidirectional lines. A second finding was that the participants who produced graphical cues in line graphs produced more arrows for process sentences compared to the participants who produced graphical cues in bar graphs. Third, the occurrence of a numerical value in the accompanying sentence resulted in more point marks on the graph proper and more horizontal projection lines compared to the absence of an explicit value in the sentence. Finally, the participants who produced graphical cues in smooth line graphs produced more projection lines than the participants who produced cues in straight line graphs. It is likely that straight line graphs with sharp corners convey information about specific numerical values more efficiently compared to smooth line graphs.

Participants' systematic production of graphical cues for bringing certain aspects of states and processes to the foreground over the others shows that time-domain graphs are interpreted not only as visualizations of domain values but also as visualizations of states and processes, and possibly they are interpreted as hints to causes and effects in the domain of discourse [7] [8]. The graph-as-data-visualization conception of graph design tools, in the recent state of the art, underestimates the potential use of graphs in reasoning and communication, thus leading to a lack of research in designing graphical cues. The results of the experiment suggest that a systematic analysis of graphical cues has the potential to fill this gap in graph design research.

## 4 Conclusion and Future Work

Among many factors that underlie the competent uses of a graph, graphical cues are design elements that facilitate reasoning as well as communication by statistical graphs. The findings in this small-scale study suggest the following guidelines for the design of graphical cues in time-domain graphs. Larger-scale usability studies would provide a complete list of design guidelines for cued graphs.

1. Processes (e.g., fall, rise, fluctuate) should be highlighted by arrows whereas durative states (e.g., remain) should be highlighted by nondirectional/bidirectional lines.
2. A line graph (instead of a bar graph) should be used to emphasize a process.
3. The emphasis on explicit numerical values in the graph can be facilitated by the use of point marks and projection lines.
4. Smooth line graphs should be avoided if the designer aims to highlight specific information on the graph.

Additional guidelines, such as 'the arrows should not be drawn on top of the graph proper but near it' (cf. Figure 1) would help the end user to produce more usable graphs. These guidelines provide the basis not only for designing graphical cues but also for the graphical tools that provide end users with the toolkits for designing graphical cues. The future work should address the following topics. The reported experiment investigates graphical cues from a production perspective. In particular, the participants were free in their producing graphical cues in contrast to being provided an inventory of graphical items for their selection, cf. [14]. The future work should analyze participants' choice given an inventory of graphical cues in addition to addressing the interpretation of graphical cues on graphs from a comprehension perspective. A relevant research topic is the interaction between dynamic gestures, different types of static and dynamic graphical cues and referring expressions in spoken communication environments. The analysis of alphanumerical annotations is a complementary research topic to the analysis of graphical cues in written communication environments and in reasoning.

**Acknowledgments.** I thank Christopher Habel for his valuable comments and suggestions in the initial drafts. Thanks METU HCI Research and Application Laboratory for providing technical support during the experiments. I also thank three anonymous reviewers for their helpful comments.

## References

1. Kosslyn, S.M.: Understanding Charts and Graphs. Applied Cognitive Psychology 3, 185–226 (1989)
2. Lowe, R.K.: Multimedia Learning of Meteorology. In: Mayer, R.E. (ed.) The Cambridge Handbook of Multimedia Learning, pp. 429–446. Cambridge University Press, New York (2005)
3. Attneave, F.: Some Informational Aspects of Visual Perception. Psychological Review 61, 183–193 (1954)
4. Freeman, H.: Shape Description via the Use of Critical Points. Pattern Recognition 10, 159–166 (1978)
5. Feldman, J., Singh, M.: Information along Contours and Object Boundaries. Psychological Review 112, 243–252 (2005)
6. Lowe, R.K., Boucheix, J.-M.: Cueing Complex Animations: Does Direction of Attention Foster Learning Processes? Learning and Instruction 21, 650–663 (2011)
7. Acarturk, C., Habel, C., Cagiltay, K.: Multimodal Comprehension of Graphics with Textual Annotations: The Role of Graphical Means Relating Annotations and Graph Lines. In: Stapleton, G., Howse, J., Lee, J. (eds.) Diagrams 2008. LNCS (LNAI), vol. 5223, pp. 335–343. Springer, Heidelberg (2008)
8. Habel, C., Acartürk, C.: Causal Inference in Graph-Text Constellations: Designing Verbally Annotated Graphs. Tsinghua Science and Technology 16, 7–12 (2011)
9. Tversky, B., Zacks, J., Lee, P., Heiser, J.: Lines, Blobs, Crosses and Arrows: Diagrammatic Communication with Schematic Figures. In: Anderson, M., Cheng, P., Haarslev, V. (eds.) Diagrams 2000. LNCS (LNAI), vol. 1889, pp. 221–230. Springer, Heidelberg (2000)
10. Heiser, J., Tversky, B.: Arrows in Comprehending and Producing Mechanical Diagrams. Cognitive Science 30, 581–592 (2006)
11. Kong, N., Agrawala, M.: Perceptual Interpretation of Ink Annotations on Line Charts. In: Proceedings of the 22nd ACM Symposium on User Interface Software and Technology, pp. 233–236. ACM, New York (2009)
12. Shah, P., Mayer, R.E., Hegarty, M.: Graphs as Aids to Knowledge Construction: Signaling Techniques for Guiding the Process of Graph Comprehension. Journal of Educational Psychology 91, 690–702 (1999)
13. Mautone, P.D., Mayer, R.E.: Cognitive Aids for Guiding Graph Comprehension. Journal of Educational Psychology 99, 640–652 (2007)
14. Tversky, B., Lee, P.U.: Pictorial and Verbal Tools for Conveying Routes. In: Freksa, C., Mark, D.M. (eds.) COSIT 1999. LNCS, vol. 1661, pp. 51–64. Springer, Heidelberg (1999)

# Enriching Indented Pixel Tree Plots with Node-Oriented Quantitative, Categorical, Relational, and Time-Series Data

Michael Burch, Michael Raschke, Miriam Greis, and Daniel Weiskopf

VISUS, University of Stuttgart

**Abstract.** Indented Pixel Tree Plots are useful for an overview of large and deep hierarchical data. As a major benefit, these plots scale to pixel or even subpixel resolution, still clearly visualizing the hierarchical structures and substructures in a redundant-free representation. Consequently, there is display space available that may be used to show additional information such as enlarged or filtered subregions, details-on-demand, or control panels. In this paper, we demonstrate how this compact indented diagram can be enriched with additional data associated with both leaf and inner nodes of the hierarchy. To this end, we support quantitative, categorical, relational, and time-series data. By such a combination, exploration and analysis of visual patterns and anomalies on different levels of hierarchical granularity are possible in a static diagram. Furthermore, interactive features such as expanding/collapsing of subhierarchies, horizontal/vertical distortions, zooming in/out, or details-on-demand are integrated to allow the user to inspect the data from different viewpoints. The usefulness of the enriched diagrams is illustrated by applying them to file system data where single software constructs are hierarchically organized. Here, we focus on quantitative, categorical, and relational data attached to the nodes of the hierarchy. In a second case study, we demonstrate how evolving water level data of rivers in Germany can be represented by our plots.

**Keywords:** Hierarchy visualization, data visualization, pixel-based techniques, time-varying data.

## 1 Introduction

Hierarchical datasets occur in many application domains, including software development processes, where a complete software system is structured into packages, directories, subdirectories, files, classes, and methods or functions. Similarly, the file system of a user's workspace is hierarchically organized in the form of directories and subdirectories. Apart from their structural organization, these hierarchical entities always have a certain size and fall into a certain category given by the file type—resulting in quantitative and categorical data attached to the nodes of the hierarchy.

In today's typical software projects, we have to deal with very large systems reaching hierarchies that consist of several thousands of elements. If we want to explore an even more fine-granular substructure, e.g., on source-code level, we soon reach sizes

P. Cox, B. Plimmer, and P. Rodgers (Eds.): Diagrams 2012, LNAI 7352, pp. 102–116, 2012.

of several million hierarchically organized elements, leading to a huge dataset that cannot be explored and analyzed by inspecting the raw data. Consequently, visualization in the form of a diagram, a chart, or a plot is required to support the viewer in obtaining an overview and additional insights about the dataset. Tapping the full potential of computer-generated visualization, interactive features may be applied to gain further insights that might not have been uncovered by the static diagram only. Here, we follow the Visual Information Seeking Mantra: Overview first, zoom and filter, then details-on-demand [22].

In particular, our requirements for the static diagram and the accompanying interactive features are:

- Display of the structure of large hierarchical data
- Additional, simultaneous display of information associated with the vertices of the hierarchy on different levels of granularity
- Support for comparisons of such attached information between hierarchical elements

To achieve these goals, we combine the Indented Pixel Tree Plots proposed by Burch et al. [5]—which intrinsically support large-data tree visualization—with the visual representation of several kinds of additional data associated with hierarchical elements: quantitative, categorical, relational, and time-series data. By such a combination, an exploration of the additional data on different levels of hierarchical granularity is easily possible.

The combination of hierarchical data with additionally attached node-oriented data leads to three variants of diagrams:

- **Indented Bar Diagram:** Quantitative as well as categorical data types can be explored by adding bar charts and color coding to the indented plot.
- **Indented Timeline Diagram:** Time-varying quantitative data are shown by color-coded timelines aligned with the indented plot.
- **Indented Matrix Diagram:** Weighted relational data can be analyzed in both vertical and horizontal directions by adding a matrix diagram to two indented plots at the vertical and horizontal axes.

All three variants of these diagrams are scalable pixel-based and even subpixel-based representations using overplotting or aggregation of the compressed single data points. Therefore, all these diagrams support large-data exploration on different levels of granularity of visual representation. Our visualization techniques provide several interactive features: subhierarchies can be collapsed and expanded, the plot can be distorted horizontally as well as vertically, subregions and subintervals can be selected, and we allow details-on-demand, data annotations, and the like.

The usefulness of the enriched Indented Pixel Tree Plots is illustrated by exploring hierarchical data from large file systems and time-varying water levels of rivers in Germany. The insights gained by our interactive visualization tool are demonstrated in two corresponding case studies.

## 2 Related Work

Visualizing hierarchical data has been in the focus of many researchers throughout the years, as reviewed in the surveys by Jürgensmann and Schulz [19], [20]. Many visual metaphors have been invented, developed, and enhanced, leading to a variety of diagrams for hierarchical data.

Node-link diagrams are a good and most popular choice when displaying hierarchical organizations in an intuitive way. Burch et al. [4] used eye tracking to evaluate node-link visual metaphors for tree diagrams. As a major result of this study, it is advantageous if the root node is placed on top and the subordinate elements on layers below depending on their depth in the hierarchy, as also used in early work by Eades [11] and also by Reingold and Tilford [18].

In this work, we address the issue of displaying additional information attached to a tree diagram, supporting a viewer to inspect datasets on different levels of hierarchical granularity in a static diagram and serving as a good overview of the whole dataset by using aggregation and overplotting. Examples of this line of research include the display of timelines (of dynamic quantitative data) attached to edges of an orthogonal node-link tree diagram [7], the visual encoding of transaction data in Timeline Trees [2], or heatmap representations attached to dendrograms [12] used in bioinformatics. The drawback of these approaches is the limited scalability in the number of vertices and that additional data cannot be attached to leaf and to inner vertices at the same time.

Being visually most related to indented plots, layered icicle plots [14] exploit the principle of vertically stacked bars growing from top to bottom, leading to diagrams that look very similar to icicles in nature. A drawback of layered icicles is the fact that much ink is used for inner vertices and borders are needed to separate sibling nodes. Most important in the context of this paper, inner nodes in icicle plots have no representative elements on the outline of the tree. Therefore, data associated with inner vertices is difficult to represent. Similarly, interactive visualization tools that use radial layered icicle plots—such as Information Slices [1], Sunburst [23], or InterRing [26]—may represent additional node-associated data but radially growing visual elements are difficult to judge and to compare [10].

Treemaps [21] are based on the principle of nested boxes. This approach generates space-filling hierarchy representations and allows us to additionally encode quantitative data (in the form of the box sizes) and categorical data (by color). Following this principle, inner vertices typically cover the sum of the areas of all their child vertices. Therefore, it is difficult to attach other kinds of information that does not add up when traversing the hierarchy. In particular, treemaps soon reach their limitations when trying to attach relational data and lead to visual clutter caused by many link crossings, e.g. in the ArcTrees visualization [17], in the Trees in a Treemap approach [8], or by overlaying graph links on treemaps [13]. Attaching dynamic data to treemap representations is in focus of the TennisViewer tool [15] and the Contrast Spiral Treemap [24]. The biggest problem when combining time-varying data with treemap representations is the difficulty that a viewer has when simultaneously comparing the data in subintervals. Time-varying relations in information hierarchies are an even harder visualization challenge and have been investigated by Burch et al. [3] in their recent work on parallel edge splatting.

Most of the aforementioned tree diagrams enriched with additional node-oriented data are useful for their respective tasks and application domains. However, they lack scalability in either the hierarchy visualization part or attached data visualization part. Furthermore, the efficiency of using the given display space when reflecting the hierarchical structure varies between the visualization techniques [16]. For example, treemaps may be used to represent large hierarchies. However, attaching relational data is difficult and may cause visual clutter. Attaching quantitative and categorical data is difficult to interpret on different levels of granularity, and attaching time-series data is hard to compare in subintervals because treemap boxes do not allow aligned timelines for all levels of a hierarchy. Node-link diagrams and icicle plots do not scale for large hierarchies because of the much ink used for drawing the link information. These also do not allow the attachment of data to inner nodes because only leaf nodes are represented at the outline of the tree. Furthermore, separation lines are additionally needed in treemaps and icicle plots to visually differ between sibling nodes.

We address these issues and drawbacks by using Indented Pixel Tree Plots as the basis for hierarchy visualization [5] and corresponding tree browsing [6]. In particular, Indented Pixel Tree Plots naturally lend themselves to scalability in the number of nodes and depth of the hierarchy because of their pixel-based drawing approach. This paper adds scalability in terms of visualization of additional data associated with inner and leaf nodes.

## 3 Data Model and Indented Metaphor

To display hierarchical structures, we use the indented visual metaphor that is popular in graphical file browsers and pretty printing of source code. The greatest benefits of indentation are the facts that as little ink as possible is needed to represent the hierarchical data and that the diagrams can be scaled down to pixel or even subpixel size. Hence, they scale up to large hierachical datasets. The diagrams are mapped to one-dimensional zigzag curves and, thus, provide much free space to attach additional visual information. Another advantage is the fact that each vertex—leaf and inner vertex alike—has a representative element on the one-dimesional axis and, hence, allows data comparisons on different levels of hierarchical granularity.

### 3.1 Data Model

We model a hierarchy as an ordered pair

$$H = (V_I, E_I)$$

where $V_I$ denotes the set of vertices and $E_I \subset V_I \times V_I$ the set of directed edges, i.e., the parent–child relationships that we call inclusion edges to distinguish them from another kind of relations called adjacency edges. One vertex is designated the root vertex of the hierarchy.

Quantitative data attached to the hierarchy can be modeled by a function

$$f_q : V_I \longrightarrow \mathbb{R}$$

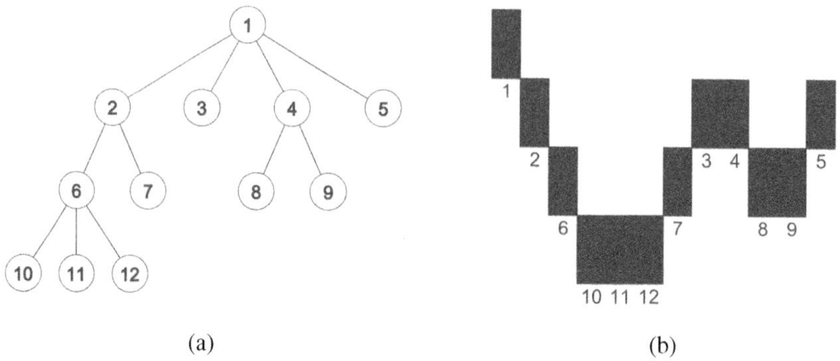


**Fig. 1.** Comparsion of node-link diagram and indented plot for the same hierarchy. (a) Node-link diagram: explicit link structures are used to express parent–child relationships. (b) Indented plot: hierarchical structure is illustrated by indenting vertically.

mapping each vertex to a real-valued number. Additional categorical data is described by a function

$$f_c : V_I \longrightarrow T$$

where $T$ expresses a set of different categories. Dynamic data is modeled by a function

$$f_d : V_I \longrightarrow (a_i)$$

i.e., a sequence of real-valued numbers $a_i \in \mathbb{R}$ or categorical data $a_i \in T$. The index $i \in \{1,\ldots,N\}$ corresponds to the time for which the quantitative or categorical data was recorded. Relations between hierarchy elements can be modeled as a graph $G = (V_I, E_A)$ where $V_I$ contains the same vertices as in the hierarchy and $E_A$ are adjacency edges expressing weighted relations between pairs of vertices, i.e. $E_A \subseteq V_I \times V_I$. A function

$$g_q : E_A \longrightarrow \mathbb{R}$$

attaches real-valued numbers to each adjacency edge, leading to a weighted directed graph in information hierarchies. A similar function

$$g_c : E_A \longrightarrow T$$

may be used to attach a category to each edge, leading to a graph with edge attributes.

### 3.2 Indented Plot Generation

Here, we briefly review the idea and structure of the original Indented Pixel Tree Plots. For details, we refer to the paper by Burch et al. [5]. Indented plots can be drawn easily, without caring for the visual space required for the subhierarchies, because they are depicted step-by-step from left to right. The elements of a hierarchical structure are processed one-by-one with depth-first traversal. For each vertex, we draw a vertical line of some fixed length and some fixed width, where the $y$-position on screen only depends on the depth of the corresponding vertex in the hierarchy and the $x$-position depends linearly on the index number of the vertex (the same index as from the traversal order in the depth-first strategy). To visually cluster leaf vertices for aesthetically pleasing diagrams, the hierarchy may be ordered in a way as to group leaf vertices on each hierarchy level to the right.

Figure 1 shows an example of a hierarchy visualized as a node-link diagram (a) and as an indented plot (b). To transform the data to the indented plot, the link information is removed, leading to a plot using as little ink as possible for displaying the data. The root node of each subhierarchy is placed to the left of all its corresponding subordinate nodes.

### 3.3 Visual Interpretation of an Indented Plot

Indented Pixel Tree Plots are new and, thus, have to be learned to be read. However, the user study by Burch et al. [5] showed that the learning curve is not steep: participants were able to understand the indented visual metaphor within a few minutes time, by just reading a short tutorial. The comparative evaluation with traditional node-link diagrams showed that reading, understanding, exploring, and deriving insights from an indented plot is easy and effective.

The indented plot visually encodes subordinate nodes to the right and vertically below their parent nodes recursively in all hierarchy levels. This simple visual structure can be exploited for efficient task solution strategies. For example, the parent node $X$ of some other node $Y$ is quickly identified by just looking for the closest vertical graphical primitive to the left in the indented plot that lies exactly one step above the node $Y$ in the vertical direction. As another example, the least common ancestor of a group of vertices $v_1, \dots, v_n$ can be found by starting at the right hand side of the plot and inspecting the lowest peak that is to the left of the leftmost vertex of $v_1, \dots, v_n$ and that is still above all of the highest peaks between any two subsequent vertices of $v_1, \dots, v_n$ in the plot.

## 4 Combination with Additional Node-Associated Data

There are many scenarios where additional data in hierarchical structures exists and there is a need for a diagram that shows the hierarchy in combination with that data at one glance. In this section, we will describe how certain, relevant data types can be combined with hierarchical structures in a scalable way with the goal to obtain a good overview of the data and to analyze it on different levels of hierarchical granularity.

### 4.1 Indented Bar Diagram

Quantitative data may be represented as aligned bar charts or histograms, as evaluated by Cleveland and McGill [10], because the single lengths can be judged and compared very accurately and reliably by the human visual system. Categorical data is typically represented by color or shape, see Card and Mackinlay [9]. In our diagrams, we use color coding because shape representation is not feasible for pixel-based or even subpixel-based visualization.

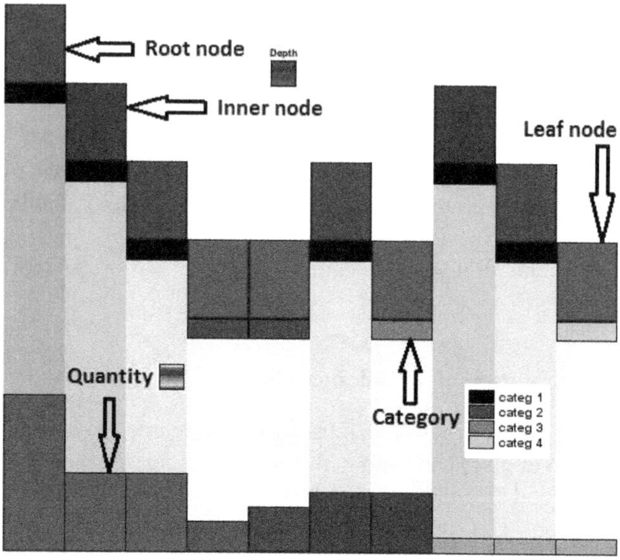


**Fig. 2.** Example of an Indented Bar Diagram: Combination of hierarchical data with quantitative as well as categorical data in a single diagram

The Indented Bar Diagram is composed of an Indented Pixel Tree Plot for representing the hierarchical organization of the data elements and of a bar chart for visually encoding the attached quantitative information about each data element. The category of each element is displayed by a color-coded rectangle aligned below each corresponding vertex in the indented outline, see also Figure 2. This color coding can be used to derive and compare the categories of the hierarchy and subhierarchy elements. Grayscale is used to additionally link the hierarchical organization with the corresponding data elements and to obtain the impression of a single diagram as a whole and not of two separated parts. The gray values are linearly adapted to the depth of each element $v \in V_I$ in the hierarchy.

All color tables can be changed interactively and independently for each represented data type: hierarchy depth, quantitative data, and categorical data. Relational and time-series data either consist of single quantitative data points, i.e., the weights of the graph or of single categorical data points.

Figure 2 shows an Indented Bar Diagram for a hierarchical dataset containing 10 elements at a maximal depth of 3. Here, a blue-to-red color gradient is chosen for encoding the depths in the tree—the redder, the deeper in the hierarchy. Gray bars connect the indented plot on top with the bar chart at the bottom, where the gray values also reflect the depth of the hierarchy element. If a categorical data for the hierarchy elements exists, this is drawn as shown in the color legend in Figure 2. Quantitative data for each hierarchy element is displayed as a bar chart and additionally color coded by using the vegetation color scale, see the color legend in Figure 2. By using the same horizontal axis as reference (e.g., for zero values), the quantitative values are easy to compare, just

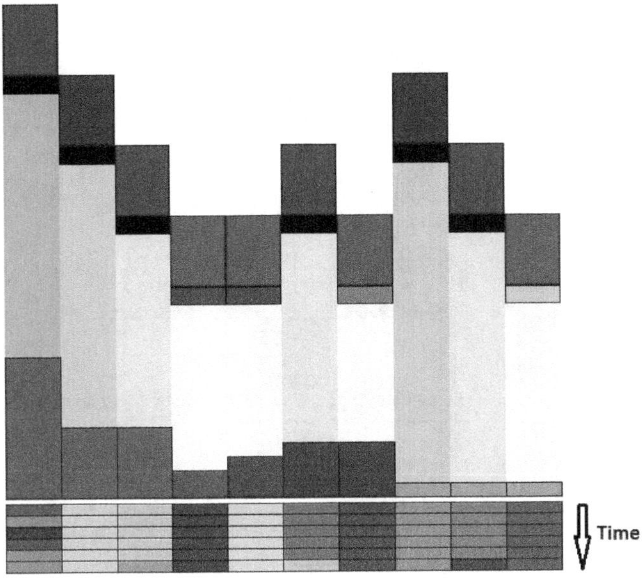


**Fig. 3.** Example of an Indented Timeline Diagram: Combination of hierarchical data with quantitative as well as categorical data in a single diagram. Additionally, the evolution of either quantitative or categorical data is shown below the Indented Bar Diagram as an aligned timeline representation supporting comparisons of subintervals.

like in a regular bar chart. Here, color is used to improve the perception of differences in the quantitative data. One can easily see, for example, that the leftmost bar (the one for the root node) is highest and red.

### 4.2 Indented Timeline Diagram

Time-varying quantitative data may be visualized by a static time-to-space mapping instead of a time-to-time mapping as commonly used in animations. We choose a static visualization because it reduces cognitive efforts and preserves a viewer's mental map. Drawbacks of animated diagrams are discussed in more detail by Tversky et al. [25].

The Indented Timeline Diagram attaches a timeline representation to each hierarchy element in an aligned way and hence, shows the time-varying data in a static diagram with the goal to support a viewer when comparing the dynamic data in different subintervals and also on different levels of hierarchical granularity. By using a time-to-space mapping the evolving data can easily be explored for dynamic phenomena such as trends, countertrends, temporal shifts, periodicities, stagnation, or anomalies.

Figure 3 illustrates the idea for the small example of 10 hierarchical elements and a time-varying quantitative dataset containing 6 timesteps. For example, the quantitative data for the rightmost element in the diagram is not changing over time, which can be seen by the constant red color for all time steps. This fact also holds for some other elements, whereas the timeline for the root node at the leftmost position of the diagram shows heavy fluctuations over time. It starts with a red color and then changes to green, blue, and green again.

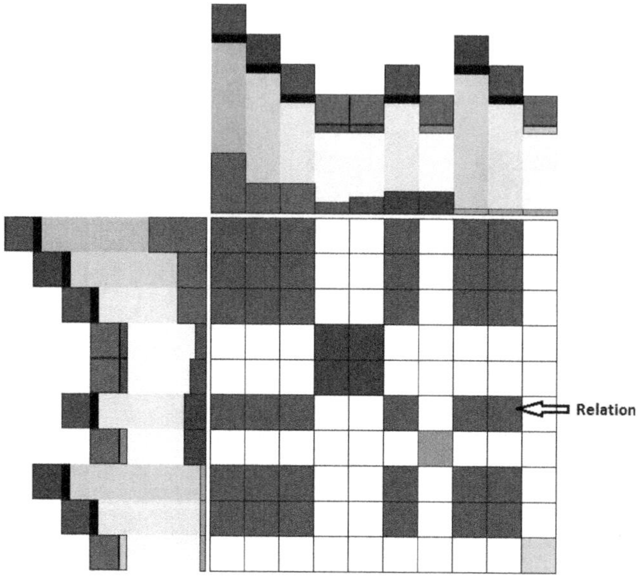


**Fig. 4.** Example of an Indented Matrix Diagram: Combination of hierarchical data with relational data in one single diagram

## 4.3 Indented Matrix Diagram

A more complex but also easy-to-understand representation is given by the Indented Matrix Diagram. In this scenario, a weighted relational dataset is visualized where the single related elements are additionally hierarchically organized. The hierarchy has to be represented twice, vertically as well as horizontally in the indented metaphor and the weighted directed graph data is shown as a matrix representation that benefits from reduction of visual clutter for dense graphs, i.e., graphs with very many edges.

In Figure 4, such an Indented Matrix Diagram is shown for a hierarchy with 10 elements. Strongly related subhierarchies can easily be uncovered by inspecting the indented plot on top and to the left combined with the color-coded matrix representation in-between the indented plots. In the figure, we can easily derive from the matrix cell colored in yellow that the strongest relation is a relation of the rightmost element with itself, a so-called self-edge. The small blue colored blocks indicate that there is a small clique among several elements and the hierarchical belonging can also be tracked easily.

## 4.4 Interactive Features

A static diagram serves as a good overview of a huge dataset but we also support a viewer by interactive features to tap the full potential of computer-generated visualizations. Following the Visual Information Seeking Mantra—Overview first, zoom and filter, then details-on-demand [22]—we provide the following interactive features:

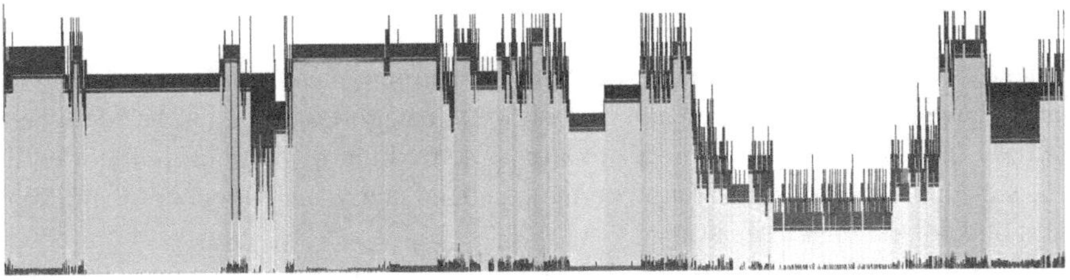

**Fig. 5.** An Indented Bar Diagram of a directory structure containing 36,856 files and subdirectories represented as a combination of hierarchical data with quantitative and categorical data

- **Expanding/Collapsing of Subhierarchies:** Clicking on a node in the indented plot leads to a collapsed subhierarchy of the corresponding subordinate nodes and an aggregation of the quantitative or categorical data. Several aggregation modi exist for generating the new values such as maximum, sum, or average value. Clicking on a collapsed node leads to an expansion of the subtree again.
- **Horizontal/Vertical Distortions:** Since indented plots scale for large hierarchical datasets and allow still readable pixel or subpixel representations, we support the user by horizontal and vertical distortion techniques that create new space for displaying additional information.
- **Separate Color Codings:** All components of the combined diagrams can be attached to a separate color coding to avoid misinterpretations that would be caused by using similar color schemes for different components.
- **Weight Filtering:** Quantitative data can be filtered for values in a given interval, categorical data for several categories, relational data for vertex groups, time-varying data for subintervals, and hierarchical data for certain subhierarchies. Also a logarithm function can be applied to reduce the visual dynamic range between high and low values.
- **Textual Search:** If the hierarchical data contains elements with an additional descriptive information, i.e., labels, the user can apply textual search to highlight these interesting elements.
- **Details-on-Demand:** Hovering the mouse cursor over any part of any component of the combined plots shows a detail-on-demand information if one is present, i.e., the value of the focused matrix cell for example with its corresponding vertical and horizontal hierarchy elements.

## 5 Case Study

To illustrate the usefulness and the scalability of the enriched plots, we apply them to file system data containing several thousand hierarchically organized elements to which a quantitative and categorical data type may be attached. Furthermore, relations among the elements exist, generating a dense weighted and directed graph, and also an evolution over time may be possible for the quantitative as well as the categorical data. In a second application example, we explore dynamic water level data from the river system in Germany for trends, countertrends, periodicities, temporal shifts, and anomalies.

## 5.1 File Systems

Typically, file systems can become very large containing many subdirectories and files that all have different sizes and fall into different categories given by the file type for example. To get an overview of all this data, some kind of diagram is required that shows the hierarchical organization on the one hand and visually encodes the different types of data on the other hand.

Figure 5 shows an Indented Bar Diagram of a directory structure containing 36,856 files (leaf nodes) and subdirectories (inner nodes) at a maximal hierarchy depth of 17. We use an additional blue-to-red color gradient besides the vertical indentation to visually strengthen the depth of a node in the hierarchy. File and directory sizes are displayed as an aligned bar chart right below each corresponding element in the hierarchy by using a vegetation color scale (i.e. blue and green color codings mean low values and

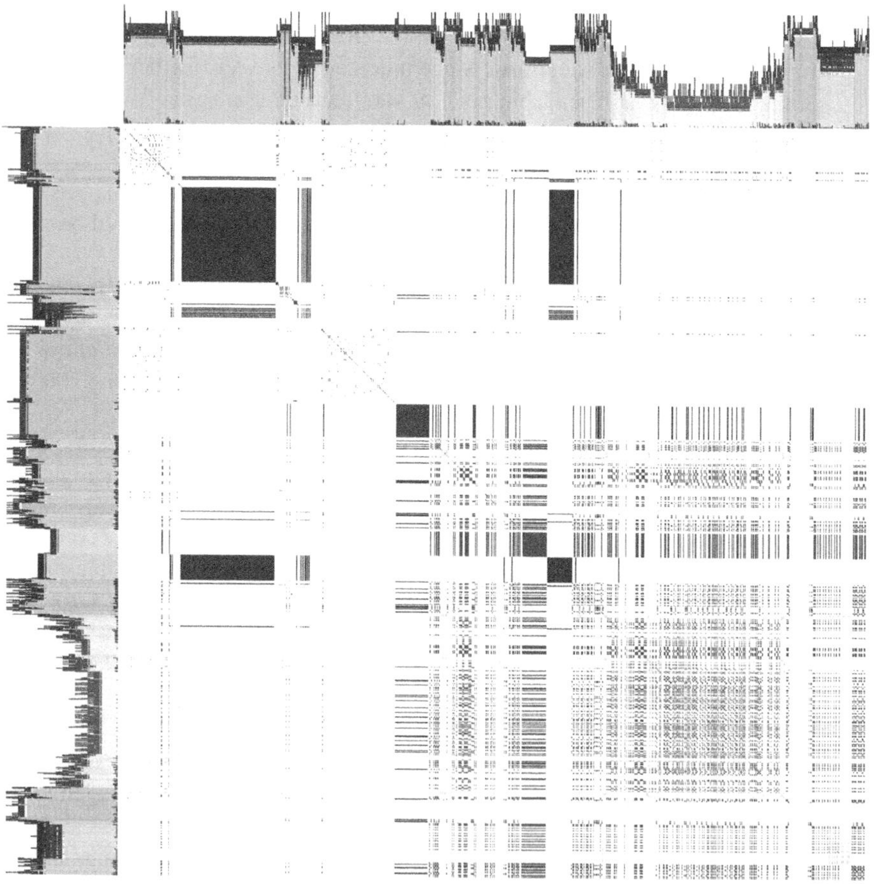

**Fig. 6.** An Indented Matrix Diagram of a directory structure containing 36,856 files and subdirectories and more than 1,350,000,000 weighted directed relations in a compressed and aggregated view serving as an overview representation

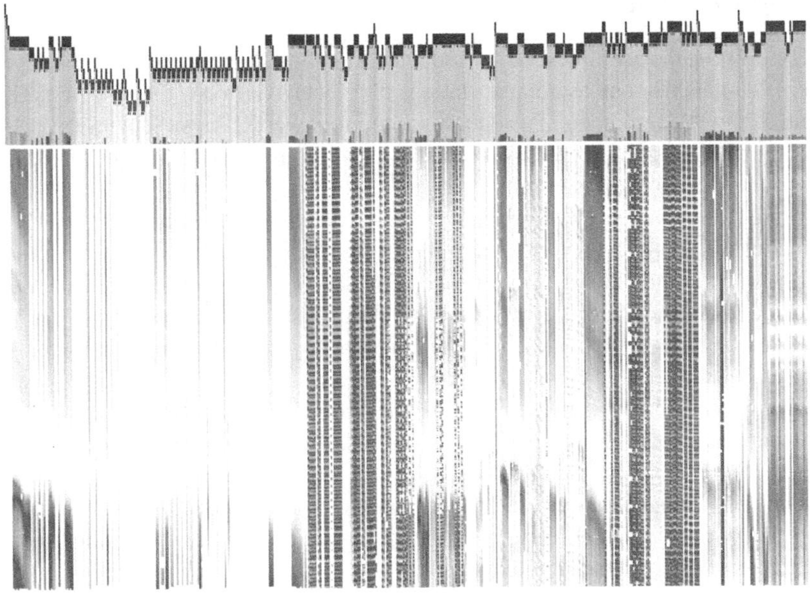

**Fig. 7.** An Indented Timeline Diagram represents several measurement stations for water level data of rivers in Germany as a hierarchical system and shows the evolution of the water levels over time for 768 points in time

red and yellow color codings encode high values). We additionally apply a logarithm function to the values to keep the differences between maximum and minimum values small. Each file falls into a certain category given by the type of the file in this application example. This information is visually encoded to the lower end of each graphical primitive in the indented plot of the corresponding hierarchical element.

From this diagram, we already obtain many insights by just inspecting the static picture. First of all, the deepest level of the hierarchy structure is visualized in the right hand part of the diagram, which can be visually uncovered by the red color coding and the vertically indented outline. If we additionally inspect the corresponding bar chart right below in this subregion, only blue colored and flat bars can be detected, indicating that the files located there are not that large compared to many others in the diagram.

Figure 6 shows which files are related to each other with respect to the same file type. Here, 1,350,000,000 weighted directed relations are displayed in a compressed and aggregated matrix representation allowing a good overview for the hierarchical structure, the categorical file types, the quantitative file and directory sizes, and weighted relations between the hierarchy elements. This overview representation can be used as a starting point for further exploration and analysis tasks. Following the Visual Information Seeking Mantra [22], we can inspect the data piece-by-piece, supported by the interactive features provided by the visualization tool, until we finally understand the data.

### 5.2 Dynamic Water Levels

To demonstrate the usefulness of our Indented Timeline Diagram, we visualize dynamic water level data from measurement stations of rivers in Germany that are hierarchically organized by the confluents. The data was acquired over 32 days (September 2010 plus 2 days) and a measurement was recorded every hour, generating a sequence of quantitative data over 768 points in time.

Figure 7 demonstrates trends, countertrends, temporal shifts, periodic behaviors, and anomalies of this time-varying dataset. For example, the periodic behavior in many of the timelines is the natural consequence of low and high tides of the North Sea and Baltic Sea. Measurement stations close to these Seas and the mouths of rivers into these Seas consequently show this phenomenon. It can be visually uncovered by the alternating yellow to red color in several timelines.

On the left hand side of the diagram, one can see a temporal shift pattern, i.e., a flood wave is moving downward the river. A detail-on-demand request shows that this phenomenon belongs to measurement stations along the river Rhine. There are many more visual features and dynamic patterns in the dataset visible by just inspecting the static diagram. Interactive features of the visualization tool can be applied to further analyze and explore these large datasets.

## 6 Conclusion and Future Work

We have demonstrated how hierarchically organized data with additionally attached different kinds of data—such as quantitative, categorical, relational, or time-series data—can be represented visually in a single static diagram serving as an overview representation as a starting point for further and more detailed exploration tasks.

In particular, we have combined Indented Pixel Tree Plots with bar diagrams, timeline diagrams, and matrix diagrams, although combinations with other node-oriented plots are conceivable as well. Our approach scales for large hierarchical datasets, uses as little ink as possible to display the hierarchy, and allows data to be attached for all vertices—leaf and inner vertices alike.

Apart from the static diagrams, several interactive features can be applied to explore the data from different viewpoints and also on different levels of hierarchical granularity. We have shown the usefulness of the diagrams by visually exploring a large file system with several thousand elements and additionally attached data types. Furthermore, in a second application scenario, we have demonstrated how time-varying water level data can be represented in a hierarchically oranized river system to allow for a good overview and to better compare the dynamic data in different time intervals side-by-side for trends, countertrends, temporal shifts, periodicities, and anomalies.

For future work, we plan to conduct a user study to evaluate the enriched visualization techniques. Furthermore, we want to apply it to datasets from different application domains, e.g., biological data where the hierarchical organization of organisms and species plays a crucial role.

## References


1. Andrews, K., Heidegger, H.: Information Slices: Visualising and Exploring Large Hierarchies Using Cascading, Semi-Circular Discs. In: Proceedings of the IEEE Symposium on Information Visualization, pp. 9–12 (1998)
2. Burch, M., Beck, F., Diehl, S.: Timeline Trees: Visualizing Sequences of Transactions in Information Hierarchies. In: Proceedings of Advanced Visual Interfaces, pp. 75–82 (2008)
3. Burch, M., Vehlow, C., Beck, F., Diehl, S., Weiskopf, D.: Parallel Edge Splatting for Scalable Dynamic Graph Visualization. IEEE Transactions on Visualization and Computer Graphics 17(12), 2344–2353 (2011)
4. Burch, M., Heinrich, J., Konevtsova, N., Höferlin, M., Weiskopf, D.: Evaluation of Traditional, Orthogonal, and Radial Tree Diagrams by an Eye Tracking Study. IEEE Transactions on Visualization and Computer Graphics 17(12), 2440–2448 (2011)
5. Burch, M., Raschke, M., Weiskopf, D.: Indented Pixel Tree Plots. In: Bebis, G., Boyle, R., Parvin, B., Koracin, D., Chung, R., Hammoud, R., Hussain, M., Kar-Han, T., Crawfis, R., Thalmann, D., Kao, D., Avila, L. (eds.) ISVC 2010. LNCS, vol. 6453, pp. 338–349. Springer, Heidelberg (2010)
6. Burch, M., Schmauder, H., Weiskopf, D.: Indented Pixel Tree Browser for Exploring Huge Hierarchies. In: Proceedings of the International Symposium on Visual Computing, pp. 301–312 (2011)
7. Burch, M., Weiskopf, D.: Visualizing Dynamic Quantitative Data in Hierarchies. In: Proceedings of International Conference on Information Visualization Theory and Applications, pp. 177–186 (2011)
8. Burch, M., Diehl, S.: Trees in a Treemap: Visualizing Multiple Hierarchies. In: Proceedings of Visualization and Data Analysis, pp. 224–235 (2006)
9. Card, S.K., Mackinlay, J.: The Structure of the Information Visualization Design Space. In: Proceedings of the IEEE Symposium on Information Visualization, pp. 92–99 (1997)
10. Cleveland, W.S., McGill, R.: Graphical Perception: Theory, Experimentation, and Application to the Development of Graphical Methods. Journal of the American Statistical Association 79(387), 531–554 (1984)
11. Eades, P.: Drawing Free Trees. Bulletin of the Institute for Combinatorics and its Applications 5, 10–36 (1992)
12. Eisen, M., Spellman, P., Brown, P., Botstein, D.: Cluster Analysis and Display of Genome-Wide Expression Patterns. Proceedings of the National Academy of Sciences 95, 14863–14868 (1998)
13. Fekete, J.-D., Wang, D., Dang, N., Aris, A., Plaisant, C.: Overlaying Graph Links on Treemaps. In: Proceedings of the IEEE Symposium on Information Visualization, Poster Compendium, pp. 82–83 (2003)
14. Kruskal, J., Landwehr, J.: Icicle Plots: Better Displays for Hierarchical Clustering. The American Statistician 37(2), 162–168 (1983)
15. Jin, L., Banks, D.C.: TennisViewer: A Browser for Competition Trees. IEEE Computer Graphics and Applications 17(4), 63–65 (1997)
16. McGuffin, M.J., Robert, J.M.: Quantifying the Space-Efficiency of 2D Graphical Representations of Trees. Information Visualization 9(2), 115–140 (2009)
17. Neumann, P., Schlechtweg, S., Carpendale, S.: ArcTrees: Visualizing Relations in Hierarchical Data. In: Proceedings of Eurographics IEEE VGTC Symposium on Visualization, pp. 53–60 (2005)
18. Reingold, E.M., Tilford, J.S.: Tidier Drawings of Trees. IEEE Transactions on Software Engineering 7, 223–228 (1981)

19. Jürgensmann, S., Schulz, H.-J.: A Visual Survey of Tree Visualization. In: Proceedings of the IEEE Conference on Information Visualization, Poster Compendium (2010)
20. Schulz, H.-J.: Treevis.net: A Tree Visualization Reference. IEEE Computer Graphics and Applications 31(6), 11–15 (2011)
21. Shneiderman, B.: Tree Visualization with Tree-Maps: 2-D Space-Filling Approach. ACM Transactions on Graphics 11(1), 92–99 (1992)
22. Shneiderman, B.: The Eyes Have It: A Task by Data Type Taxonomy for Information Visualizations. In: Proceedings of the IEEE Symposium on Visual Languages, pp. 336–343 (1996)
23. Stasko, J.T., Zhang, E.: Focus+Context Display and Navigation Techniques for Enhancing Radial, Space-Filling Hierarchy Visualizations. In: Proceedings of the IEEE Symposium on Information Visualization, pp. 57–65 (2000)
24. Tu, Y., Shen, H.-W.: Visualizing Changes of Hierarchical Data Using Treemaps. IEEE Transactions on Visualization and Computer Graphics 13(6), 1286–1293 (2007)
25. Tversky, B., Bauer Morrison, J., Bétrancourt, M.: Animation: Can it Facilitate? International Journal of Human-Computer Studies 57(4), 247–262 (2002)
26. Yang, J., Ward, M.O., Rundensteiner, E.A., Patro, A.: InterRing: A Visual Interface for Navigating and Manipulating Hierarchies. Information Visualization 2(1), 16–30 (2003)

# Interpreting Effect Size Estimates through Graphic Analysis of Raw Data Distributions

Michael T. Bradley[1], Andrew Brand[2], and A. Luke MacNeill[1]

[1] University of New Brunswick
Department of Psychology
P.O. Box 5050
Saint John, NB E2L 4L5 Canada
Bradley@unb.ca
[2] Kings College
Institute of Psychiatry
P.O. Box 77
De Crespigney Park
London SE5 8AF, GB

**Abstract.** Effect size estimates are altered by many factors, including, and perhaps most importantly, the shapes of compared distributions. There have been many long time advocates of the necessity of graphing raw data to truly understand analysis. Though they were and remain correct, there is little evidence in the published literature in psychology that their recommendations have been followed. This paper argues their case, but with the advantage of the recent emphasis on effect sizes promoted by, amongst others, the American Psychological Association publication guide. Unlike Null Hypothesis Statistical Testing (NHST), effect size estimates are not robust to distributional deviations from normality. As a consequence of effect size sensitivity to distributional distortions from normality, it is all the more important to understand the qualities of the distributions from which estimates are derived. In this paper, we consider and simulate cases where graphical analyses reveal distortion in effect size estimates, and in doing so highlight the value of graphing data to interpret effect size estimates.

## 1 Introduction

Graphic approaches to understanding Social Science data lead to insights [1]. Cohen [2], echoing the advice given by Tukey [3], suggested that researchers should attempt to understand their raw data through graphic representation. Beyond the compelling visual examples modelled for interpreting confidence intervals by Cumming and Finch [4], there is not strong evidence that his advice has been followed with regularity. Perhaps there is reluctance on the part of researchers, reviewers, and editors to learn and consider a perceived myriad of techniques when they feel comfortable with an approved set of methods associated with null hypothesis statistical tests (NHST). Furthermore, computational aspects of NHST have been so

P. Cox, B. Plimmer, and P. Rodgers (Eds.): Diagrams 2012, LNAI 7352, pp. 117–123, 2012.

routinized that data analysis to some could seem like a matter of simply entering the data. Historically, an unintended consequence of Box [5] and earlier researchers in documenting the robustness of t and F tests to violations of normality may have also contributed. The general message from their studies is that t and F tests are so robust that researchers need not concern themselves with the distribution shape of their raw data. Therefore, the researcher is presented, on the one hand, with robust techniques that are well laid out, as versus, on the other hand, techniques which offer a learning curve perceived as steep.

Things may be ripe for change. Years of criticism of NHST has led to a greater emphasis on effect size measures [6], [7], [8]. Moving this approach into prominence could stimulate recognition of the value of graphic approaches, as we will try to demonstrate in this paper through the use of simulated and empirical data sets. As mentioned, many have tried before to guide researchers towards a more graphic approach, but in this current attempt, we have two advantages. One is from the emphasis on effect size, and the second comes through the benefit of hindsight. With hindsight, it is arguable that a graphic approach should 1) emphasize simplicity, 2) illustrate common or highly probable examples, and 3) link examples to known statistical techniques and descriptors. By simplicity, we mean raw data graphic approaches that are not overwhelming, and to which the majority of psychologists have been exposed at some point in their education. There are at least 39 major probability distributions [9]. That number alone can be intimidating. The potential number of moments for any distribution could perhaps be even more intimidating, since, in theory, it is the number of measures sampled minus one. However, by simply concentrating on the normal distribution and three major deviations reflected in variance, skew, and kurtosis, we argue that many cases in the social sciences are covered to the extent that most researchers will see value in graphing. In certain specialized areas (e.g., reaction time measurement), exploration beyond our presentation will be and has been undertaken. Means, variance, skew, and kurtosis are within the realm of training typical for social scientists, and they are very revealing about the structure of raw data for subsequent analysis. Virtually all researchers are intimately familiar with the $1^{st}$ and $2^{nd}$ moments, the mean and variance, respectively, and they have at the least a passing familiarity with the $3^{rd}$ and $4^{th}$ moments, skew and kurtosis. These moments are readily comprehended visually and are often focused on in Finance courses as key to describing stock market activity. All distributions can be at least partially understood by these moments.

The alleged robust nature of NHST can be counterproductive when considering effect sizes [5]. Effect sizes are meant to be accurate estimates of the size of a phenomenon, and the more accurate and precise the estimate the better. It turns out, however, that effect size estimates are not robust to the very distortions to which a statistical significance test supposedly is. Brand, Bradley, Best, and Stoica, [10], [11], [12] have spent some effort detailing when effect sizes may or may not be accurate reflections of the intended measure, but perhaps the most important situation arises with deviations from the normal distribution. These deviations are most readily apparent from graphs. It is evident with graphs that effect sizes depend very much on the underlying distribution assumptions, as we intend to show. Consider variance:

Standardized estimates of mean differences are based on estimates of variability, and, as a consequence of graphing, researchers may pay attention to variability.

## 2 Three Examples of Effect Size Sensitivity

**Variance Manipulation.** In the following illustration, the initial or control distribution was conceptualized as a distribution of 38 measures with a mean of 10 and a standard deviation of 2. A hypothetical manipulation created sets of distributions of 38 measures with means of 10.4, 11 and 11.6. These values correspond to Cohen's effect sizes of .2, .4 and .8. Standard deviations for each of the means in the second set of distributions ranged from .5 to 4. Effect sizes were calculated between the standard reference distribution and each of the manipulated distributions. The proportional differences remained approximately the same across different effect sizes, so one set of effect size numbers covers all cases. The effect sizes were reduced by 36% from the pooled estimates with the largest SD to an increase of 36% with the smallest SD. Accurate measurement is a hallmark of science but it may be difficult to obtain with not only error in measurement of means but also error in estimating variability.

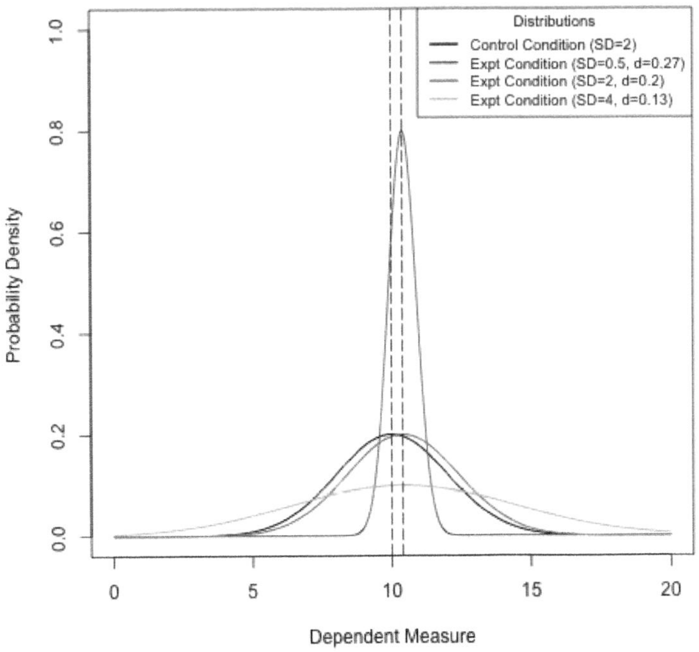


**Fig. 1.** A normal control distribution with mean = 10, SD = 2 compared with three distributions with a mean of 10.4 (d = .2) and SDs of .5, 2 and 4. The vertical dotted lines show the placement of the means.

Examination of the graphs in figure 1 show what is happening with robustness and a potential inaccuracy of standardized effect sizes. With graphs it is obvious that the differences in effect size are from the increase or decrease in variance. On the one hand, increased variability may, with traditional inference testing, result in a failure to obtain statistically significant results, whereas, with a decrease, not only is there an increase in the probability of statistical significance, but also the reported effect size may be exaggerated. We use "may be exaggerated" because analysis does not stop with the graphing of the data. Theory or past findings also matter. A decrease in variance could be legitimate if the manipulation does actually shrink variance. For example, nitrous oxide makes virtually all people laugh continuously during its application. On the other hand, it could be an artifact. For example, many measurement scales have a limited range and result in a compression of variance.

**Scales and Measurement.** To understand potential measurement effects, it is worth considering the data sets presented in figure 2. For example, Likert scales may have only five, seven, or 10 points, and manipulations that move participants' ratings unidirectionally away from a midpoint are almost certain to create skewed, leptokurtic distributions with restricted variability. The graphs presented in figure 2 are based on ten point scales, which seemed reasonable at the time of creation. However examination of the three graphs together may raise the questions as to whether or not the scales were nuanced enough to adequately discriminate amongst participants. Panel A shows love ratings amongst university-aged individuals. Ratings of love average 8.4 and are skewed and leptokurtic. Ratings of security, with a mean of 8.1 and similar levels of skew and kurtosis, follow the same pattern (see Panel B).

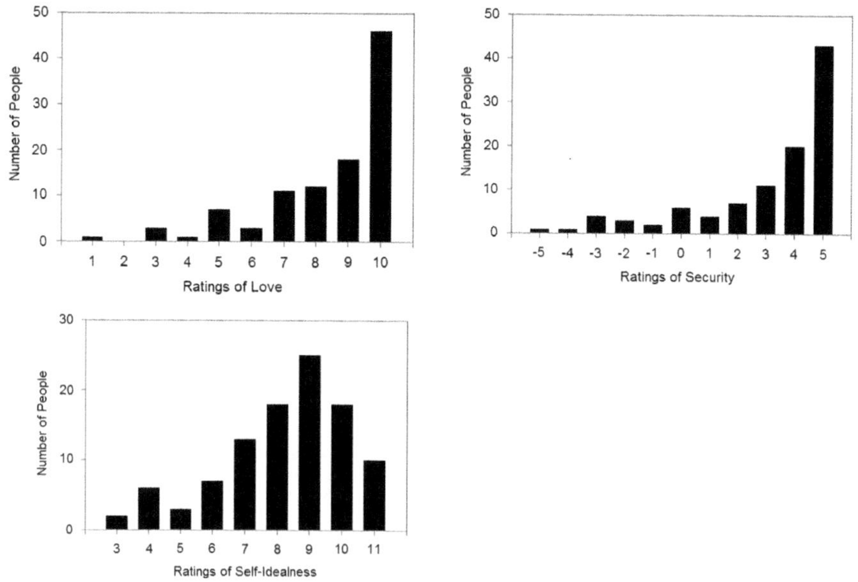


**Fig. 2.** Three raw data graphs showing different levels of deviation from normality to inform interpretation. Ratings of love (Panel A), ratings of security (Panel B), and ratings of self-idealness (Panel C) are presented.

The graphs revealed a further point of interest. There was actually an error in that the idealness scale went to 11. Graphing made this completely evident and showed the scale scores had to be reduced by 9%. With that reduction the mean was 7.5 and was different from the love mean. Thus graphing revealed two things: a clear error, and, even without the error, a distribution difference that reflected a less intense appraisal of self idealness than ratings of love. Overall, with a look at the graphs, a more complicated appreciation of data is gained in comparison to the simple analysis of means. The surprise was with idealness. This rating was an estimation of how ideal each individual estimated themselves for their particular partner (see Panel C). The distribution mean was 8.2 but the distribution was mesokurtic (approaching normal) and only mildly skewed. Thus individuals were rating themselves as less than ideal even though they were intensely in love. This could be interpreted as modesty in estimating one's own impact on another.

It is worth noting that Anscombe [13] had some time ago encouraged graphic analysis of raw data for the same distributional reasons we discuss. Anscombe [13] presented four distributions that visually were radically different from each other but shared equal means and variances. That paper, at the time, presented a compelling argument for graphic understanding of data, but it may not have had the impact it deserved for two reasons. It was written before the emphasis on effect size [6], and Anscombe [13] did not manipulate variances. With variance free to vary and effect sizes, as not only prominent metrics, but also demonstratively sensitive to variance manipulations perhaps there will be greater appreciation of this type of work.

**Bimodal Distribution.** Perhaps the most compelling case for the value of graphs could occur with bimodal distributions. In the following illustration, the beginning distribution approximates normality, whereas the manipulated distribution approximates a bimodal distribution. The means of the two distributions are the same for the simulation. Under this circumstance, a traditional F test discovers no statistically significant difference, and the effect size approaches 0. An analysis with no reference to graphs or higher moments of the distribution would suggest that nothing happened. However, examination of variability and kurtosis reveal distributions that differ from each other in important and informative ways. This simulation could model the evaluation of a politician. The initial description may be relatively neutral, and present a sincere, honest, established individual who has a family and is interested in serving the ordinary citizens of the country to the best of her/his ability. In figure 3, the ratings are represented by the unimodal normal curve depicted with the solid line. After the initial rating, the politician, in the North American context, could be identified with the contentious issue of gun control. The issue is potentially divisive enough to create a bimodal distribution which, in this case, is depicted with the dotted lines. The increase in variance and the increase in platykurtosis associated with the bimodal distribution, so clearly illustrated in figure 3, indicate that there are at least two groups reacting to this particular issue.

Examination of figure 3 makes it obvious that knowledge could be furthered by identifying two groups of responders, perhaps right wing and more centrist voters. The logical follow-up would be to create a more complex design. Before such an observation is trivialized, because we know some factors in this particular example, it should be considered that similar distribution changes can occur in drug research, and in the evaluation of art, movies and products.

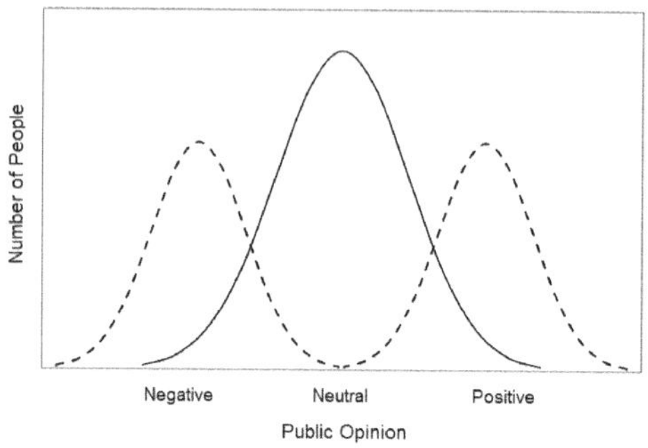


**Fig. 3.** Graph representation of a nonsignificant, 0 effect size comparison where the manipulated distribution is bimodal and obviously different from the original normal distribution

# 3 Conclusion

The human condition is complex, and the perpetuation of testing one mean against another without the thorough graphic examination of the distributions, as we have tried to present, needlessly limits the potential of Social science. At the most problematic level, a researcher may simply input data into a program, obtain a significance value, and then fail to look at the data distributions. The hesitation to further examine the data can range from a lack of realization of potential insights to be gained, to a misunderstanding of the conventions of research. Such conventions suggest that once a significance test fails there is little to be done with the data collected. That view may hold for well developed and understood areas, but it is arguably not the case for relatively underdeveloped areas of science. Furthermore, we argue, through our examples, that an appreciation of raw data distributions gained from graphing is a necessary adjunct to understanding effect size estimates, since these estimates are very sensitive to various and common departures from normality. This holds at the micro level of one's particular area and informs theory and measurement practice. At the macro level, involving general reading in one's discipline, or in new areas, and with important findings, it is necessary to have at least the trust, if not the actual graph or some form of evidence, that the author considered the raw data distribution form.

# References

1. Wainer, H., Velleman, P.F.: Statistical graphics: Mapping the pathways of science. Annual Review of Psychology 52, 305–335 (2001)
2. Cohen, J.: The Earth is round (<.05}. American Psychologist 49(12), 997–1003 (1994)
3. Tukey, J.W.: The Future of Data Analysis. The Annals of Mathematical Statistics 33, 1–67 (1962)
4. Cumming, G., Finch, S.: Inference by eye: Confidence intervals and how to read pictures of data. American Psychologist 60(2), 170–180 (2005)

5. Box, G.E.P.: Non-normality and tests on variances. Biometrika 40(3/4), 318–335 (1953)
6. Wilkinson, L., Task Force on Statistical Inference: Statistical Methods in Psychology Journals: Guidelines and Explanations. American Psychologist 54(8), 594–604 (1999)
7. American Psychological Association: Publication manual of the American Psychological Association, 5th edn., Washington, DC (2001)
8. American Psychological Association: Publication manual of the American Psychological Association, 6th edn., Washington, DC (2010)
9. Evans, M., Hastings, N., Peacock, B.: Statistical distributions, 2nd edn. John Wiley and Sons (1993)
10. Brand, A., Bradley, M.T., Best, L., Stoica, G.: Accuracy of effect size estimates from published psychological research. Perceptual and Motor Skills 106, 645–649 (2008)
11. Brand, A., Bradley, M.T., Best, L., Stoica, G.: Multiple Trials May Yield Exaggerated Effect Size Estimates. Journal of General Psychology 138(1), 1–11 (2011)
12. Brand, A., Bradley, M.T., Best, L., Stoica, G.: Accuracy of Effect Size Estimates from Published Psychological Experiments Involving Multiple Trials. Journal of General Psychology 138(4), 281–291 (2011)
13. Anscombe, F.J.: Graphs in statistical analysis. American Statistician 27, 17–21 (1973)

# Psychological Evidence of Mental Segmentation in Table Reading

Takeshi Sugio[1], Atsushi Shimojima[1], and Yasuhiro Katagiri[2]

[1] Doshisha University, 1-3 Tatara-Miyakodani, Kyotanabe, Kyoto, Japan
[2] Future University Hakodate, 116-2 Kamedanakano, Hakodate, Hokkaido, Japan

**Abstract.** How we organize elements when reading a table was examined in a psychological experiment using a modified spatial-cuing paradigm. Table-like stimuli consisting of 16 square elements arranged in a four-by-four matrix form were used. Participants were instructed to discriminate whether the presented stimuli could be read as containing either one element or two elements in accordance with the induced reading direction. The results showed that when two elements were presented along with the induced direction, it was easier to read as such than when two elements were presented orthogonal to the induced direction. Although there was no contour line in the stimuli, participants were able to mentally segment and organize them into global units lying in the particular direction, which was instrumental to reading the tables efficiently.

**Keywords:** object-based attention, table reading, global unit.

## 1 Introduction

Empirical studies on the comprehension of information graphics can be divided into several broad categories according to what aspect of the comprehension process they are interested in. One traditional line of research is concerned with the *estimation* process of the relative magnitudes of individual graphical objects, such as the lines, bars, and pie slices in statistical charts [1,2]. Active research is also conducted on the *interpretation* process, where the features obtained in the estimation process are translated into meaningful information on a represented domain. Typical subjects are the translation of a steeper line in a line graph to a faster pace of change [3] and of a "descendant staircase" in a bar graph to a decreasing trend [4] as well as other interpretation processes yielding what we later discuss as "higher level" information. In addition, there is an emerging body of research shedding light on a more fundamental aspect of graphics comprehension, namely, the *attention* process, where graphical objects to be estimated and interpreted are segregated out from other candidate objects. Several researchers (e.g., [5,6]) started investigating how people allocate attention to different locations in information graphics by using participants' eye movements and fixations as clues.

Strictly speaking, however, the issue of attentional shifts and focusing is different from the issue of attentional *content*—the order and durations in which different locations are attended to do not determine what exact graphical objects are segregated during that process. It is this latter issue that the present work is concerned with. We borrowed a spatial-cuing experimental paradigm [7] used in standard attention research

P. Cox, B. Plimmer, and P. Rodgers (Eds.): Diagrams 2012, LNAI 7352, pp. 124–131, 2012.

and applied it to investigate the working of our attentional system over a typical tabular representation (See Fig. 1 in section 2).

A dedicated investigation on attentional content is also important for at least two reasons. The first reason is related to the phenomenon of *object-based attention*. Roelfsema and his colleagues [8,9] proposed computational and neurological models of the "bounded activation" operation in Ullman's sense [10], whose function is to define coherent units of regions in an unarticulated visual scene so that further operations can be applied selectively to the activated regions. "Object-based attention," actively investigated by Duncan [11] and Kramer and Jacobson [12], largely overlaps with the operation of bounded activation.

Given that we are endowed with this ability, the content of our attention is not a simple function of the center of attention as indicated by the fixation center. Attention may spread over various graphical elements in and near the center, grouping them into a single "global unit" to which estimation and interpretation operations are applied. Multiple global units may be bounded in this manner, even forming a hierarchical structure. What objects are bounded and how they are organized in the attentional process therefore need dedicated examination.

The second reason is not particular to graphics comprehension research but related more generally to the functional significance of object-based attention. Object-based attention has been a central topic in the psychological research of visual attention, and its existence has been abundantly evidenced. When it comes to its *functions*, however, much still remains unclear. Why do we need to bound our attention into a particular unit of regions rather than just radiating it from a certain center?

One plausible explanation is that the units of regions we bound our attention into strongly affect the kinds of information we read off from a scene, and graphics comprehension is a quintessential example of the reading process affected in that way. Researchers and designers have long noted that information graphics can express "higher level" information as well as "lower level" information (e.g., [13,14,4,15]), and what higher level information we read off depends heavily on what units of regions we attend to. Taking the case of a scatter plot, for example, we usually read off the strength of correlation between two variables by attending to the "cloud" formed by multiple dots as a unit of interpretation [16]. We usually read off the existence of two dominant subfamilies from a family tree by attending to the relevant sub-trees as units of interpretation. Likewise, a membership table usually lets us read off the number of groups to which more than two members belong only when we attend to rows (or columns) as units of interpretation. In two experiments using the modified spatial-cuing paradigm, we attempt to demonstrate the involvement of object-based attention in the reading of information in the table. By connecting research on object-based attention to graphics comprehension research in this way, we can expect to identify one concrete function of object-based attention.

## 2 Experiment

It has been shown that object-based attentional effects can be separated from space-based effects by controlling the distances among cued locations and target locations

[7]. We used the experimental paradigm developed by [7] to show the degree of the involvement of the object-based attention that underlies our ability to read tables.

We conducted a preliminary experiment by using a detection task, which was used to detect whether a target (filled square) was present in the table-like stimulus. We found that the mean reaction time in the valid-cue condition was faster than those in invalid-cue conditions. A significant difference was not observed between two invalid-cue conditions (within-object and between-object), which suggested that participants were able to extend their attentional focus to entire stimuli regardless of the reading direction instructed in the detection task. On the basis of this preliminary result, we investigated whether such attentional enhancement can also be found for more complex task like discrimination.

## 2.1 Method

*Participants, Apparatus, and Stimuli.* The participants were 15 undergraduate students, of which 11 were females. All had normal or corrected-to-normal vision and were paid for their participation. The experiment was programmed and conducted using Psychtoolbox (version 3.0.8) running on a MATLAB (R2007b) platform on an IBM-compatible PC [17]. A response pad (Cedrus RB-530) was used to collect data. All stimuli were presented on a 17-in. color monitor (85-Hz refresh rate) at a viewing distance of about 57 cm. Participants' heads were immobilized with a chin rest. The experiment took place in a dark room.

The fixation cross, letters (labels and instructions used to induce the reading direction of the stimuli), and element squares (including targets, filled squares) were gray, and the cue was the color white. The fixation cross was a plus sign (+), which subtended about $0.5 \times 0.5$ deg in a visual angle. Row labels, column labels, and the letters G (**G**roup) and P (**P**erson) were the same size as the fixation. Each square subtended about $2 \times 2$ deg with a stroke of approximately 2.0 deg, and the distances between two squares were 1.3 deg in both directions.

*Design and Procedure.* Before beginning the experiment, several examples of the table-like stimuli were presented to participants with descriptions on how the stimuli were to be comprehended in accordance with the reading direction. Each trial started with a preview display (16 squares in $4 \times 4$ arrangement with the instruction letter at the center) for 1,000 ms. Participants were instructed to remember the instructed reading direction of the target display. Then, the cue (white) was presented at one of the four corners for 100 ms. The cue was replaced with the fixation display (16 squares with a fixation cross at the center) for another 200 ms. The target (filled square or squares) was presented, and it overlapped this fixation display until there was a response from pressing a button or 2,000 ms elapsed with no response. The task was to discriminate the table-like stimuli on the basis of the number of people that each group had ("G") or the number of groups that each person belonged to ("P"). Figure 1 illustrates how the identical stimuli can be interpreted differently according to the instructed reading direction. Participants pressed the assigned button (left or right button of the response pad) for each number ("one" or "two") as rapidly and as accurately as possible. A feedback beep was presented for errors. The next trial began after a 500-ms blank intertrial interval. Participants

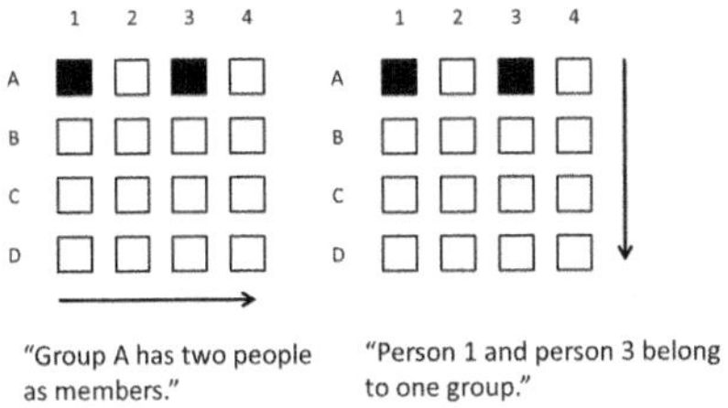


**Fig. 1.** Difference in the interpretation of the table according to the reading direction

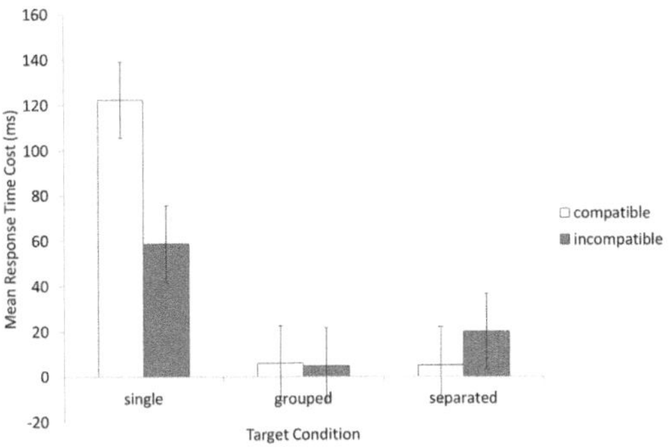


**Fig. 2.** Mean response time costs of invalid-cuing for correct trials between target type and instruction compatibility. Error bars represent 95% within-confidence intervals [18].

were instructed to maintain fixation on the center throughout each trial. The order was randomized for each participant. There were 1,600 trials in all, and they were divided into 16 blocks of 100 trials each. A self-paced rest was allowed between blocks.

One of the targets always appeared at the cued location for the valid condition (76% of the trials). For the single target conditions, the target appeared either at the cued location (valid cue) or at the uncued location (invalid cue, either compatible or incompatible with the instructed reading direction). For two-target conditions, the spatial relation between two elements was manipulated (grouped or separated).

## 2.2 Results

Two separate cost-benefit analyses were performed. One was based on the difference in response times between valid and invalid-cuing conditions, which indicates the degree of the effectiveness of spatial cuing at a particular location in the table-like stimuli. The other was based on the difference in response times between single-target and two-target conditions, which may reflect the additional time necessary for processing an element in the stimuli.

First, a two-way repeated-measures ANOVA was performed for the difference in median response times for the valid and invalid conditions (Figure 2). The two within-

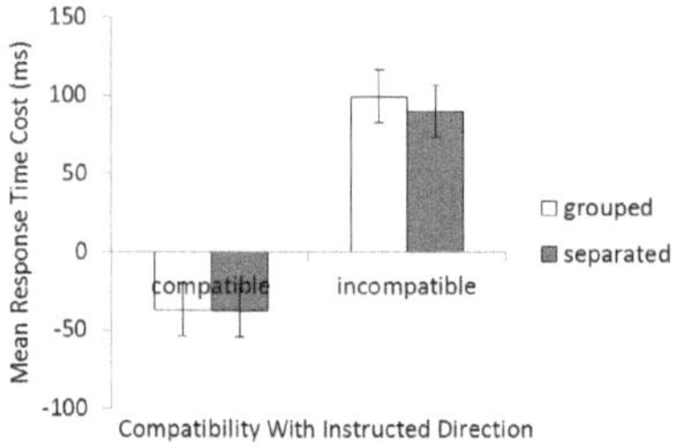


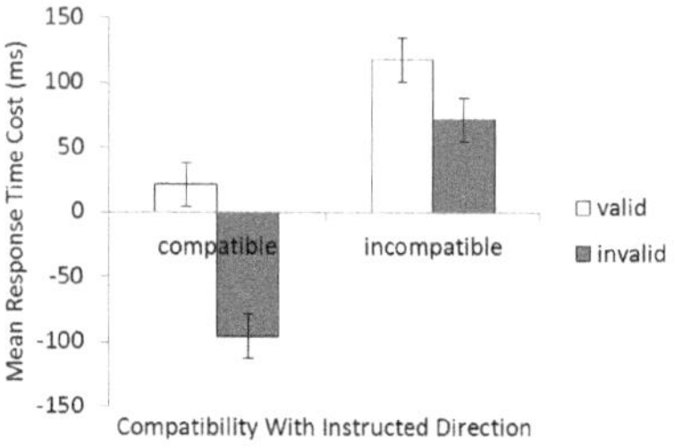


**Fig. 3.** Mean response time costs of processing an additional target for correct trials between target relation and instruction compatibility (left) and between cue validity and instruction compatibility (right). Error bars represent 95% within-confidence intervals [18].

participant factors were target type (single target, grouped targets, or separated targets) and compatibility with instruction (compatible or incompatible). The main effect of the target type and the interaction between the target type and compatibility were significant; $F(2, 28) = 60.619$, $p < .001$, $F(2, 18) = 14.963$, $p < .001$. The main effect of compatibility was marginally significant; $F(1, 14) = 4.580$, $p = .050$. A post-hoc analysis using Shaffer's method revealed significant differences between a single target and grouped targets and between a single target and separated targets (all $p < .05$). Simple main effects were analyzed for the interaction. The simple main effect of a target type was significant in both compatible and incompatible conditions (all $p < .05$). In comparison, the simple main effect of an instruction-compatibility was significant in a single-target condition only ($p < .05$).

Next, a three-way repeated-measures ANOVA was performed for the difference in median response time for single-target and two-target conditions. The three within-participant factors were the distance between two targets (grouped or separated), cue validity (valid or invalid), and compatibility with an instruction (compatible or incompatible). All main effects and two interactions (target relation × instruction compatibility, cue validity × instruction compatibility) were significant; $F(1, 14) = 4.924$, $p = .044$, $F(1, 14) = 86.143$, $p < .001$, $F(1, 14) = 184.999$, $p < .001$, $F(1, 14) = 5.213$, $p = .039$, $F(1, 14) = 26.401$, $p < .001$ (Figure 3). The simple main effect for the interaction between target relation and instruction compatibility showed that when the target location was along the same reading direction with the cue, there was no difference in discrimination time between grouped and separated targets. In comparison, when the target location was on the orthogonal reading direction to the cue, discriminating grouped targets as each belonging to different group (or person) required more time than with separated targets (100 – 90 ms). The simple main effects of instruction compatibility were significant for both grouped and separated targets ($p < .05$). Furthermore, all simple main effects for the interaction between cue validity and instruction compatibility were significant ($p < .05$).

## 2.3 Discussion

The interaction between target type and instruction compatibility showed quite a discrepant result with previous studies [7]. Since the task was to discriminate whether the presented stimuli could be read as containing only one element or two elements for each group (or person), participants had no need to continue the search for another object when they successfully found two elements along the induced reading direction. However, when there was only one element within the object that participants were attending to, they had to continue the search to confirm their responses, and they also had to maintain which object had been already attended to in their working memory.

Since the exogenous cue automatically drew attention to the cued location, the information reading process must have begun at the cued location. It is plausible that the automatic allocation of attention to the cued location lead to inefficient processing when sufficient information for making a decision could not be obtained there. Attending to a specific element may have activated an incorrect response in such trials, and participants had to inhibit the incorrect response.

A considerable amount of difference in the costs of processing additional elements between two instruction-compatibility conditions was observed. This result indicates that multiple targets appearing within the cued mental object were easily discriminated as two independent elements irrespective of the proximity between them. In other words, the spatial resolution of the attended mental object was better than that of an unattended one.

In comparison, when two targets were presented at locations orthogonal to the induced reading direction, the discrimination time was longer than in the single-target condition, which suggests that each mental object was searched sequentially. However, considering that the cost for discriminating grouped targets was larger than that of separate targets, the search result of each uncued object seemed to be maintained in a degraded manner, namely, it was a cost to discriminate two consecutive objects as two distinct objects. Since there was no contour line presented in the stimuli, it was demanding to maintain multiple unattended objects in working memory. As indicated from the simple main effect of instruction compatibility for a valid cue, the cost of processing an additional element was rather small, suggesting the enhanced discrimination process within the attended object.

These results suggest that space-based and object-based attention may not have contributed to the efficient processing of information in a simple additive manner. Rather, space-based attention seems to have dominated object-based attention when both are directed at the same area in the stimuli. The function of object-based attention is not to further highlight particular portions in the stimuli but to bound sub-areas within it. The nature of the top-down knowledge that served to segment mental objects remains to be investigated. Some sort of rule-based processing, such as syntactic processing, might be one of the possibilities [19].

## 3 Conclusion

Combined with a preliminary experiment, the experiment indicates our ability to keep the entire array of a table activated for feature detection while maintaining a particular segmentation of the array into a set of rows or columns depending on the task at hand. Contrast our finding to a more simple-minded possibility that task-dependent segments (rows or columns) are segregated sequentially as the comprehension of the table proceeds from one segment to another. Our results indicate a more parallel attentive mechanism, where a global unit in the display, along with its sub-units, is maintained simultaneously. The existence of this hierarchical structure was evidenced by the facilitation of target-detection within the global unit that is comparable to the facilitation within a rectangle observed in [7] (the preliminary experiment) and the facilitation of target-discrimination within each sub-unit regardless of which particular sub-unit the cue had fallen into (the main experiment).

Here, the segregation of the global unit appears to be space-based, so the spot light metaphor is appropriate here. Yet, the lighting is not plain but rather divided, with individual divisions allocated to different task-dependent sub-units. Although our experiments were confined to a table, we can easily imagine that this strategy of global spot-lighting and local segmentation is extended to portions of node-link graphs, maps, statistical charts, or any information graphics of which the segmentation of non-well-defined objects is functionally important. This has a deep implication on the way we extract information from information graphics, especially because it implies that multiple sub-units can be maintained in parallel so far as they are within the coverage of a spot light, aka, space-based attention to a superior global unit. Our results also have an important implication on the functional significance of object-based attention: it serves to segment the area segregated by space-based attention. Here, the role of object-based attention is only segmentation (not initial segregation), but it is informed of the demand of the reading task at hand. This task-sensitivity appears to be an important feature of object-based attention, apparently missing from space-based attention.

## References


1. Cleveland, W.S., McGill, R.: An experiment in graphical perception. International Journal of Man-Machine Studies 25(5), 491–500 (1986)
2. Spence, I.: Visual psychophysics of simple graphical elements. Journal of Experimental Psychology: Human Perception and Performance 16(4), 683–692 (1990)
3. Gattis, M., Holyoak, K.J.: Mapping conceptual to spatial relations in visual reasoning. Journal of Experimental Psychology 22(1), 231–239 (1996)
4. Pinker, S.: A theory of graph comprehension. In: Aritificial Intelligence and the Future of Testing, pp. 73–126. L. Erlbaum Associates (1990)
5. Carpenter, P., Shah, P.: A model of the perceptual and conceptual processes in graph comprehension. Journal of Experimental Psychology: Applied 4(2), 75–100 (1998)
6. Peebles, D., Cheng, P.C.: Modeling the effect of task and graphical representation on response latency in a graph reading task. Human Factors 45(1), 28–46 (2003)
7. Egly, R., Driver, J., Rafal, R.D.: Shifting visual attention between objects and locations: Evidence from normal and parietal lesion subjects. Journal of Experimental Psychology: General 123(3), 161–177 (1994)
8. Roelfsema, P.R., Lamme, V.A.F., Spekreijse, H.: The implementation of visual routines. Vision Research 40, 1385–1411 (2000)
9. Roelfsema, P.R.: Cortical algorithms for perceptual grouping. Annual Review of Neuroscience 29, 203–227 (2006)
10. Ullman, S.: Visual routines. Cognition 18, 97–159 (1984)
11. Duncan, J.: Selective attention and the organization of visual information. Journal of Experimental Psychology: General 113(4), 501–517 (1984)

12. Kramer, A.F., Jacobson, A.: Perceptual organization and focused attention: the role of objects and proximity in visual processing. Perception & Psychophysics 50(3), 267–284 (1991)
13. Bertin, J.: Semiology of Graphics: Diagrams, Networks, Maps. The University of Wisconsin Press, Madison (1973)
14. Tufte, E.R.: Envisioning information. Graphics Press, Cheshire (1990)
15. Cleveland, W.S.: The Elements of Graphing Data. Hobart Press, Summit (1994)
16. Kosslyn, S.M.: Elements of Graph Design. W. H. Freeman and Company, New York (1994)
17. Brainard, D.H.: The psychophysics toolbox. Spatial Vision 10(4), 433–436 (1997)
18. Masson, M.E.J., Loftus, G.R.: Using confidence intervals for graphically based data interpretation. Canadian Journal of Experimental Psychology 57(3), 202–220 (2003)
19. Jansen, A.R., Marriott, K., Yelland, G.W.: Comprehension of algebraic expressions by experienced users of mathematics. The Quarterly Journal of Experimental Psychology 56A(1), 3–30 (2003)

# Proof-Theoretical Investigation of Venn Diagrams: A Logic Translation and Free Rides

Ryo Takemura

Nihon University, Japan
takemura@abelard.flet.keio.ac.jp

**Abstract.** In the literature on diagrammatic reasoning, Venn diagrams are abstractly formalized in terms of minimal regions. In view of the cognitive process to recognize Venn diagrams, we modify slightly the formalization by distinguishing conjunctive, negative, and disjunctive regions among possible regions in Venn diagrams. Then we study a logic translation of the Venn diagrammatic system with the aim of investigating how our inference rules are rendered to resolution calculus. We further investigate the free ride property of the Venn diagrammatic system. Free ride is one of the most basic properties of diagrammatic systems and it is mainly discussed in cognitive science literature as an account of the inferential efficacy of diagrams. The soundness of our translation shows that a free ride occurs between the Venn diagrammatic system and resolution calculus. Furthermore, our translation provides a more in-depth analysis of the free ride. In particular, we calculate how many pieces of information are obtained in the manipulation of Venn diagrams.

## 1 Introduction

Venn and Euler diagrams are two of the most basic diagrams for logical reasoning, originally introduced to illustrate syllogisms. These diagrams have been studied since the 1990s from both a mathematical and formal logic viewpoint, and have been applied, beyond simple syllogisms, to various areas such as knowledge representation and ontologies.

In the literature on diagrammatic reasoning, a Venn diagram is abstractly defined as a set of *minimal regions* (refer to Howse et al. [4] and to [11] for a survey).

For example, the Venn diagram, $\mathcal{V}$, in Fig. 1 is specified as a diagram in which "the region inside $A$, but outside $B$ and $C$ (denoted by $A\overline{B}\overline{C}$) is empty," and "the region inside $A$ and $C$, but outside $B$ ($A\overline{B}C$) is empty." Thus, $\mathcal{V}$ is abstractly formalized as the set of shaded minimal regions $\{A\overline{B}\overline{C}, A\overline{B}C\}$.

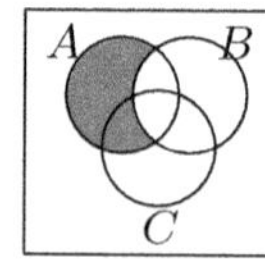


**Fig. 1.** Venn diagram, $\mathcal{V}$

Although we can read from this $\mathcal{V}$, "the region inside $A$, but outside $B$ ($A\overline{B}$) is empty," such a (non minimal) region is generally defined as the union of minimal regions, i.e., $A\overline{B}\overline{C} \cup A\overline{B}C$, and hence, only minimal regions are mathematically

P. Cox, B. Plimmer, and P. Rodgers (Eds.): Diagrams 2012, LNAI 7352, pp. 132–147, 2012.

sufficient for the abstract formalization of Venn diagrams. Based on this formalization of Venn diagrams, Euler diagrams are formalized by considering shaded regions of Venn diagrams as "missing" regions (cf. [4]). Thus, both Venn and Euler diagrams are abstractly formalized in terms of minimal regions, and we call this framework a "region-based" framework.

Based on the above formalization of Venn and Euler diagrams, their influence on human reasoning, particularly syllogistic reasoning, is discussed in the cognitive science literature by Gurr-Lee-Stenning [3] and Shimojima [9], amongst others. From a cognitive viewpoint, it is often observed that Venn diagrams are harder to handle in actual reasoning than Euler diagrams. However, from the mathematical formalization of Venn and Euler diagrams in the region-based framework, it is difficult to derive any differences between them especially in their cognitive complexities [3].

In contrast to studies using the region-based framework, Mineshima-Okada-Takemura [5] introduced another framework to formalize Euler diagrams. The framework is based on the idea that, from a cognitive viewpoint, the most basic and essential component of Euler diagrams is topological (inclusion and exclusion) relations between circles. Thus, in this framework, Euler diagrams and manipulations thereof are formalized in terms of the topological relations. The inference rules are designed to be as natural as possible to reflect intuitive manipulations of Euler diagrams. We describe this approach being "relation-based."

Along similar lines, if we consider processes to recognize Venn diagrams, it seems that the existing mathematical formalization in terms of minimal regions is somewhat inadequate. For example, in the diagram $\mathcal{V}$ in Fig. 1, in order to read "There is nothing that is $A$ but not $B$," i.e., "All $A$ are $B$," it is not sufficient to grasp the emptiness of the minimal regions $A\overline{B}\overline{C}$ and $A\overline{B}C$ discretely. Instead it is necessary to grasp the emptiness of the union of these regions as a whole. Thus, it is considered appropriate for the formalization of Venn diagrams to take into account the unions of certain minimal regions. Based on this idea, we modify slightly the region-based formalization of Venn diagrams. Of the possible regions in Venn diagrams, we distinguish conjunctive, negative, and disjunctive regions. In particular, each conjunctive region denotes "Things that are $X_1$ and ... and $X_n$, but not $Y_1$ or ... or $Y_m$." As the most basic system, we are mainly concerned with a formal Venn diagrammatic system based on conjunctive regions in Section 2, and we define an abstract Venn diagram as a set of shaded conjunctive regions.

In Section 3, we investigate a proof theory of the Venn diagrammatic system. In particular, we study a logic translation of the system with the aim of investigating how our Venn diagrammatic inference rules are rendered to resolution calculus. The logic translation enables us various further investigation of the Venn diagrammatic system. We investigate, among others, the free ride property of our system in Section 4. Free ride occurs when by adding a certain piece of information to a diagram, the resulting diagram somehow comes to represent pieces of information not contained in the given diagram or in the original piece of information. Shimojima [9] analyzed its semantic conditions within the framework of Barwise-Seligman's channel theory [1].

In contrast to the semantic investigation, Takemura [12] introduced a proof-theoretical framework to analyze the free ride through logic translations. It was shown that the soundness of a translation between two logical systems implies the occurrence of free ride (in the sense of [9,1]) between the systems. Based on the result, we show that free ride occurs between our Venn diagrammatic system and resolution calculus in Section 4. Our logic translation gives us a more in-depth analysis of free ride. We are able to investigate what kind of and how many inference steps are required in terms of the translated logical system, to derive freely obtained pieces of information in the manipulation of diagrams. In particular, we investigate the number of pieces of information, that we call "free rides," occurring in an application of a Venn diagrammatic inference rule, which serves to estimate recognition steps for the free rides.

Since our formalization and analysis of the free ride is based on the general method of logic translation in the proof theory, they are also applicable to other appropriately defined diagrammatic systems.

In Section 5, based on the analysis given in Section 4 and that given in [12], we discuss a difference between the Venn and Euler diagrammatic systems with respect to free rides.

## 2 Venn Diagrammatic System

In this section, we recall the syntax, semantics, and inference rules for Venn diagrams. See, e.g., Howse et al. [4], Shin [10] for detailed and formal descriptions thereof. In particular, in Section 2.1 we introduce a classification among regions in Venn diagrams, namely conjunctive, negative, and disjunctive regions. Inference rules, described in Section 2.2, are those defined in [6].

### 2.1 Venn Diagrams

A (concrete) **Venn diagram** (denoted by $\mathcal{V}, \mathcal{V}_1, \mathcal{V}_2, \dots$) consists of a finite number of named circles (i.e., simple closed curves) on a plane $\mathbb{R}^2$ enclosed by a boundary rectangle that satisfies the *partial-overlapping* condition, i.e., all possible intersections of circles must occur.

A **minimal region** (denoted by $z, z_1, z_2, \dots$) is a part of the plane that lies inside some of the circles and outside the remaining circles of the diagram. (Although our "minimal region" may be more appropriately called a **zone** as in [4], we prefer to retain our terminology in this paper.) Thus, if $\mathcal{L}$ is the set of names of circles in a Venn diagram $\mathcal{V}$, each minimal region is specified using the names in $\mathcal{L}$ as $X_1 \dots X_n \overline{Y_1} \dots \overline{Y_m}$, where $\{X_1, \dots, X_n, Y_1, \dots, Y_m\} = \mathcal{L}$, and $X_1, \dots, X_n$ are the names of the circles enclosing the region, $Y_1, \dots, Y_m$ are the names of the circles outside of which the region lies.

A (general) **region** is the union of certain minimal regions.

*Example 1.* The following Venn diagram $\mathcal{V}$ consists of two circles $A$ and $B$, which have four minimal regions: $z_1$ is inside $A$, but outside $B$ (and is denoted by $A\overline{B}$); $z_2$ is inside both $A$ and $B$ ($AB$); $z_3$ is outside $A$, but inside $B$ ($\overline{A}B$); and $z_4$ is outside both $A$ and $B$ ($\overline{A}\overline{B}$).

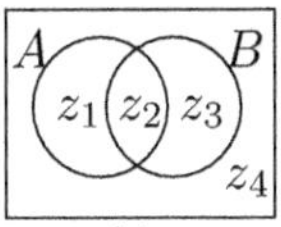


$\mathcal{V}$

In general, we denote by $r$ a sequence, say $X_1 \dots X_n \overline{Y_1} \dots \overline{Y_m}$, of names and overlined names of circles, in which we ignore the order and repetition of elements, i.e., we regard $r$ as a set. Then, by $|r|$ we denote the set of names $\{X_1, \dots, X_n, Y_1, \dots, Y_m\}$.

For any sequences $r$ and $r'$, the concatenation thereof is denoted by $rr'$.

Of the regions in a given Venn diagram, we distinguish conjunctive, negative, and disjunctive regions.

**Definition 1 (Conjunctive region).** A **conjunctive region** $r$ of a Venn diagram $\mathcal{V}$ is a region of $\mathcal{V}$ that is specified by the names of circles in $\mathcal{V}$ as follows. For $n \geq 1, m \geq 0$,

$$X_1 \dots X_n \overline{Y_1} \dots \overline{Y_m} \ .$$

A region specified by using only overlined names of the form $\overline{Y_1} \dots \overline{Y_m}$ is called a **negative region**. A region that is neither conjunctive nor negative is called a **disjunctive region**.

Observe that if $\mathcal{L}$ is the set of names of circles in $\mathcal{V}$, a conjunctive or negative region $r$ such that $|r| = \mathcal{L}$ is a usual minimal region. Note also that conjunctive, negative, and disjunctive regions are disjoint.

*Example 2.* In the diagram $\mathcal{V}$ in Example 1, the regions $z_1$ $(= A\overline{B}), z_2$ $(= AB), z_3$ $(= \overline{A}B), z_1 \cup z_2$ $(= A)$, and $z_2 \cup z_3$ $(= B)$ are all conjunctive (the first three are also minimal); regions such as $z_4$ $(= \overline{A}\overline{B})$, and $z_3 \cup z_4$ $(= \overline{A})$ are negative; while the other regions e.g., $z_1 \cup z_3$, and $z_1 \cup z_2 \cup z_3$, are disjunctive.

Each conjunctive region can naturally be read as "Things that are $X_1$ and ... and $X_n$, but not $Y_1$ or ... or $Y_m$." We are mainly concerned with conjunctive regions in our formalization of Venn diagrammatic system, which can be extended by including other regions.

Each conjunctive region may be **shaded**. In particular, a Venn diagram in which no regions are shaded is called a **primary diagram**. For a Venn diagram $\mathcal{V}$, we denote by $m(\mathcal{V})$, $shm(\mathcal{V})$, and $shc(\mathcal{V})$ the set of names of minimal regions, shaded minimal regions, and shaded conjunctive regions, respectively of $\mathcal{V}$.

An **abstract Venn diagram** is defined as the set of shaded conjunctive and/or negative, disjunctive regions. Instead of giving a formal definition of the abstract syntax here, we explain it through our translation into a resolution calculus in Section 3.2.

Although we do not deal explicitly with points and the linking thereof in this paper, we may technically regard points (with linking) in Venn diagrams as special circles that do not contain or overlap any other circles. See [6].

We define the semantics of Venn diagrams in the same way as Howse et al. [4]. A **model** is a pair $M = (U, I)$, where $U$ is a non-empty set called the universe, and $I$ is an interpretation function that assigns to each circle a subset of $U$. The interpretation function $I$ is naturally extended to interpret conjunctive regions as follows. For any conjunctive region $r$, specified as $X_1 \ldots X_n \overline{Y_1} \ldots \overline{Y_m}$, the interpretation $I(r)$ is defined by $I(r) = I(X_1) \cap \cdots \cap I(X_n) \cap \overline{I(Y_1)} \cap \cdots \cap \overline{I(Y_m)}$, where $\overline{I(Y_j)}$ is the complement of the set $I(Y_j)$. $M = (U, I)$ is a model of the Venn diagram $\mathcal{V}$, denoted as $M \models \mathcal{V}$, if each shaded conjunctive region is interpreted as the empty set, i.e., $\bigcup_{r \in shc(\mathcal{V})} I(r) = \emptyset$.

## 2.2 Venn Diagrammatic Inference System

We review the most basic Venn diagrammatic inference system $\mathsf{V}$ from Mineshima-Okada-Takemura [6]. See [4,10], for example, for a formal description of rules.

**Definition 2 (Inference rules for $\mathsf{V}$).**

Axiom Any primary diagram (i.e., a diagram in which no regions are shaded) is an axiom.

Introduction of a circle Let $\mathcal{V}$ be a Venn diagram that does not contain a circle $A$. Then, to obtain the conclusion diagram $\mathcal{V}+A$, add circle $A$ to $\mathcal{V}$ observing the partial-overlapping rule, i.e., each conjunctive or negative region (as well as shaded one) of $\mathcal{V}$ is split into two regions.

Superposition of diagrams Let $\mathcal{V}_1$ and $\mathcal{V}_2$ be Venn diagrams that contain the same circles. Then, to obtain the conclusion diagram $\mathcal{V}_1 + \mathcal{V}_2$, construct a primary diagram with the same circles as $\mathcal{V}_1$. Then, shade all minimal regions that are shaded in $\mathcal{V}_1$ or $\mathcal{V}_2$.

Erasure of shading Let $\mathcal{V}$ be a Venn diagram in which a minimal region $z$ is shaded. Then, to obtain the conclusion diagram $\mathcal{V} - \{z\}$, erase the shading of $z$.

Erasure of a circle Let $\mathcal{V}$ be a Venn diagram that contains a circle $A$. Then, to obtain the conclusion diagram $\mathcal{V} - A$, circle $A$ must be erased from $\mathcal{V}$ and any shading remaining in only a part of a minimal region after the erasure of $A$ should also be erased.

The notion of *Venn diagrammatic proof* is defined inductively as tree structures consisting of the above inference rules (see Example 3 below). Note in particular that each leaf of a proof-tree is a premise diagram or a primary diagram (i.e., an axiom).

*Example 3.* Fig. 2 is a Venn diagrammatic proof of $\mathcal{V}_3$ from premises $\mathcal{V}_1$ and $\mathcal{V}_2$.

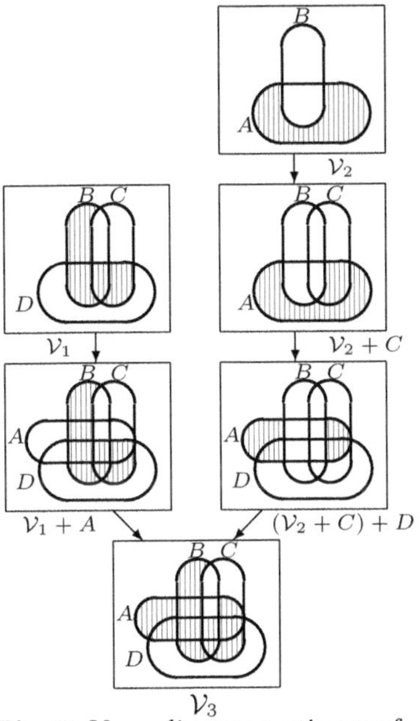


**Fig. 2.** Venn diagrammatic proof

# 3 Translation of Venn Diagrammatic System

In this section, after a brief review of the resolution calculus in Section 3.1, we define a translation of the Venn diagrammatic inference system V into resolution calculus in Section 3.2. In particular, we translate the rule for Superposition (instead of the rule for Erasure of a circle as in [6]) into a combination of the resolution principle.

## 3.1 Resolution Calculus

Propositional and first-order resolution were introduced by Robinson "for use as a basic theoretical instrument of the computer theorem-proving program" [7]. The "resolution principle" is very powerful in that it forms by itself a complete system of propositional and first-order logic. The efficiency of the rule makes it easy to implement decision procedures involved in establishing the provability of given formulas. The resolution principle gives the theoretical basis of logic programming such as Prolog. For details refer to [2], for example.

A **literal** is either an atom $A$ or its negation $\overline{A}$. In the context of resolution, we use the "overbar" symbol for negation instead of the usual $\neg$. We denote literals by $L, L_1, L_2, \ldots$. If $L$ is a literal of the form $\overline{A}$, then $\overline{L}$ denotes the unnegated literal $A$. A **clause** is a finite set of literals, and is denoted by $r, x, y, z, \ldots$. If a clause $r$ is $\{L_1, \ldots, L_n\}$, it is usually expressed as $L_1 \cdots L_n$. A set of clauses is called a **clause set**, and is denoted by $\Gamma, \Delta, \Sigma, \ldots$. In particular, the singleton $\{r\}$ of a clause $r$ is often denoted by $r$. Let $r_1$ and $r_2$ be clauses such that $L \in r_1$

and $\bar{L} \in r_2$. The **resolution principle** is defined by the following operation to derive the clause $(r_1 \setminus \{L\}) \cup (r_2 \setminus \{\bar{L}\})$, called the **resolvent** of $r_1$ and $r_2$:

$$\frac{r_1 \qquad r_2}{(r_1 \setminus \{L\}) \cup (r_2 \setminus \{\bar{L}\})}\ res, L \quad .$$

A **resolution derivation** is then defined as the process of deriving a clause $r$ from a given clause set $\Gamma$ by applications of the resolution principle.

For any non-empty clause set $\Gamma = \{r_1, \ldots, r_m\}$, we denote by $d(\Gamma)$, the disjunctive normal form (DNF) formula $(\wedge r_1) \vee \cdots \vee (\wedge r_m)$.

To investigate the translation of the Venn diagrammatic system V, we introduce a calculus VR over clause sets based on the resolution principle.

**Definition 3** (VR). The resolution calculus for Venn diagrams VR consists of the following rules over clause sets.

$$\frac{\{rL\} \quad \{r\bar{L}\}}{\{r\}}\ res, L \qquad \frac{\Gamma}{\Gamma L}\ intro, L \qquad \frac{\Gamma_1 \quad \cdots \quad \Gamma_n}{\Gamma_1 \cup \cdots \cup \Gamma_n}\ sup \qquad \frac{\Gamma \cup \Delta}{\Gamma}\ er$$

Here, $\Gamma L$ (or equivalently $L\Gamma$) signifies $\{r_1 L, \ldots, r_n L\}$ when $\Gamma = \{r_1, \ldots, r_n\}$. We assume the empty clause $\emptyset$ is an axiom of VR.

Note that the above resolution principle in VR is slightly restricted, since $r$ is shared in the premises. However, by using the *intro* rule, the usual resolution principle is easily simulated by the above rule.

By $Res(\Gamma)$, we denote the set of clauses that is derivable from a clause set $\Gamma$ by applications of only the above resolution principle in VR.

Instead of proving the soundness of VR, we explain the inference rules of VR in terms of those of natural deduction under our interpretation of clauses. We interpret each clause set, say $\Gamma = \{r_1, \ldots, r_n\}$, as the negation of the disjunctive normal form formula $\neg d(\Gamma) = \neg((\wedge r_1) \vee \cdots \vee (\wedge r_n))$ (cf. the semantics of Venn diagrams). In particular, the empty clause set $\emptyset$ is interpreted as the constant $\top$ which expresses the truth. Then, using this interpretation and the usual equivalence of formulas, the *intro* rule corresponds to the disjunction introduction rule of natural deduction, since, when $\Gamma = \{r_1, \ldots, r_n\}$, $\neg((\wedge r_1) \vee \cdots \vee (\wedge r_n))$ implies $\neg((\wedge r_1) \vee \cdots \vee (\wedge r_n)) \vee \neg L$, which is equivalent to $\neg((\wedge r_1 \wedge L) \vee \cdots \vee (\wedge r_n \wedge L))$. The *sup* rule corresponds to the $n$-ary conjunction introduction rule since $\neg d(\Gamma_1), \ldots, \neg d(\Gamma_n)$ implies $\neg d(\Gamma_1) \wedge \cdots \wedge \neg d(\Gamma_n)$, i.e., $\neg(d(\Gamma_1) \vee \cdots \vee d(\Gamma_n))$. The *er* rule corresponds to the conjunction elimination rule since $\neg(d(\Gamma) \vee d(\Delta))$, which is equivalent to $\neg d(\Gamma) \wedge \neg d(\Delta)$, implies $\neg d(\Gamma)$.

### 3.2 Translation of V

We present a translation of the Venn diagrammatic inference system V into the resolution calculus VR. Our translation here is different from and more suitable than that in [6] for comparing free rides in Venn and Euler diagrammatic systems.

Translation between logical systems is one of the basic methods in proof theory, and can be applied to various systems for a variety of reasons. Compared with semantic interpretation, a logic translation reveals various properties of inference rules in a logical system. For example, it enables us to study logical connectives and inference rules of the original system in terms of those of the translated system.

In general, diagrams correspond to formulas in symbolic logic, and diagram manipulations correspond to applications of inference rules in a certain logical system. In particular, our Venn diagrams abstractly specified in terms of conjunctive regions correspond to clause sets, i.e., disjunctive normal form formulas, and the inference rules of the Venn diagrammatic system V correspond to inference rules over clause sets.

We first define a translation of a Venn diagram into a clause set.

**Definition 4 (Translation of Venn diagrams).** Let $\mathcal{V}$ be a Venn diagram such that $shc(\mathcal{V}) = \{r_1, \ldots, r_n\}$.

- Each shaded conjunctive region $r_i$ such that $X_1 \ldots X_n \overline{Y_1} \ldots \overline{Y_m}$ is translated into a clause $r_i^\bullet$ such that $X_1 \ldots X_n \overline{Y_1} \ldots \overline{Y_m}$.
- The Venn diagram $\mathcal{V}$ is translated into the clause set $\mathcal{V}^\bullet = \{r_1^\bullet, \ldots, r_n^\bullet\}$. In particular, each primary diagram is translated as $\emptyset$.

To avoid notational complications, we denote simply by $r$ (instead of $r^\bullet$) the translation of a region $r$.

*Example 4.* Venn diagrams $\mathcal{V}_1$ and $\mathcal{V}_2$ in Example 3 are translated as follows: $\mathcal{V}_1^\bullet = \{\overline{B}CD, BCD, B\overline{C}D, B\overline{C}\overline{D}, CD, BD, B\overline{C}\}$ and $\mathcal{V}_2^\bullet = \{A\overline{B}\}$.

Based on the above translation of Venn diagrams into clause sets, inference rules for V are translated into resolution calculus VR.

**Definition 5 (Translation of V).** Rules for V are translated as follows.

Introduction of a circle: The conclusion diagram $\mathcal{V} + A$ is specified in terms of the set of shaded conjunctive regions as follows:

$$shc(\mathcal{V} + A) = shc(\mathcal{V}) \cup \{rA \mid r \in shc(\mathcal{V})\} \cup \{r\overline{A} \mid r \in shc(\mathcal{V})\}.$$

Then this rule is translated as follows.

$$\dfrac{\mathcal{V}^\bullet \quad \dfrac{\mathcal{V}^\bullet}{A\mathcal{V}^\bullet}\,intro, A \quad \dfrac{\mathcal{V}^\bullet}{\mathcal{V}^\bullet\overline{A}}\,intro, \overline{A}}{\mathcal{V}^\bullet \cup A\mathcal{V}^\bullet \cup \mathcal{V}^\bullet\overline{A}}\,sup$$

Superposition of diagrams: The conclusion diagram $\mathcal{V}_1 + \mathcal{V}_2$ is specified in terms of the set of shaded conjunctive regions as follows. Let $n$ be the length, i.e., the number of names of each minimal region of $\mathcal{V}_1$, which is uniquely determined. Let $\Gamma_0 = \{r \mid rL \in shm(\mathcal{V}_1), r\overline{L} \in shm(\mathcal{V}_2)\}$ and $\Gamma_i = \{r \mid rL, r\overline{L} \in \Gamma_{i-1}\}$.

$$shc(\mathcal{V}_1 + \mathcal{V}_2) = shc(\mathcal{V}_1) \cup shc(\mathcal{V}_2) \cup \textstyle\bigcup_{i=0}^{n} \Gamma_i$$

Then this rule is translated as shown below.

$$\frac{\mathcal{V}_1^\bullet \quad \mathcal{V}_2^\bullet \quad Res(\mathcal{V}_1^\bullet \cup \mathcal{V}_2^\bullet)}{\mathcal{V}_1^\bullet \cup \mathcal{V}_2^\bullet \cup Res(\mathcal{V}_1^\bullet \cup \mathcal{V}_2^\bullet)}\ sup$$

Here, $Res(\mathcal{V}_1^\bullet \cup \mathcal{V}_2^\bullet)$ denotes the set of clauses derived from $\mathcal{V}_1^\bullet \cup \mathcal{V}_2^\bullet$ by applying the resolution principle.

**Erasure of shading** The conclusion diagram $\mathcal{V} - \{z\}$ is specified as follows:

$$shc(\mathcal{V} - \{z\}) = shc(\mathcal{V}) \setminus (\{z\} \cup \{r \in shc(\mathcal{V}) \mid \exists r'.rr' = z\}) .$$

Then this rule is translated as shown below. Let $\mathcal{V}^\bullet$ be the disjoint union $\{z\} \cup \Gamma_1 \cup \Gamma_2$, where for all $r \in \Gamma_1$, there exists $r'$ such that $rr' = z$, and for all $x \in \Gamma_2$, there is no $r'$ such that $xr' = z$.

$$\frac{\{z\} \cup \Gamma_1 \cup \Gamma_2}{\Gamma_2}\ er$$

**Erasure of a circle** The conclusion diagram $\mathcal{V} - A$ is specified as follows:

$$shc(\mathcal{V} - A) = shc(\mathcal{V}) \setminus \{r \in shc(\mathcal{V}) \mid A \in |r|\} .$$

Then this rule is translated as shown below. Let $\mathcal{V}^\bullet = \Gamma_1 \cup \Gamma_2$ in which $A \in |r|$ for all $r \in \Gamma_1$, and $A \notin |x|$ for all $x \in \Gamma_2$.

$$\frac{\Gamma_1 \cup \Gamma_2}{\Gamma_2}\ er$$

*Example 5.* The proof in Fig. 2 for Example 3 is translated as $\pi$ in Fig. 3.

$$\pi_1 \left\{ \frac{\mathcal{V}_1^\bullet \quad \dfrac{\mathcal{V}_1^\bullet}{\mathcal{V}_1^\bullet A}\ intro, A \quad \dfrac{\mathcal{V}_1^\bullet}{\mathcal{V}_1^\bullet \overline{A}}\ intro, \overline{A}}{(\mathcal{V}_1 + A)^\bullet = \mathcal{V}_1^\bullet \cup \left\{ \begin{array}{l} A\overline{B}CD,\ ABCD,\ AB\overline{C}D,\ AB\overline{C}\,\overline{D},\ ACD,\ ABD,\ AB\overline{C}, \\ \overline{A}\overline{B}CD,\ \overline{A}BCD,\ \overline{A}B\overline{C}D,\ \overline{A}B\overline{C}\,\overline{D},\ \overline{A}CD,\ \overline{A}BD,\ \overline{A}B\overline{C} \end{array} \right\}}\ sup \right.$$

Here, $\mathcal{V}_1^\bullet = \{\overline{B}CD, BCD, B\overline{C}D, B\overline{C}\,\overline{D}, CD, BD, B\overline{C}\}$

$$\pi_2 \left\{ \frac{\dfrac{\mathcal{V}_2^\bullet = \{A\overline{B}\} \quad \dfrac{\mathcal{V}_2^\bullet}{\mathcal{V}_2^\bullet C}\ intro, C \quad \dfrac{\mathcal{V}_2^\bullet}{\mathcal{V}_2^\bullet \overline{C}}\ intro, \overline{C}}{(\mathcal{V}_2 + C)^\bullet = \mathcal{V}_2^\bullet \cup \{A\overline{B}C, A\overline{B}\,\overline{C}\}}\ sup \quad \dfrac{(\mathcal{V}_2 + C)^\bullet}{(\mathcal{V}_2 + C)^\bullet D}\ intro, D \quad \dfrac{(\mathcal{V}_2 + C)^\bullet}{(\mathcal{V}_2 + C)^\bullet \overline{D}}\ intro, \overline{D}}{((\mathcal{V}_2 + C) + D)^\bullet = (\mathcal{V}_2 + C)^\bullet \cup \{A\overline{B}CD, A\overline{B}C\overline{D},\ A\overline{B}\,\overline{C}D, A\overline{B}\,\overline{C}\,\overline{D},\ A\overline{B}D, A\overline{B}\,\overline{D}\}}\ sup \right.$$

$$\pi \left\{ \frac{\begin{array}{c} \vdots\ \pi_1 \\ (\mathcal{V}_1{+}A)^\bullet \end{array} \quad \begin{array}{c} \vdots\ \pi_2 \\ ((\mathcal{V}_2{+}C){+}D)^\bullet \end{array} \quad \dfrac{\dfrac{(\mathcal{V}_1 A)^\bullet}{AB\overline{C}D} \quad \dfrac{(\mathcal{V}_2 CD)^\bullet}{A\overline{B}\,\overline{C}D}}{A\overline{C}D} \quad \dfrac{\dfrac{(\mathcal{V}_1 A)^\bullet}{AB\overline{C}\,\overline{D}} \quad \dfrac{(\mathcal{V}_2 CD)^\bullet}{A\overline{B}\,\overline{C}\,\overline{D}}}{A\overline{C}\,\overline{D}} \quad \dfrac{\dfrac{(\mathcal{V}_1 A)^\bullet}{ABD} \quad \dfrac{(\mathcal{V}_2 CD)^\bullet}{A\overline{B}D}}{AD} \quad \dfrac{\dfrac{(\mathcal{V}_1 A)^\bullet}{AB\overline{C}} \quad \dfrac{(\mathcal{V}_2 CD)^\bullet}{A\overline{B}\,\overline{C}}}{A\overline{C}}}{(\mathcal{V}_1 + A)^\bullet \cup ((\mathcal{V}_2 + C) + D)^\bullet \cup \{A\overline{C}D, A\overline{C}\,\overline{D}, AD, A\overline{C}\}}\ sup \right.$$

**Fig. 3.** Translation of the Venn diagrammatic proof in Fig. 2

In order to justify the above translation, we show the following lemma.

**Lemma 1.** *Let $r$ be a region of a given Venn diagram $\mathcal{V}$, and let $X$ be a circle of $\mathcal{V}$ such that $X \notin |r|$. $r \in shc(\mathcal{V})$ if and only if $rX, r\overline{X} \in shc(\mathcal{V})$.*

*Proof.* $\Leftarrow$) When $rX, r\overline{X} \in shc(\mathcal{V})$, we have $r = rX \cup r\overline{X} \in shc(\mathcal{V})$.
$\Rightarrow$) Let $r \in shc(\mathcal{V})$. Since $X \notin |r|$, $r$ is still shaded after the erasure of circle $X$, i.e., $r \in shc(\mathcal{V} - X)$. Then, since each region is split into two parts by the introduction of $X$ to $\mathcal{V} - X$, we have $rX, r\overline{X} \in shc((\mathcal{V} - X) + X)$.

Assume to the contrary that $rX \notin shc(\mathcal{V})$ or $r\overline{X} \notin shc(\mathcal{V})$. If $rX \notin shc(\mathcal{V})$, since unshaded regions are kept by the erasure and introduction of circle $X$, we have $rX \notin shc((\mathcal{V} - X) + X)$, which contradicts $rX \in shc((\mathcal{V} - X) + X)$. The same applies to the case $r\overline{X} \notin shc(\mathcal{V})$. Therefore, we have $rX, r\overline{X} \in shc(\mathcal{V})$. ■

By the above Lemma 1, we have the following lemma.

**Lemma 2.** *Let $\mathcal{V}$ be a Venn diagram, and $r$ be a region of $\mathcal{V}$. $r \in shc(\mathcal{V})$ if and only if $r \in Res(shm(\mathcal{V}))$.*

*Proof.* $\Leftarrow$) Let $r \in Res(shm(\mathcal{V}))$. We show $r \in shc(\mathcal{V})$ by induction on the length $n$ of a given derivation of $r$. The base step $n = 0$ is immediate, since we have $r \in shm(\mathcal{V}) \subseteq shc(\mathcal{V})$. In the induction step $n > 0$, $r$ is obtained by the resolution principle from $rX$ and $r\overline{X}$ for some $X$. By the induction hypothesis, we have $rX, r\overline{X} \in shc(\mathcal{V})$, and hence, by Lemma 1, we have $r \in shc(\mathcal{V})$.
$\Rightarrow$) Let $r \in shc(\mathcal{V})$. When $r$ is a minimal region, we immediately have $r \in Res(shm(\mathcal{V}))$. When $r$ is not a minimal region, by Lemma 1, for some $X_1 \notin |r|$, we have $rX_1, r\overline{X_1} \in shc(\mathcal{V})$.

On the one hand, when $rX_1$ and $r\overline{X_1}$ are minimal regions, we have $r \in Res(shm(\mathcal{V}))$ by the resolution principle. On the other hand, when $rX_1$ and $r\overline{X_1}$ are not minimal, again by Lemma 1, for some $X_2 \notin |rX_1|$, we have $rX_1X_2, rX_1\overline{X_2} \in shc(\mathcal{V})$ and $r\overline{X_1}X_2, r\overline{X_1}\overline{X_2} \in shc(\mathcal{V})$. Since the length of a minimal region in $\mathcal{V}$ is finite, we are able to reduce these case to the case of minimal regions by repeated applications of Lemma 1. In this way, we have $r \in Res(shm(\mathcal{V}))$. ■

Based on Lemmas 1 and 2, the soundness of the translation is easily obtained.

**Theorem 1 (Soundness of translation of VR).** *If $\mathcal{V}$ is provable from $\mathcal{V}_1, \ldots, \mathcal{V}_n$ in the Venn diagrammatic system* V, *then $\mathcal{V}^\bullet$ is derivable from $\mathcal{V}_1^\bullet, \ldots, \mathcal{V}_n^\bullet$ in resolution calculus* VR.

## 4 Free Ride in Venn Diagrammatic System

We now investigate the free ride property of our Venn diagrammatic system V. In Takemura [12], a translation of Euler diagrammatic system of Mineshima-Okada-Takemura [5] into a natural deduction system is given, and it is shown that the soundness theorem of the translation implies the occurrence of free ride (in the sense of [9,1]) between the systems. In this way, free ride is formalized in the proof-theoretical framework through the logic translation. Thus, the soundness of our translation in Section 3.2 shows that free ride occurs between the Venn diagrammatic system V and the resolution calculus VR. See Takemura [12].

Our formalization of free ride in the proof-theoretical framework makes it possible to analyze free ride in more detail than that in the semantic framework. In the translation (Definition 5) of the Venn diagrammatic system V into the resolution calculus VR, it is revealed that which conjunctive regions are shaded in the conclusion diagram after the application of an inference rule. Some of such regions are not specified in the given premise diagrams or in the description of the diagrammatic operation of the rule, and they are automatically represented by the application of the rule. By slightly extending the notion of free ride of Shimojima [9], let us call the regions "free rides" those that are automatically represented in a diagram after a manipulation of given diagrams. Our free rides may also be called "emergent objects."

For example, when we carry out the operation of Superposition (cf. Definition 2) as illustrated in Fig. 2, we only need to consider shaded minimal regions in both premise diagrams; we do not need to take the other regions into account. Hence, the conjunctive regions $A\overline{C}D, A\overline{C}\overline{D}, AD, A\overline{C}$ in Fig. 3 are automatically represented by the application of Superposition, and they are free rides of this application of Superposition.

In this way, for our Venn diagrammatic system V, we are able to define the free rides of an application of each inference rule in terms of shaded conjunctive regions. In general, the free rides of an application of an inference rule are defined as the regions obtained by subtracting from the shaded conjunctive regions of the conclusion, the shaded conjunctive regions of premises and those described in the operation of the rule. Thus, Introduction of a circle, Erasure of shading, and Erasure of a circle have no free rides, and in particular, the free rides of Superposition are defined as follows.

**Definition 6 (Free rides of superposition).** In an application S of Superposition between $\mathcal{V}_1$ and $\mathcal{V}_2$, the following set of conjunctive regions are called the *free rides* of S:

$$shc(\mathcal{V}_1 + \mathcal{V}_2) \setminus \left(shc(\mathcal{V}_1) \cup shc(\mathcal{V}_2)\right) .$$

Based on the above definition of free rides, we are able to investigate what kind of and how many inference steps are required in terms of the resolution calculus, to derive free rides. We here investigate the number of free rides occurring in an application of Superposition.

First, we calculate the number of conjunctive regions $\mathsf{cr}(\mathcal{V})$ in a given diagram $\mathcal{V}$ consisting of $n$ circles as follows:

$$\mathsf{cr}(\mathcal{V}) = \sum_{i=1}^{n} {}_n\mathrm{C}_i \times 2^i - \sum_{i=1}^{n} {}_n\mathrm{C}_i .$$

Here, ${}_n\mathrm{C}_n \times 2^n$ $(= 2^n)$ is the number of usual (i.e., including negative) minimal regions of $\mathcal{V}$ (cf. truth table), and $\sum_{i=1}^{n} {}_n\mathrm{C}_i$ is the number of negative regions. Thus, for a Venn diagram $\mathcal{V}_3$ (resp. $\mathcal{V}_4$) consisting of 3 (resp. 4) circles, we have $\mathsf{cr}(\mathcal{V}_3) = 19$, i.e., $\{A, B, C, AB, AC, A\overline{B}, A\overline{C}, B\overline{C}, \overline{A}B, BC, \overline{A}C, \overline{B}C, ABC, A\overline{B}C,$

$A\overline{B}\overline{C}, AB\overline{C}, \overline{A}B\overline{C}, \overline{A}BC, \overline{A}\overline{B}C\}$ (resp. $\mathsf{cr}(\mathcal{V}_4) = 65$). (Cf. Example 2 for the case of $n = 2$.)

Next we calculate the number of free rides in the worst case, where the maximum number of free rides occur. Let $\mathcal{V}_1$ and $\mathcal{V}_2$ be Venn diagrams such that (1) $m(\mathcal{V}_1) = m(\mathcal{V}_2)$ (i.e., $\mathcal{V}_1$ and $\mathcal{V}_2$ have the same minimal regions); (2) $shc(\mathcal{V}_i) \setminus shm(\mathcal{V}_i) = \emptyset$ for each $i = 1, 2$ (i.e., the shaded regions are only minimal regions); and (3) $r \in shc(\mathcal{V}_1)$ if and only if $r \notin shc(\mathcal{V}_2)$, for any conjunctive region $r$ of $\mathcal{V}_1$ (i.e., $\mathcal{V}_1$ and $\mathcal{V}_2$ are dual diagrams with respect to shading). Thus, after an application of Superposition, all conjunctive regions are shaded in $\mathcal{V}_1 + \mathcal{V}_2$. The following Fig. 4 illustrates such $\mathcal{V}_1$ and $\mathcal{V}_2$ consisting of three circles.

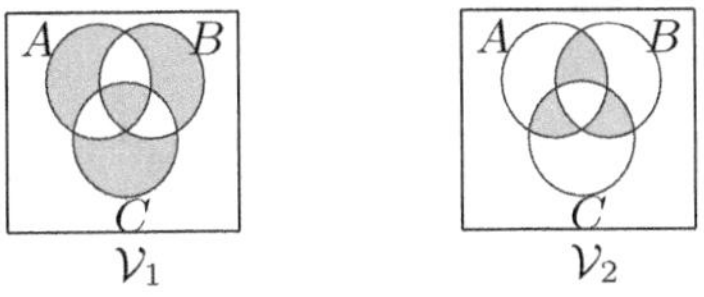


**Fig. 4.**

Then, the number of free rides in $\mathcal{V}_1 + \mathcal{V}_2$, which consists of $n$ circles, is as follows.

$$\mathsf{fr}(\mathcal{V}_1 + \mathcal{V}_2) = \mathsf{cr}(\mathcal{V}_1 + \mathcal{V}_2) - ({}_n\mathrm{C}_n \times 2^n - {}_n\mathrm{C}_n)$$

Here, ${}_n\mathrm{C}_n$ is the number of negative minimal regions, and ${}_n\mathrm{C}_n \times 2^n - {}_n\mathrm{C}_n$ is the number of conjunctive minimal regions. Thus, for example, if $\mathcal{V}_1$ and $\mathcal{V}_2$ consist of 2 (resp. 3, and 4) circles, $\mathsf{fr}(\mathcal{V}_1 + \mathcal{V}_2) = 5 - 3 = 2$ (resp. $19 - 7 = 12$ and $64 - 14 = 50$). In this way, if $n \geq 3$, we have $2^n < \mathsf{fr}(\mathcal{V}_1 + \mathcal{V}_2)$, and hence, the number of free rides increases exponentially with the number of circles $n$.

Furthermore, if we consider negative and disjunctive regions as well, the number of free rides is estimated as follows. Since a (general) region is a non-empty union of certain minimal regions, and since the number of minimal regions $\mathsf{mr}(\mathcal{V})$ of a Venn diagram consisting of $n$ circles is $2^n$, the number of regions $\mathsf{r}(\mathcal{V})$ is $2^{2^n} - 1$. Then, the number of free rides in the worst case is calculated by subtracting the number of minimal regions from the total number of regions:

$$\mathsf{fr}(\mathcal{V}_1 + \mathcal{V}_2) = \mathsf{r}(\mathcal{V}_1 + \mathcal{V}_2) - \mathsf{mr}(\mathcal{V}_1 + \mathcal{V}_2) .$$

Thus, for example, if $\mathcal{V}_1 + \mathcal{V}_2$ consists of 2 (resp. 3) circles, we have $\mathsf{fr}(\mathcal{V}_1 + \mathcal{V}_2) = 11$ (resp. $= 248$).

## 5 A Comparison of Venn Diagrams and Euler Diagrams

We try to compare Venn and Euler diagrams with respect to free rides. In Section 5.1, we briefly review the analysis, given in Takemura [12], on free rides in our Euler diagrammatic system. Then, in Section 5.2, we discuss differences between the Venn and Euler diagrammatic systems, and discuss our future work.

### 5.1 Free Rides in Euler Diagrammatic System

While the most basic and essential components of Venn diagrams are regions and shading of them, those of Euler diagrams are topological (inclusion and exclusion) relations between circles and points. Based on this idea, Mineshima-Okada-Takemura [6] distinguished a "region-based" framework for Venn diagrams, where a diagram is specified in terms of shaded regions, and a "relation-based" framework for Euler diagrams, where a diagram is specified in terms of inclusion and exclusion relations. Then Mineshima-Okada-Takemura investigated a relation-based Euler diagrammatic inference system in [5].

This inference system comprises two kinds of inference rules: Deletion and Unification. Essentially, Deletion allows us to delete a diagrammatic object from a given Euler diagram. Unification allows us to combine two Euler diagrams into one diagram, where the semantic information is equivalent to the conjunction of the two original diagrams. To characterize intuitive manipulations on diagrams as formal inference rules, the unification rules are defined by requiring that one of the unified diagrams be a *minimal diagram* that consists of two objects (circles and points). Each inference rule is described in terms of relations by specifying (i) premise diagrams; (ii) the constraints that the premise diagrams should satisfy; and (iii) diagrammatic operations to introduce a new object into, or to rearrange a configuration of objects of, one of the premise diagrams. See [5,12] for a detailed description.

The following diagrammatic proof on the left in Fig. 5 is an application of our unification rule, in which, to obtain the unified diagram $\mathcal{D} + \alpha$, circle $A$ in the minimal diagram $\alpha$ is added to diagram $\mathcal{D}$ so that "$A$ is inside $B$" holds.

While our region-based Venn diagrammatic system is naturally translated into resolution calculus as seen in Section 3.2, the relation-based Euler diagrammatic system is translated into a natural deduction system (cf. [6,12]). An Euler diagram specified in terms of relations corresponds to a conjunction of implicational formulas, and an inference rule of our Euler diagrammatic system corresponds to natural deduction inference rules associated with the implicational connective. For example, the Euler diagram $\mathcal{D}$ (resp. $\alpha$) on the left in Fig. 5 is translated into the conjunction of implicational formulas

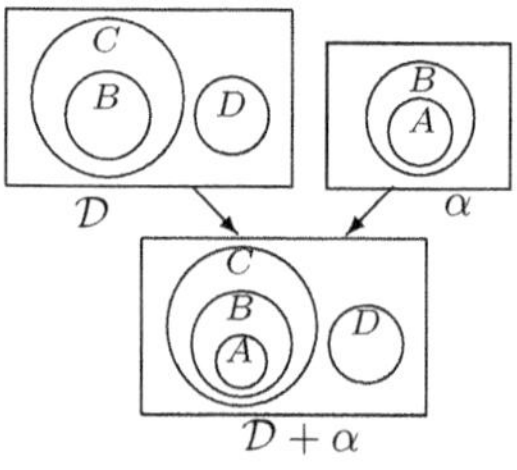


$$\dfrac{\mathcal{D}^\circ \quad \alpha^\circ \quad \dfrac{\dfrac{\dfrac{[A]^1 \quad \genfrac{}{}{0pt}{}{\alpha^\circ}{A\to B}}{B} \quad \dfrac{\mathcal{D}^\circ}{B\to C}}{C}}{A\to C}\,1 \quad \dfrac{\dfrac{\dfrac{[A]^1 \quad \genfrac{}{}{0pt}{}{\alpha^\circ}{A\to B}}{B} \quad \dfrac{\mathcal{D}^\circ}{B\to \neg D}}{\neg D}}{A\to \neg D}\,1}{\mathcal{D}^\circ \wedge \alpha^\circ \wedge \underbrace{(A\to C)\wedge(A\to\neg D)}_{\text{Free rides}}}$$

**Fig. 5.** Free rides of an Unification in the Euler diagrammatic proof that corresponds to the Venn diagrammatic proof in Fig. 2

$\mathcal{D}^{\circ} = (B \to C) \wedge (B \to \neg D) \wedge (C \to \neg D) \wedge (B \to B) \wedge (C \to C) \wedge (D \to D)$ (resp. $\alpha^{\circ} = (A \to B) \wedge (A \to A) \wedge (B \to B)$). Then the application of Unification is translated into the natural deduction proof on the right in Fig. 5, in which natural deduction rules for the conjunction are generalized to those for $n$-ary conjunction. See [12] for a detailed description.

Note that, when we carry out the operation of Unification in Fig. 5, we only need to consider the relations between $A$ and $B$, and not the other circles. Hence, the relations corresponding to $A \to C$ and $A \to \neg D$, i.e., "$A$ is inside $C$" and "$A$ is outside $D$," are automatically represented by this application of Unification, and we call these relations "free rides." In general, the free rides for an application of each Unification are defined in terms of relations that exist in the diagrams, as the relations obtained by subtracting, from the relations holding at the conclusion, the relations existing in the premises and the relations required to exist in the description of the operation of the rule (see [12]).

Let us calculate the number of free rides in the worst case. For an application of Unification between a (general) diagram $\mathcal{D}$ and a minimal diagram $\alpha$, assume that $\mathcal{D}$ contains $n$ circles $A_1, \ldots, A_n$ which are laid out in a line with respect to the inclusion relation so that $A_1$ is the smallest circle, and assume that "$B$ is inside $A_1$" holds in $\alpha$. Such a Unification $\mathcal{D} + \alpha$ is a worst case, where the maximum number of free rides occur, and it is easily seen that the number of free rides is $n-1$, i.e., the relations "$B$ is inside $A_i$" for $2 \leq i \leq n$. The application of Unification in Fig. 5 also illustrates a worst case, where $\mathcal{D}$ contains 3 circles, with a total of 2 free rides.

### 5.2 Discussion and Future Work

In order to compare Venn diagrams and Euler diagrams with respect to free rides, we consider the examples of translations given in Figs. 2 and 3, and Fig. 5.

On the one hand, in the application of Superposition of Venn diagrams in Fig. 2, free rides are the regions $A\overline{C}D, A\overline{C}\overline{D}, AD, A\overline{C}$, i.e., $\neg(A \wedge \neg C \wedge D), \neg(A \wedge \neg C \wedge \neg D), \neg(A \wedge D)$, and $\neg(A \wedge \neg C)$ as seen in Fig. 3. On the other hand, in the application of Unification of Euler diagrams in Fig. 5, free rides are $A \to C$ and $A \to \neg D$. In view of the equivalence between formulas $\neg(A \wedge \neg C)$ (resp. $\neg(A \wedge D)$) and $A \to C$ (resp. $A \to \neg D$), it is immediately seen that the Superposition of Venn diagrams has more free rides than the Unification of Euler diagrams. This generally holds, as we have already seen in the estimation of the number of free rides in Venn and Euler diagrammatic systems. In other words, an application of Superposition to Venn diagrams generally provides more pieces of information automatically than that of Unification of Euler diagrams.

However, this does not necessarily imply that Venn diagrams are more useful than Euler diagrams. In particular, from a cognitive viewpoint, it is often observed that Venn diagrams are harder to handle in actual reasoning than Euler diagrams, and there are some experimental results to support this claim, e.g., Sato-Mineshima-Takemura [8].

Although the translated inference steps in resolution calculus and natural deduction, are not simply regarded as cognitive/psychological processes to recognize free rides represented in diagrams, we may assume at least one step is required to recognize each of the free rides. Then, the number of free rides can be calculated as the number of steps required to recognize these free rides. Thus, we could be able to interpret our results as follows. On the one hand, in Venn diagrams the number of recognition steps for free rides increases exponentially with the number of circles contained in the given diagrams, whereas in Euler diagrams, it increases linearly. We consider this to be part of the reason that Euler diagrams are considered to be more tractable than Venn diagrams.

As future work, we need a more in-depth analysis of the estimation of free rides. The Venn diagrams used in our worst case analysis of the number of free rides do not correspond to Euler diagrams. It is better to evaluate free rides in an application of Superposition to Venn diagrams obtained by translating Euler diagrams. Although not fully investigated, we have made the following observations. In the simplest case where only conjunctive regions are considered, and moreover, at most three circles are contained in the given Venn diagrams (this case corresponds to the simplest one-step syllogism covering only universal sentences), the number of free rides in Superposition of such Venn diagrams is the same as that in Unification of the corresponding Euler diagrams. However, if we also consider negative or disjunctive regions, the number is greater in Superposition than that in Unification. Furthermore, in the case where the given Venn diagrams contain more than three circles, as shown in Figs. 3 and 5, the number of free rides is greater in Superposition than in Unification when considering only conjunctive regions. In view of these observations, we need to investigate how many free rides occur in the Superposition of Venn diagrams that are translated from Euler diagrams.

In our analysis of free rides, we assumed an ideal person who is able to recognize all shaded conjunctive regions fully and uniformly. However, in our actual reasoning, some regions may be more difficult to grasp than others. In fact, our conjunctive regions are not exactly equivalent to visually connected regions, which depend on a particular representation of a diagram. Our analysis may be considered as the one on the abstract or informational aspect of free ride. For the thorough analysis on free ride, we need further study on the cognitive or perceptual aspect of free ride.

## References

1. Barwise, J., Seligman, J.: Information Flow: The Logic of Distributed Systems. Cambridge University Press (1997)
2. Buss, S.R.: An Introduction to Proof Theory. In: Buss, S.R. (ed.) Handbook Proof Theory. Elsevier, Amsterdam (1998)
3. Gurr, C.A., Lee, J., Stenning, K.: Theories of diagrammatic reasoning: Distinguishing component problems. Minds and Machines 8(4), 533–557 (1998)
4. Howse, J., Stapleton, G., Taylor, J.: Spider Diagrams. LMS Journal of Computation and Mathematics 8, 145–194 (2005)

5. Mineshima, K., Okada, M., Takemura, R.: A Diagrammatic Inference System with Euler Circles. Journal of Logic, Language and Information (to appear), A preliminary version is available at: http://abelard.flet.keio.ac.jp/person/takemura/index.html
6. Mineshima, K., Okada, M., Takemura, R.: Two Types of Diagrammatic Inference Systems: Natural Deduction Style and Resolution Style. In: Goel, A.K., Jamnik, M., Narayanan, N.H. (eds.) Diagrams 2010. LNCS (LNAI), vol. 6170, pp. 99–114. Springer, Heidelberg (2010)
7. Robinson, J.A.: A Machine-Oriented Logic Based on the Resolution Principle. Journal of the ACM 12(1), 23–41 (1965)
8. Sato, Y., Mineshima, K., Takemura, R.: The Efficacy of Euler and Venn Diagrams in Deductive Reasoning: Empirical Findings. In: Goel, A.K., Jamnik, M., Narayanan, N.H. (eds.) Diagrams 2010. LNCS (LNAI), vol. 6170, pp. 6–22. Springer, Heidelberg (2010)
9. Shimojima, A.: On the Efficacy of Representation, Ph.D. thesis, Indiana University (1996)
10. Shin, S.-J.: The Logical Status of Diagrams. Cambridge University Press (1994)
11. Stapleton, G.: A survey of reasoning systems based on Euler diagrams. In: Proc. of Euler 2004. Electronic Notes in Theoretical Computer Science, vol. 134(1), pp. 127–151 (2005)
12. Takemura, R.: Proof theory for reasoning with Euler diagrams: a logic translation and normalization. Studia Logica (to appear), A preliminary version is available at: http://abelard.flet.keio.ac.jp/person/takemura/index.html

# Euler Diagram Encodings

Paolo Bottoni[1], Gennaro Costagliola[2], and Andrew Fish[3,*]

[1] Dipartimento di Informatica - "Sapienza" University of Rome, Italy
bottoni@di.uniroma1.it
[2] Dipartimento di Informatica - University of Salerno, Italy
gencos@unisa.it
[3] School of Computing, Engineering and Mathematics - University of Brighton, UK
Andrew.Fish@brighton.ac.uk

**Abstract.** Euler Diagrams are a well-known visualisation of set-based relationships, used in many application areas and at the basis of more complex notations. We propose a *static* code for concrete Euler Diagrams, which enables efficient storage (vs. storage of concrete diagrams), and transformations preserving concrete-level structure, hence the viewer's mental map. We provide the theoretical underpinnings of the encoding, examples and deductions, and an indication of their utility. For use in an interactive setting, we provide algorithms to update the code upon curve addition and removal. Independently, we show that the code identifies minimal regions, enabling the computation of the abstract zone set.

## 1 Introduction

Euler Diagrams (EDs) are a popular visualisation method for representing relationships between set-based data. They consist of a set of curves representing sets and their relationships, e.g. disjointness and containment. The same name is also used to refer to the extension, termed Euler Diagrams *with items* in [3], in which items (elements) are visualised within the regions (set intersections) determined by the curves.

From a logical perspective, EDs are a diagrammatic system for representing, and reasoning with, logic expressions. Subsequent to pioneering work by Shin [18], recent work has included producing automatic reasoning with ED logics [19], heterogeneous reasoning systems [20], combination with Conceptual Graphs [5], and more expressive diagrammatic logics, such as Spider Diagrams [13], and Constraint Diagrams [15].

From an information visualisation perspective, EDs with items are used in application domains such as network visualisation [17], resource management [3], and for display of search query results [21]. EDs with area-proportional regions to indicate relative sizes of sets are used in bio-informatics [16] and for statistical data representation [2].

When considering interaction with EDs, ED transformations (e.g. curve addition and deletion) become important. This has been investigated in several contexts, e.g. developing dual graph transformations corresponding to diagram transformations [7]. Transformations should preserve a viewer's mental map (a term provided for graphs in [6], but used more broadly), controlling variations between consecutive diagrams.

For visual languages, the concrete level refers to the drawing, whilst an abstract level captures information that is deemed semantically important. Abstract EDs consist

* Thanks to UK EPSRC for support via grant EP/J010898/1 Automatic Diagram Generation.

P. Cox, B. Plimmer, and P. Rodgers (Eds.): Diagrams 2012, LNAI 7352, pp. 148–162, 2012.

of an abstract set of zones which correspond to the set intersections to be represented as regions in the concrete ED. In [8], alternative ED abstractions were considered, and shown to be equivalent, including the view of EDs as a *building sequence* of curve additions [10]. In [3,4], efficient algorithms were provided for the online ED abstraction problem: given a concrete ED compute the associated abstract ED and update this efficiently upon curve addition and removal. This utilised intersection points created in a building sequence, using these points (or equivalently curve segments) to mark zones. Checking membership of these marked points within regions can then be used to determine the new zone set quickly. By not utilising a graph based methodology within the interactive setting, the implementation becomes simple and computations efficient.

In this paper, we complement, and abstract from, the work in [8,4], inspired by the use of codes in Knot Theory, and provide a notion of encodings of EDs. In detail, Section 2 provides preliminaries and terminology conventions. The encoding is introduced in Section 3, providing examples, theoretical underpinnings, some results and an indication of the utility of the encoding. Section 4 presents detailed algorithmic procedures for dynamical code update and in Section 5 we provide a method for obtaining the abstract diagram from a code. We discuss related avenues and conclude in Section 6.

## 2 Preliminaries

Various classes of EDs are considered in the literature, with different topological and geometric *well-formedness* conditions on curves [11]. We develop the theory of ED codes for several of these classes. We assume that all curves have unique identifiers (labels) and we do not distinguish between curves and their identifiers. An extension to multiple curves with the same label is also possible.

**Definition 1.** *A* concrete ED *is a finite set of closed curves $\mathcal{C}$ in the plane, subject to a set of conditions $W$. Let $\mathcal{E}$ denote the class of concrete EDs with $W$ requiring that there are a finite number of curve intersections, each of which is a transverse double point (exactly two arcs cross at any intersection). For any partition $(X, Y)$, where $X$ and $Y$ can be empty, of the curves $\mathcal{C}$ of $d \in \mathcal{E}$, let $R$ be the maximal set of points which are in the interior of all curves in $X$ and in the exterior of all curves in $Y$. If $R$ is non-empty, we say that it is a* concrete zone *of $d$. A* minimal region *of $d$ is a component of the plane in the complement of $\mathcal{C}$. Let $\mathcal{E}_s \subset \mathcal{E}$ denote the class for which $W$ is extended so that curves are simple (no self-intersections), and $\mathcal{E}_w \subset \mathcal{E}_s$ the further extension where all concrete zones are minimal regions.*

The class $\mathcal{E}$ agrees with conditions on knot diagrams. The class $\mathcal{E}_w$ corresponds to well-formed diagrams in [11]. The class $\mathcal{E}_s$ admits non-connected zones needed for the incremental construction of well-formed EDs (EDs in $\mathcal{E}_s$ are termed *weakly reducible* EDs in [3]), important in an interactive setting. Examples are provided in Figure 1.

The most common abstraction for EDs is the following *zone-based abstraction*, used as the specification for the ED generation problem in [11], although alternative abstractions [8] may be appropriate according to the context.

**Definition 2.** *An* abstract Euler Diagram *$d^*$ is a set of labels $L$, called* abstract curves, *together with a set of* abstract zones *which are partitions $(X, Y)$ of $L$, where $X$ and*

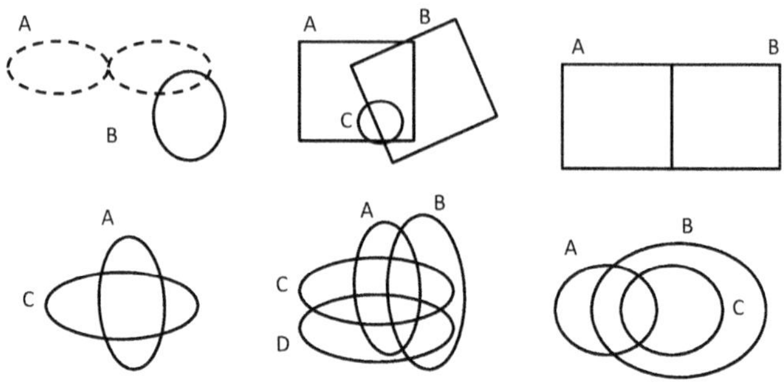


**Fig. 1.** Euler Diagrams from various classes: (top row) the left diagram is in $\mathcal{E}$ with $A$, a non-simple curve shown dashed, whilst the other two are not in $\mathcal{E}$ since the left one has a triple point whilst the right one has an infinite number of intersection points. (bottom row) all diagrams are in $\mathcal{E}_s$: the middle and right diagrams are well-formed (in $\mathcal{E}_w$). The left diagram is not in $\mathcal{E}_w$ since the zone $(A, C)$ which in inside $A$ but outside $C$ comprises of a union of two minimal regions; this diagram can occur in the incremental construction of the middle diagram.

*$Y$ can be empty, and are called the* inside *set and* outside *set, respectively. If there is a bijection between the set of abstract curves of $d^*$ and the set of curves of a concrete Euler Diagram d, which induces a bijection between the abstract zones of $d^*$ and the concrete zones of d, then $d^*$ is the* abstraction *of d, and d is a* realisation *of $d^*$.*

The concrete ED $d$ on the bottom left of Figure 1 has two curves, labelled by $A$ and $C$, decomposing the plane into six minimal regions. The four regions: outside both curves; inside just $A$; inside just $C$; inside both $A$ and $C$ are concrete zones. The abstraction of $d$ has label set $\{A, C\}$ and abstract zones: $(\emptyset, \{A, C\}), (\{A\}, \{C\}), (\{C\}, \{A\})$, and $(\{A, C\}, \emptyset)$. We abbreviate zone descriptions by removing set brackets and concatenating labels to ease reading; i.e. we write $(\emptyset, AC), (A, C), (C, A), (AC, \emptyset)$. Each of the concrete zones $(A, C)$ and $(C, A)$ is the union of two minimal regions.

In this paper, we assume that all concrete EDs have at least two contours and are *connected* (or non-nested [12]), i.e. every curve is decomposed into segments via its crossing points. The extension to nested diagrams is relatively straightforward. We introduce further concrete level concepts and terminology in the next section.

## 3 The Static Code

**Definition 3.** *Let d $\in \mathcal{E}$. An* orientation *of d is a choice of orientation for each of the curves of d. A* crossing point *of d is a point p of intersection of curves of d. Each curve c of d that has a crossing point naturally decomposes into a set of* segments, *overlapping at exactly the crossing points and with orientation induced from the orientation of c, giving rise to their* start *and end points. A sequence of segments $\pi$ is a* path *if the end of each segment is the start of the next segment; it is a* cycle *if the end of the last segment of the path is the start of the first segment and there is no sub-sequence of $\pi$ with this property. To each crossing point p of d we assign a unique* crossing number

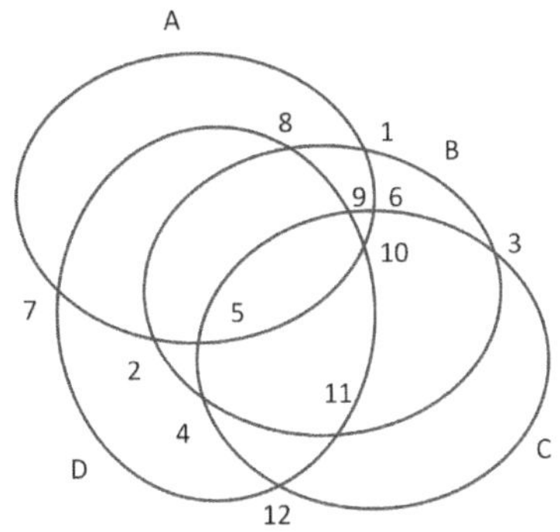


**Fig. 2.** An Euler Diagram $d \in \mathcal{E}_w$ with curves labelled and crossings numbered

$k \in \{1, \ldots, n\}$, *with* $n$ *the number of crossing points in* $d$. *Each path* $\pi$ *of segments induces a sequence of crossing numbers called the* crossing number sequence *of* $\pi$; *for a cycle we omit the final number and use parentheses to indicate that the sequence is read cyclically. For each curve* $c$ *in* $d$ *a choice of orientation of the curve and a base point give rise to a unique (up to reversal and cyclic permutation) crossing number sequence, called the* base code *of* $c$. *A* base code *of* $d$ *is a set of base codes for the curves of* $d$.

The ED $d \in \mathcal{E}_w$, shown in Figure 2, has four curves, labelled $A, B, C$ and $D$, determining 14 concrete zones, each of which is a minimal region. The crossing points have been numbered uniquely with numbers from 1 to $n = 12$. We adopted the convention of assigning numbers following the incremental construction of $d$ (adding curves $A, B, C$ and $D$ in turn), orienting the curves clockwise, and choosing an arbitrary start point on each curve. The curve $D$ decomposes into a cycle of six segments: 7-8, 8-9, 9-10, 10-11, 11-12, and 12-7. The crossing number sequence for the cycle $\pi_D$ (for curve $D$) is (7 8 9 10 11 12); to identify curve $D$ explicitly we write $D$:(7 8 9 10 11 12). The base code of $d$ is: $\{A$:(1 6 10 5 2 7), $B$:(1 3 11 4 2 8), $C$:(4 5 9 6 3 12), $D$:(7 8 9 10 11 12)$\}$.

The base code of a diagram $d$ does not encode enough information to identify its abstract zone set. For example, if the segment 7-1 of $A$ passed along the bottom of the diagram instead of along the top (e.g. enclosing point 12 within $A$ instead of point 8), then the base code remains the same, but there would be a different zone set.

**Definition 4.** *Let* $d \in \mathcal{E}$ *with curve set* $\mathcal{C}$, *and let* $s$ *be a segment of curve* $c$ *of* $d$ *with start point* $p_1$ *and end point* $p_2$. *The* containing curve set *of* $s$ *in* $d$, *denoted* $K(s, d)$, *is the set of all curves* $\mathcal{C}' \subseteq \mathcal{C}$ *whose interior contains* $s$, *with the possible exception of* $p_1$ *and* $p_2$. *We write* $K(s)$ *for* $K(s, d)$ *when no ambiguity arises. A cycle* $\pi$ *of crossing numbers is* annotated *if each consecutive pair of numbers (read cyclically) corresponding to a segment* $s$ *is annotated by* $K(s)$; *we present this annotation as a concatenation of the curves in* $K(s)$ *forming a subscript of the crossing number of the end point of* $s$. *A* static code *of a curve* $c$ *of* $d$, *denoted* $Stat(c)$, *is an annotated base code of the curve, and a* static code *of* $d$, *denoted* $Stat(d)$, *is a set of annotated base codes for the curves of* $d$.

To simplify language we refer to the (annotated) crossing number sequence of a segment as an (annotated) segment. For $d$ in Figure 2, $B$:1 3 denotes the segment $s_1$ of $B$ from curve $A$ to curve $C$ outside all other curves, and so $K(s_1) = \emptyset$. Also $B$:4 2 denotes the segment $s_2$ of $B$ from curve $C$ to curve $A$ inside the curve $D$, and so $K(s_2) = \{D\}$. We omit instances of $\emptyset$ in the annotations, so we write $B$:4 $2_D$ for the annotated segment $s_2$, and $B$:1 3, instead of $B$:1 $3_\emptyset$, for the annotated segment $s_1$. The *static code* for $d$ is given by: $\{A$:(1 $6_B$ $10_{BC}$ $5_{BCD}$ $2_{BD}$ $7_D$), $B$:($1_A$ 3 $11_C$ $4_{CD}$ $2_D$ $8_{AD}$), $C$:($4_D$ $5_{BD}$ $9_{ABD}$ $6_{AB}$ $3_B$ 12), $D$:(7 $8_A$ $9_{AB}$ $10_{ABC}$ $11_{BC}$ $12_C$)$\}$. The code fragment 1 $6_B$ $10_{BC}$ in $Stat(A)$ indicates that segment 1-6 is inside curve $B$ only, and 6-10 is inside curves $B$ and $C$ but no others. The annotated segment 12 $4_D$ in $Stat(C)$ has 12 at the end of $Stat(C)$ and $4_D$ at the start, highlighting the cyclic nature of the code. Algorithm 1 computes the static code for a diagram $d$ by recording the crossing information for each curve $c$, together with the containing set information for each segment of $c$. Here $n$ is the total number of crossings in $d$, $n_i$ is the number of crossings in a single traversal of curve $c_i$, $m_i^j$ is the $j$-th crossing number in $Stat(c_i)$, and $\%n$ denotes modulo $n$. Theorem 1 presents some properties of codes.

---

**Algorithm 1. GetStaticCode**$(d)$

---

**Input**: A concrete Euler Diagram $d \in \mathcal{E}$ with $n$ crossing points.
**Output**: A static code for $d$.

```
Assign a unique (crossing number) from 1 to n to each crossing point in d.
forall curves c_i in d do
    choose a base point p_i on c_i and an orientation of c_i.
    record in Stat(c_i) the crossing numbers met when traversing c_i once from p_i
    according to the chosen orientation.
    forall segments (m_i^j, m_i^((j+1)%n)) in Stat(c_i) do
        associate with m_i^((j+1)%n) the set of all curves containing the segment of c_i
        between the crossing points numbered m_i^j and m_i^((j+1)%n)
return (Stat(d) = {Stat(c_i) : c_i ∈ d})
```

---

**Theorem 1.** *Let $d \in \mathcal{E}_S$. Then the following properties hold for $Stat(d)$:*

1. *Each crossing number $k$ in $Stat(d)$ appears exactly twice, once in $Stat(c)$ and once in $Stat(c')$, where $c$ and $c'$ are distinct curves in $d$.*
2. *Suppose that there is exactly one segment $s$ from $Stat(c_i)$ for which curve $c_j \neq c_i$ is in $K(s)$. Then there is a segment $s'$ from $Stat(c_j)$ for which curve $c_i$ is in $K(s')$.*
3. *If curve $c_j$ is in $K(s)$ for every segment $s$ of $Stat(c_i)$ for some curve $c_i$ then curve $c_j$ contains curve $c_i$ in $d$.*

*Proof.* When constructing the static code of $d$, each crossing point is met twice, once for each curve involved in the traversed crossing. So 1 holds. If curve $c_j$ is in exactly one $K(s)$ with $s$ a segment from $Stat(c_i)$ for $c_i \neq c_j$, then $c_j$ contains exactly one segment of $c_i$. Either (i) $c_j$ crosses $c_i$ at the two endpoints of the segment $s$, and so $Stat(c_j)$ has some segment $s'$ with $c_i \in K(s')$ (left of Figure 3), or (ii) $c_j$ does not intersect with $c_i$ and so $c_i$ must lie in the interior of $c_j$ (right of Figure 3). Since $d$ is connected, there is

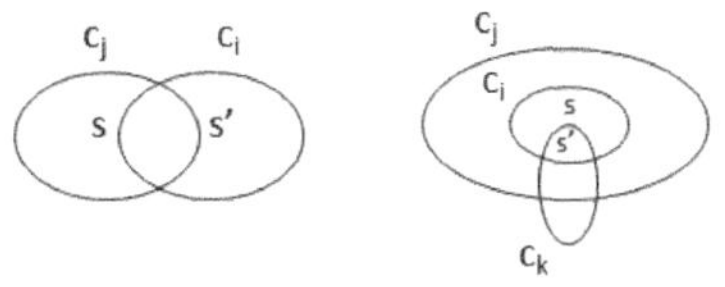


**Fig. 3.** Two relations between $c_i$ and $c_j$ in $\mathcal{E}_s$

another curve $c_k$ that intersects $c_i$. Thus $c_j$ appears more than once in curve $c_i$'s code since $c_i$ is in the interior of $c_j$ and it consists of more than one segment. This contradicts the hypothesis. So 2 holds. As the presence of a curve $c_j$ in $K(s)$ for every segment $s$ of $Stat(c_i)$ means that every segment of $c_i$ is within the interior of $c_i$, 3 holds. □

The encoding encapsulates more information than the abstract diagram. Figure 4 shows three examples of the construction of $Venn(4)$, the Venn diagram on four curves with all of the $2^4 = 16$ possible zones present. For the final diagrams in the top and middle row, $V_1$ and $V_2$, there is a bijection on the curve sets that induces a bijection on the static codes (up to renumbering and reversal) given by exchanging the roles of $B$ and $D$. The final diagram in the bottom row, $V_3$, is a distinct concrete realisation of $Venn(4)$ as it has the property that the zone $(D, ABC)$ is not topologically adjacent to the outside zone $(\emptyset, ABCD)$, whilst for $V_1$ and $V_2$ each of the zones $(A, BCD), (B, ACD), (C, ABD)$ and $(D, ABC)$, which are inside exactly one curve, are topologically adjacent to the outside zone $(\emptyset, ABCD)$. In fact, no zone inside $D$ in $V_3$ is topologically adjacent to the outside zone. This property can be observed directly from the static code by checking if there is a segment $s$ in a cycle from a curve with $K(s) = \emptyset$. Since properties such as this cannot be distinguished at the (usual) abstract level, and the abstract zone set can be recovered from the static code, we have:

**Observation 1** *For $d \in \mathcal{E}$, $Stat(d)$ is more informative than the abstraction of $d$.*

The static code of an ED can be updated upon the addition (deletion) of curves, enabling an incremental construction. Deleting a curve $c$ is straightforward: (i) delete $c$ from every containing set in which it appears; (ii) delete every instance of a crossing number appearing in $Stat(c)$; (iii) delete $Stat(c)$. For example, deleting curve $D$ from diagram $d$ in Figure 2 gives the diagram $d_2$ in Figure 5. The method above gives $Stat(d_2)$ as $\{A{:}(1\ 6_B\ 5_{BC}\ 2_B), B{:}(1_A\ 3\ 4_C\ 2), C{:}(4\ 5_B\ 6_{AB}\ 3_B)\}$, the static code for the diagram $d_2$. The effects on the static code of the addition of a curve are much more complicated, and we present a detailed algorithmic implementation.

## 4 Incremental Curve Addition

Let $d \in \mathcal{E}_s$ be a concrete diagram and $c'$ a concrete curve not in $d$, s.t. the addition of $c'$ to $d$ yields $d' \in \mathcal{E}_s$. Algorithm 1 provides $Stat(d)$, the static code of $d$. Recomputing $Stat(d')$ from scratch may be inefficient as the number of crossing points increases and code comparison is hindered since their numbering can differ significantly. Algorithm 2 takes $d$, $Stat(d)$ and $c'$ as input and computes $Stat(d')$ incrementally, enabling an easy comparison of codes for $d$ and $d'$. We utilise the following parameterised operations.

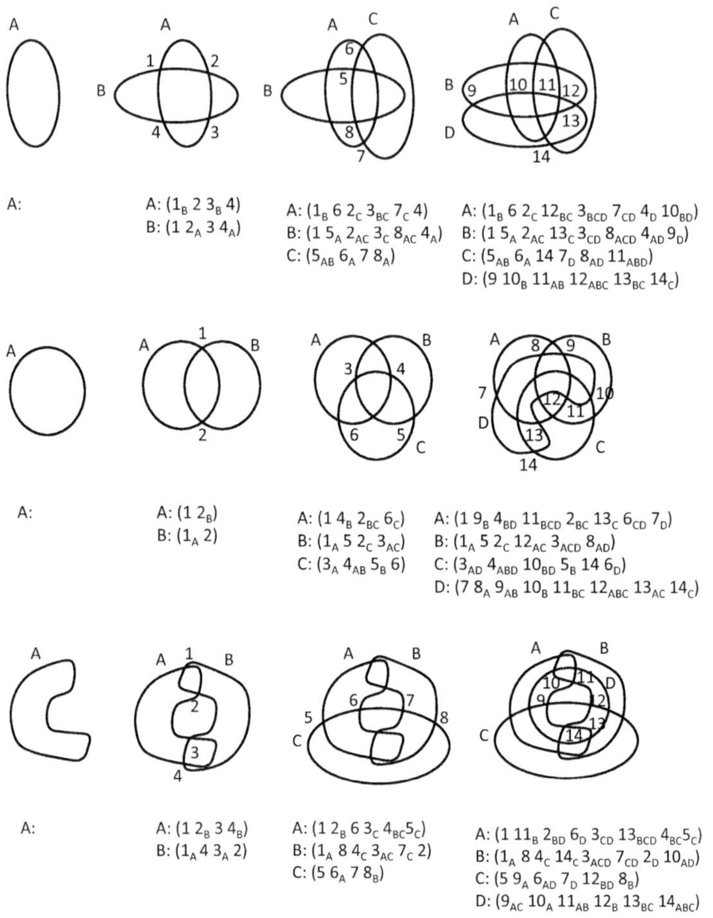


**Fig. 4.** Incremental constructions and static codes of Venn(4). The final diagrams in the first two rows have equivalent codes, not equivalent to the code of the final diagram in the bottom row.

**Definition 5.** *Let $c(d)$ denote a curve $c$ of $d \in \mathcal{E}_s$, and $c'$ a curve not in $d$, s.t. the addition of $c'$ to $d$ yields $d' \in \mathcal{E}_s$. Let $s(d)$ denote a segment of any curve in $d$, $s(k,d)$ a segment of curve $k$ in $d$, and $\hat{s}(k,d)$ the annotated crossing number sequence of $s(k,d)$. Let $y$ denote a sequence of crossing numbers, $x_1$ and $x_2$ denote annotated crossing number sequences, and $Set(x_1)$ the set of unannotated crossing numbers in $x_1$. Then:*

**intersect**$(c', s(d))$ returns the crossing number sequence that $c'$ generates along $s(d)$, ordered according to the orientation of $s(d)$;

**contained**$(s(F), c(G))$ with $F, G \in \{d, d'\}$, returns $true$ if $c(G) \in K(s(F))$; i.e. if $s(F)$ is contained in the interior of $c(G)$.

**split**$(\hat{s}(k,d), c', y)$ where $\hat{s}(k,d) = l_S\ m_T$ and $y = y^1 \dots y^p = intersect(c', s(k,d))$, returns the string $l_S\ y^1_{V^1} \dots y^p{}_{V^p}\ m_{V^{p+1}}$ which is the decomposition of $l_S\ m_T$ due

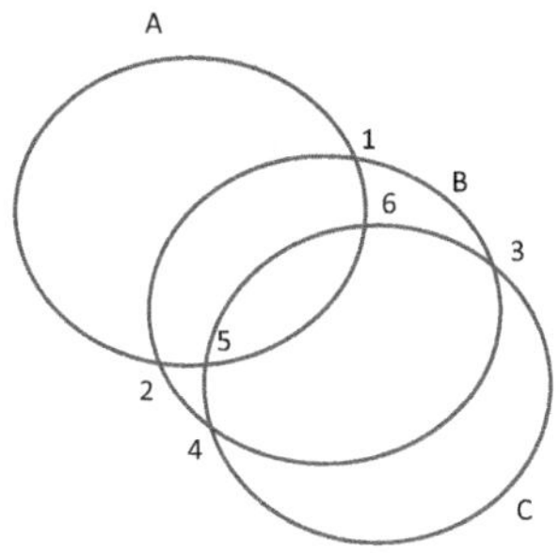


**Fig. 5.** The effect of the deletion of curve $D$ from the diagram $d$ in Figure 2

to the addition of $c'$; for $i \in \{1, \ldots, p+1\}$, $V^i = T \cup \{c'\}$ if *contained*($s_i(k, d')$, $c'(d')) = true$, where $s_i(k, d')$ is the segment of curve $k$ in $d'$ between the crossing points $y^{i-1}$ and $y^i$, taking $y^0 = l$ and $y^{p+1} = m$. Otherwise, $V^i = T$.

**join**$(c(d), x_1, x_2)$ with $x_1$ and $x_2$ annotated crossing number sequences that are subsequences of $Stat(c)$ in $Stat(d)$, returns an annotated crossing number sequence ordered with respect to the orientation of $c$, constructed as follows: form the set $u$ = $Set(x_1) \cup Set(x_2)$ of all crossing numbers from $x_1$ or $x_2$; for each $l \in u$, if $l$ appears in both $x_1$ and $x_2$ then annotate $l$ with the union of the annotations of $l$ in $x_1$ and $x_2$, otherwise annotate $l$ with the annotation that was present in either $x_1$ or $x_2$; order $u$ according to the orientation of $c$.

**encode**$(c(G), y, d)$ with $G \in \{d, d'\}$, $y$ the sequence of crossing numbers of curve $c = c(G)$, returns $Stat(c)$ by (i) sorting $y$ w.r.t. the orientation of $c$ and (ii) annotating, with $K(s)$, each consecutive pair of crossing numbers (read cyclically) corresponding to a segment $s = s(c, G)$.

Figure 6 illustrates the use of these operations. The segment $s = s(C, d)$ lies between the points numbered 3 and 4, having $\hat{s} = \langle 3_{AB}\ 4_B \rangle$, where $\langle - \rangle$ indicates that a sequence of crossing numbers is not to be read cyclically. Curve $D$, shown dashed, is added to $d$ generating new points numbered 7 to 12. We have $intersect(D, s) = \langle 8, 9, 10 \rangle$, $contained(s, D) = false$, but $contained(s, B) = true$. Then $split(\hat{s}, D, \langle 8, 9, 10 \rangle) = split(\langle 3_{AB}\ 4_B \rangle, D, \langle 8, 9, 10 \rangle) = \langle 3_{AB}\ 8_{BD}\ 9_B\ 10_{BD}\ 4_B \rangle$, and $join(c, \langle 6_{AD}\ 3_{ABD} \rangle, \langle 3_{AB}\ 8_{BD}\ 9_B\ 10_{BD}\ 4_B \rangle) = \langle 6_{AD}\ 3_{ABD}\ 8_{BD}\ 9_B\ 10_{BD}\ 4_B \rangle$. We have $encode(B, \{1, 2, 4, 6, 7, 11\}, d) = (1_{AD}\ 7_D\ 4\ 11_C\ 2_{CD}\ 6_{ACD})$.

Algorithm 2 specifies how to update the static code $Stat(d)$ to $Stat(d')$ when adding a new curve $c'$ to $d$. For each curve $G$ with cycle $Stat(G)$ in $Stat(d)$, the algorithm builds a new cycle $Stat(G')$ with elements resulting from inserting the new annotated crossing numbers produced by adding $c'$ to $d$. Then $Stat(d')$ is the set of these cycles, together with the new cycle for $c'$ (*lines* 13–14). To compute $Stat(G')$ from $G$, the algorithm takes each annotated crossing number sequence $a = \hat{s}(G, d)$ from $Stat(G)$ and: (i) computes $y'$ the sequence of crossing numbers that the insertion of $c'$ generates along the concrete segment $s(G, d)$, ordered according to its orientation (*line* 5); (ii) if $y'$ is not empty then $a$ is decomposed into a new sequence which is added to $Stat(G')$. The new sequence is computed by splitting $a$ with respect to the new crossings gener-

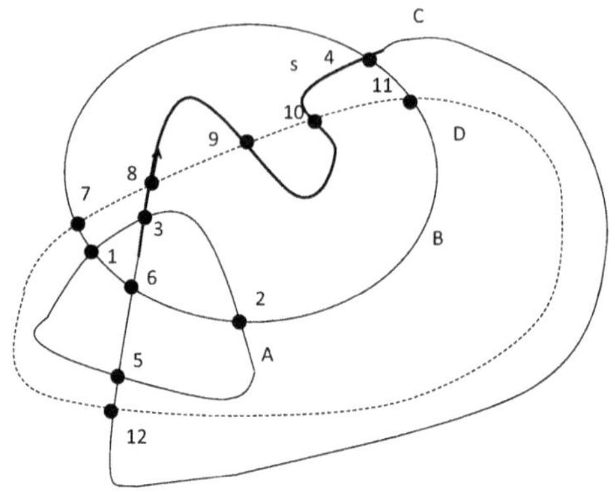


**Fig. 6.** Demonstrating operations: $s$ lies on $C$, and the curve shown dashed is the new curve $D$

---

**Algorithm 2. Incremental static curve addition**$(d, Stat(d), c')$

---

**Input**: $d \in \mathcal{E}_s$, $Stat(d)$, and a new curve $c'$ s.t. the addition of $c'$ to $d$ yields $d' \in \mathcal{E}_s$.
**Output**: A static code for $d'$.

1 $y = \emptyset$
2 **foreach** *curve* $G$ *in* $d$ **do**
3     $Stat(G') = \emptyset$
4     **foreach** $a = \hat{s}(G, d)$ *in* $Stat(G)$ *from* $Stat(d)$ **do**
5         $y' = intersect(c', s(G, d))$
6         **if** $y' \neq \emptyset$ **then**
7             $Stat(G') = join(G, Stat(G'), split(a, c', y'))$
8             $y = y \parallel y'$ where $\parallel$ is the operator concatenating two sequences
9         **else**
10             **if** *contained(*$s(G, d)$, $c'$*)* **then**
11                 update $a$ such that $K(s(G, d)) = K(s(G, d)) \cup \{c'\}$
12             $Stat(G') = join(G, Stat(G'), a)$
13     add $Stat(G')$ to $Stat(d')$
14 add $encode(c', y, d')$ to $Stat(d')$
15 **return** $(Stat(d'))$

---

ated by $c'$ ($lines$ 6–7). Since the new crossings in $y'$ are crossings in $c'$ by construction, they need to be accumulated in $y$, with accumulator denoted $||$, ($line$ 8) in order to build the cycle for $c'$ ($line$ 14); (iii) if $y'$ is empty (i.e. there are no intersections between $s(G, d)$ and the new curve $c'$), then if $s(G, d)$ is in the interior of $c'$, $a$ is updated by adding $c'$ to the containing curve set $K$ as subscript of the end crossing number of $a$. Then $a$ is added to $Stat(G')$ ($lines$ 10–12).

As an example, Table 1 shows the execution trace of Algorithm 2 upon adding curve $D$ to $d$ yielding $d'$, shown in Figure 7. Curves are oriented clockwise. We have $Stat(d) = \{Stat(A){:}1\ 2_C,\ Stat(B){:}3_C\ 4,\ Stat(C){:}1_A\ 3\ 4_B\ 2\}$. In the table we record

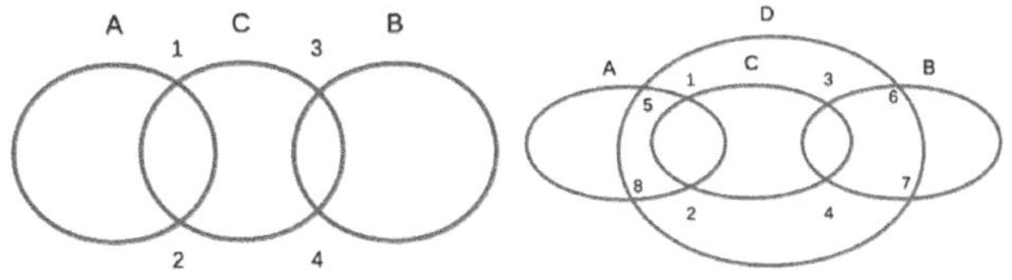


**Fig. 7.** Diagram $d'$ (right) resulting from the insertion of the curve $D$ into $d$ (left)

$split = split(a, c', y')$. Steps 1 and 2 consider the two annotated crossing number sequences $1\ 2_C$ and $2_C\ 1$ (column $a$) of the cycle $A$:$1\ 2_C$ for curve $A$ (column $G$). In Step 1, columns $y'$ and $split$ are empty since the segment 1-2 of curve $A$ does not intersect with $D$, whilst column $Stat(G')$ shows $a$ updated w.r.t. containment in $D$ ($lines$ 10–11 of the algorithm). No crossing numbers are added to the set of the crossings of $D$ (column $y$) and nothing is added to $Stat(d')$. In Step 2, the sequence resulting from intersecting segment 2-1 of curve $A$ and the curve $D$ is $y' = \langle 8, 5\rangle$ ordered w.r.t the orientation of $A$. The column $split$ shows the new sequence. The containment of the new segments only needs to be checked against the newly added curve and not against the curves in $d$, as this information is inherited from the original segment (to which they belong). Column $Stat(G')$ contains the result of joining the previously calculated (in Step 1) sequence $Stat(G') = \langle 1\ 2_{CD}\rangle$ to the annotated sequence $split = \langle 2_C\ 8_D\ 5\ 1_D\rangle$. The operation constructs the union of the crossing numbers from $Stat(G')$ and $split$, taking the union of the annotations (i.e. containing curve sets) for crossing numbers common to both $Stat(G')$ and $split$, and orders the crossing numbers w.r.t. the orientation of $A$. The column $y$ contains the sequence $\langle 8, 5\rangle$ of the calculated crossing numbers of $D$ while $Stat(d')$ is updated by adding the cycle from column $Stat(G')$. A similar description to Step 2 can be given for Step 3, but now the sequence $\langle 6, 7\rangle$ is concatenated to column $y$ to form the sequence $\langle 8, 5, 6, 7\rangle$. The remaining steps are similar to Step 1, since none of the remaining segments produces new crossings upon the addition of $D$. Finally, after ordering the sequence in $y$ with respect to the orientation of $D$ and updating the annotations (containing curve sets) for the consecutive segments (read cyclically), the cycle $(8\ 5_A\ 6\ 7_B)$ is added to $Stat(d')$. The algorithm returns: $\{A$:$(1_D\ 2_{CD}\ 8_D\ 5)$, $B$:$(3_{CD}\ 6_D\ 7\ 4_D)$, $C$:$(1_{AD}\ 3_D\ 4_{BD}\ 2_D)$, $D$:$(8\ 5_A\ 6\ 7_B)\}$.

**Table 1.** Trace of the execution of Algorithm 2 when adding $D$ to $d$ to yield $d'$ in Figure 7

| $Step$ | $G$ | $a$ | $y'$ | $split$ | $Stat(G')$ | $y$ | $Stat(d')$ |
|---|---|---|---|---|---|---|---|
| 1 | $A$ | $1\ 2_C$ | | | $1\ 2_{CD}$ | | |
| 2 | $A$ | $2_C\ 1$ | 8, 5 | $2_C\ 8_D\ 5\ 1_D$ | $1_D\ 2_{CD}\ 8_D\ 5$ | 8, 5 | added $Stat(G')$ |
| 3 | $B$ | $3_C\ 4$ | 6, 7 | $3_C\ 6_D\ 7\ 4_D$ | $3_{CD}\ 6_D\ 7\ 4_D$ | 8, 5, 6, 7 | |
| 4 | $B$ | $4\ 3_C$ | | | $3_{CD}\ 6_D\ 7\ 4_D$ | | added $Stat(G')$ |
| 5 | $C$ | $1_A\ 3$ | | | $1_A\ 3_D$ | | |
| 6 | $C$ | $3\ 4_B$ | | | $1_A\ 3_D\ 4_{BD}$ | | |
| 7 | $C$ | $4_B\ 2$ | | | $1_A\ 3_D\ 4_{BD}\ 2_D$ | | |
| 8 | $C$ | $2\ 1_A$ | | | $1_{AD}\ 3_D\ 4_{BD}\ 2_D$ | | added $Stat(G')$ |

## 5 Encoding Zones

We present a means of encoding the zones of a diagram $d \in \mathcal{E}_s$ by viewing each zone as a union of minimal regions; we encode the set of minimal regions for each zone by indicating the cycle of segments around its boundary. We use a minor modification (without changing information content) of the static code to have direct access to the curve that the segment belongs to, as well as those containing it. We show how to compute the set of all abstract zones of $d$ from a code.

**Definition 6.** *Let d $\in \mathcal{E}$ be a Euler Diagram with curve set $\mathcal{C}$. The* boundary *of a minimal region $mr$ of d is a cycle, up to orientation reversal of the segments, such that there are no segments in the interior of the region bounded by the cycle*[1]*. A zone $z$ of d can be viewed as a union of a set of minimal regions $mr_1, \ldots, mr_k$, and a segment s* lies on the boundary of $z$ *if s lies on the boundary of one of the $m_i$.*

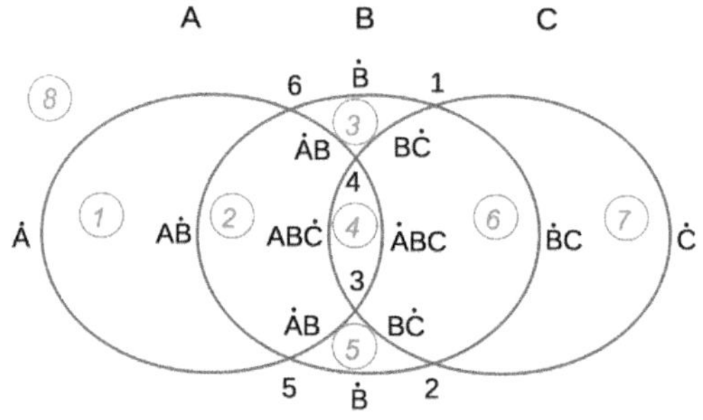


**Fig. 8.** An ED $d \in \mathcal{E}_s$ with segment annotation extended to include the curve which the segment belongs to (shown dotted). The circled numbers indicate the minimal regions.

The concrete diagram $d \in \mathcal{E}_s$ in Figure 8 is not well formed (i.e. $d \notin \mathcal{E}_w$), having 8 minimal regions, indicated by circled numbers, and 7 zones. In order, the minimal regions belong to the following zones: $(A, BC)$, $(AB, C)$, $(B, AC)$, $(ABC, \emptyset)$, $(B, AC)$, $(BC, A)$, $(C, AB)$, $(\emptyset, ABC)$. Zone $(B, AC)$ is comprised of two minimal regions, numbered 3 and 5, whilst the others are comprised of a single minimal region.

We modify the static code by extending the annotated set of curves that contain segments to include the curve that the segment belongs to, distinguished by a dot, and we omit the set notation in the examples and figures. For example, with all the curves oriented clockwise in Figure 8, we have $Stat^*(d)$ is: $Stat^*(A)$= $(6_{\dot{A}}\ 4_{\dot{A}B}\ 3_{\dot{A}BC}\ 5_{\dot{A}B})$, $Stat^*(B)$= $(6_{A\dot{B}}\ 1_{\dot{B}}\ 2_{\dot{B}C}\ 5_{\dot{B}})$, $Stat^*(C)$= $(1_{B\dot{C}} 2_{\dot{C}}\ 3_{B\dot{C}}\ 4_{AB\dot{C}})$.

**Definition 7.** *Let s be a segment of curve c in diagram d $\in \mathcal{E}$, with curve set $\mathcal{C}(d)$. The* extended containing curve set *of s in d, denoted $\dot{K}(s, d)$, is the set $K(s) \cup \{\dot{c}\}$ where $\dot{c}$ is used to indicate the distinguished element c in the set. We abbreviate $\dot{K}(s, d)$ by $\dot{K}(s)$ when no ambiguity arises. Define $Stat^\bullet(F)$ to be $Stat(F)$ with $K(s)$ replaced by $\dot{K}(s)$ throughout, for $F = d$ or $F \in \mathcal{C}(d)$.*

[1] For the outer cycle around the unbounded face this is the interior of the unbounded region.

This extension allows us to specify the boundaries of minimal regions by forming a cycle using segments from different curve's codes (possibly reversing orientation of individual segments.) For example, in Figure 8, boundaries of minimal regions are encoded as: $Stat^\bullet(mr_1)$=($6_{\dot{A}}\ 5_{A\dot{B}}$), $Stat^\bullet(mr_2)$=($4_{\dot{A}B}\ 3_{AB\dot{C}}\ 5_{\dot{A}B}\ 6_{A\dot{B}}$), $Stat^\bullet(mr_3)$=($1_{\dot{B}}\ 4_{B\dot{C}}\ 6_{\dot{A}B}$), $Stat^\bullet(mr_4)$=($3_{\dot{A}BC}\ 4_{AB\dot{C}}$), $Stat^\bullet(mr_5)$=($5_{\dot{B}}\ 3_{\dot{A}B}\ 2_{B\dot{C}}$), $Stat^\bullet(mr_6)$ = ($2_{\dot{B}C}\ 3_{B\dot{C}}\ 4_{\dot{A}BC}\ 1_{B\dot{C}}$), $Stat^\bullet(mr_7)$=($2_{\dot{C}}\ 1_{\dot{B}C}$), $Stat^\bullet(mr_8)$=($1_{\dot{B}}\ 2_{\dot{C}}\ 5_{\dot{B}}\ 6_{\dot{A}}$). Since zone $(B, AC)$ is composed of the minimal regions 3 and 5, its encoding is the set $\{(1_{\dot{B}}\ 4_{B\dot{C}}\ 6_{\dot{A}B}), (5_{\dot{B}}\ 3_{\dot{A}B}\ 2_{B\dot{C}})\}$=$\{Stat^\bullet(mr_3), Stat^\bullet(mr_5)\}$. The other zones are encoded as singleton sets containing the corresponding minimal region code.

**Definition 8.** *Let* $d \in \mathcal{E}_s$ *and* $mr$ *be a minimal region of* $d$*. Then* $Stat^\bullet(mr)$ *is defined as the path* $(x^1_{S_1}\ x^2_{S_2} \dots x^n_{S_n})$ *where each segment* $s_i$*, determined by* $\hat{s}_i = x^i x^{(i+1)\%n}$*, for* $1 \le i \le n$*, is a segment of the boundary of* $mr$ *and is annotated by the extended containing curve set* $S_i = \dot{K}(s_i)$*. Moreover,* $Stat^\bullet(z) = \{Stat^\bullet(mr_1), \dots, Stat^\bullet(mr_k)\}$*, where zone* $z$ *is a union of minimal regions* $mr_1, \dots, mr_k$*.*

**Theorem 2.** *Let* $d \in \mathcal{E}_s$*. The conditions (1) for* $2 \le i \le n+1$*,* $\exists c \in \mathcal{C}(d)$ *for which the code fragment* $x^{i-1}_{S_{i-1}}\ x^i_{S_i}$ *(with indices taken modulo* $n$*), or its reverse, occurs in* $Stat^\bullet(c)$*, (2) the* $x^i$ *are all distinct and (3)* $\exists i$ *with* $1 \le i \le n$*, s.t. for* $1 \le j \le n$ *either (i)* $set(S_i) = set(S_j)$ *or (ii)* $S_j = \{\dot{A}_j\} \cup set(S_i)$*, with* $A_j \in \mathcal{C}(d) \setminus set(S_i)$[2] *hold iff there is a minimal region* $mr$ *of* $d$ *for which* $Stat^\bullet(mr) = (x^1_{S_1}\ x^2_{S_2} \dots x^n_{S_n})$*.*

In the case of Theorem 2 we say that $S_i$ is a *seed* of the minimal region $mr$. The proof of this theorem utilises results presented after, in Lemmata 1- 3.

*Proof.* The above construction of a cycle $G = (x^1_{S_1}\ x^2_{S_2} \dots x^n_{S_n})$ is well defined since each code fragment arises from a segment of *d*, and it forms a cycle by definition (no self-intersections due to condition 2). We show that each corresponding segment of $d$ bounds the same minimal region (and hence the same zone). Firstly, by Lemma 1, $G$ cannot have other curves crossing it, and it cannot traverse the same curve for two consecutive segments. Therefore, at every crossing point that the cycle reaches, it must turn onto the adjacent segment to the right or left and not pass over other curves. Now, the extended containing curve set of a segment indicates the pair of zones that a segment bounds (Observation 2), and we can identify the curves in this set which are not part of the extended containing curve set of the seed (Lemma 2). These results are used to show that all sequences of three segments of $G$ bound a common zone (Lemma 3). Thus the cycle must bound a minimal region of a single zone. The result follows. □

As an example, the code of the boundary of minimal region 2 in Figure 8 is given by $Stat^\bullet(mr_2) = 4_{\dot{A}B}\ 3_{\dot{C}AB}\ 5_{\dot{A}B}\ 6_{\dot{B}A}$. Formally, the *seed* is $S_1 = \{\dot{A}, B\}$, $S_2 = \{\dot{C}, A, B\}$, $S_3 = \{\dot{A}, B\}$ and $S_4 = \{\dot{B}, A\}$. Thus $set(S_3) = set(S_4) = set(S_1) = \{A, B\}$, and $S_2 = \{\dot{C}, A, B\} = \{\dot{C}\} \cup set(S_1)$.

**Lemma 1.** *Given any pair of consecutive segments* $s_{j-2}$ *(with* $\hat{s}_{j-2} = x^{j-2}x^{j-1}$*) and* $s_{j-1}$ *(with* $\hat{s}_{j-1} = x^{j-1}x^j$*) in G, no curve* $c \in C$ *can cross this segment pair at* $x^{j-1}$*. Hence no two consecutive segments of G can have the same curve-label.*

[2] The $set$ operator takes all of the elements of $S_j$ forgetting the distinguished role of one curve, returning the set of curves containing a segment together with the curve owning the segment.

*Proof.* Suppose curve $A$ crosses the pair at $x^{j-1}$. Then $A$ is in exactly one of $K(s_{j-2})$ and $K(s_{j-1})$. Then condition 3, in the hypothesis of Theorem 2, implies that one of the two segments must belong to the curve $A$, contradicting the fact that the curve $A$ crossed the segment pair. Furthermore, since no curve crosses a consecutive pair of these segments, if we had two consecutive segments in $G$ belonging to the same curve then the crossing point $x^{j-1}$ would be a tangential point of the diagram, which is not allowed due to the choice of ED class under consideration (i.e. $d \in \mathcal{E}_s$). □

**Observation 2** *Let $s_j$ be any segment of d, and $K_j = K(s_j)$. If $\dot{K}(s_j) = \{\dot{A}_j\} \cup K_j$, then $s_j$ bounds exactly the two zones $\{(K_j, \mathcal{C} \setminus K_j), (K_j \cup A_j, \mathcal{C} \setminus (K_j \cup A_j))\}$.*

**Lemma 2.** *Suppose that $S_i$ is a seed, and the segment $s_i$ belongs to curve $c_i$. Then condition 3 ensures that every other segment $s_j$, belonging to curve $c_j$, in the cycle either has: $K(s_i) = K(s_j)$ and $c_i = c_j$; $K(s_i) \setminus \{c_j\} = K(s_j) \setminus \{c_i\}$; or $K(s_j) = K(s_i) \cup \{c_i\}$ and $c_j \notin K(s_i)$. Furthermore, if there is a pair of consecutive segments, $s_{j-2}$ (with $\hat{s}_{j-2} = x^{j-2}x^{j-1}$) and $s_{j-1}$ (with $\hat{s}_{j-1} = x^{j-1}x^{j}$) in G, such that $c_k$ is not in $K(s_{j-2})$ or in $K(s_{j-1})$ then $c_k \notin S_i = \dot{K}(s_i)$.*

**Lemma 3.** *All sequences in the cycle must lie on the boundary of a common zone.*

*Proof.* We show that if there are 3 consecutive segments in the constructed cycle that do not all bound a common zone then condition 3 is violated. Let $s_1, s_2$ and $s_3$ be 3 consecutive segments in the cycle s.t. $s_1$ and $s_3$ do not bound a common zone. By construction of the cycle $s_1$ and $s_2$ share a common zone, and so do $s_1$ and $s_2$. We have $s_j$ has $\dot{K}(s_j) = \{\dot{A}_j\} \cup S_j$, for each $j \in \{1, 2, 3\}$. By Lemma 1, $A_1 \neq A_2$ and $A_2 \neq A_3$. The seed $S_i$ bounds the two zones $(K_i, \mathcal{C} \setminus K_i)$ and $(K_i \cup A_i, \mathcal{C} \setminus (K_i \cup A_i))$ by Observation 2, whilst $s_1$ bounds $(K_1, \mathcal{C} \setminus K_1)$ and $(K_1 \cup A_1, \mathcal{C} \setminus (K_1 \cup A_1))$ and $s_3$ bounds $(K_3, \mathcal{C} \setminus K_3)$ and $(K_3 \cup A_3, \mathcal{C} \setminus (K_3 \cup A_3))$. Moreover, the conditions on the seed imply that it bounds a zone in common with every segment in $G$. Lemma 2 relates conditions on the seed to conditions on the other segments, and indicates that any curve not appearing in any consecutive pair of segments cannot be in the extended containing curve set of the seed. Hence, since $s_1$ and $s_3$ do not bound a common zone, and each segment bounds 2 zones determined by the curve the segment belongs to, $\dot{K}(s_1)$ and $\dot{K}(s_3)$ must differ by such an extent that they cannot both relate to the seed as is required. Thus, there cannot be a seed, contradicting the hypothesis. □

**Theorem 3.** *Let $d \in \mathcal{E}_s$. Then the abstract zone set of d can be computed from $Stat^\bullet(d)$.*

*Proof.* Each minimal region $mr_1$ of $d$ is part of a single concrete zone $z_1$. The abstract zone corresponding to $z_1$ can be computed from $Stat^\bullet(mr_1)$ as $(Set(S_i), \mathcal{C}(d) \setminus Set(S_i))$, where $S_i$ is a seed of $mr_1$. In fact, it can also be computed as the union of all $K(s)$ for $s$ a segment of the cycle for the boundary of $mr_1$. Computing the abstract zone for each minimal region in the diagram (obtained by enumerating through the relevant cycles in the code) yields the result. □

For example, the two codes $(1_{\dot{B}}\ 4_{B\dot{C}}\ 6_{\dot{A}B})$ and $(5_{\dot{B}}\ 3_{\dot{A}B}\ 2_{B\dot{C}})$ for minimal regions 3 and 5 in Figure 8, respectively, are not equivalent (i.e. they are not the same up to cyclic permutation and reversal) but they share the same seed $S = \{\dot{B}\}$. They both give rise to the same abstract zone $(\{B\},\{A,C\})$. The set of minimal region codes $\{(1_{\dot{B}}\ 4_{B\dot{C}}\ 6_{\dot{A}B}), (5_{\dot{B}}\ 3_{\dot{A}B}\ 2_{B\dot{C}})\ \}$ with seed $S$ is said to *encode* the zone $(\{B\},\{A,C\})$.

## 6 Discussion and Conclusions

We have developed the foundations for a theory of ED codes, providing access to the specification of features of concrete diagrams in addition to the usual zone-based abstraction. This finer control over diagram specification has potential for major benefits in multiple areas, and opens up many avenues for future work. For example, identifying properties of the encoding relative to the use of variants of the well-formedness condition will be useful in ED interpretation; integration with [4] will develop the theory whilst providing access to existing algorithms and libraries. The encoding will also benefit the automatic generation of EDs; foundational works [2,11] have led to much research activity on relaxing various forms of the well-formedness conditions and improving layout aesthetics, and the encodings will permit a finer specification of the desired layout than is currently used. Storing marked points, as in [4], rather than segments is likely to improve efficiency, but we chose to consider segments in the static code in order to simplify comprehension of the abstraction by humans (allowing the future option of utilising control points).

In [22], a methodology was provided which takes a set of polygons (i.e. regions determined by sets of non-overlapping curves) and outputs a set of non-overlapping polygons (essentially the boundary of the zones), enabling the computation of polygon operations such as union, intersection, difference and clipping. They use a graph based structure and their algorithm "corrects" input containing degeneracies (e.g. concurrency). Since analysing topological relationships between contours may be sufficient to compute the information required to reconstruct the abstract diagram, from a purely topological viewpoint, certain cases of singularities or degeneracies (e.g. concurrency or multiple points) may be deemed irrelevant. However, our long term goal is to relax the wellformedness conditions, developing codes which enable the identification and use of such singularities, permitting their use within wider contexts.

In [1], the Grünbaum encoding is shown to uniquely identify simple (wellformed) Venn diagrams which are monotone and polar symmetric; their codes rely upon numbering the curves (adopting conventions based on the curve segment in certain faces to fix the choices) and recording the sequence of curve numbers as one traverses a given curve. Our ED encoding is more generic, applying to a wider class of diagrams, and is based of numbering crossings rather than curves. Investigating connections between these encoding variations is an interesting line of future investigation.

We considered connected EDs, but the extension to nested diagrams [12], where one diagram is embedded in the zone of another, is straightforward. This generalises to disconnectable EDs considered in the context of diagram generation, where deleting a curve disconnects the diagram [9]; investigating relationships with ED codes is a natural step that should benefit the ED generation programme. The notion of encoding EDs was inspired by the codes in Knot Theory. For knot diagrams, one encapsulates "over" and "under" information of curves at crossings by a signed Gauss code [14], whereas we capture the containment relationship for EDs, which is the key information in this case.

## References

1. Cao, T., Mamakani, K., Ruskey, F.: Symmetric Monotone Venn Diagrams with Seven Curves. In: Boldi, P. (ed.) FUN 2010. LNCS, vol. 6099, pp. 331–342. Springer, Heidelberg (2010)
2. Chow, S.C.: Generating and Drawing Area-Proportional Euler and Venn Diagrams. PhD thesis, University of Victoria (2007)
3. Cordasco, G., De Chiara, R., Fish, A.: Interactive visual classification with Euler diagrams. In: Proc. VL/HCC 2009, pp. 185–192. IEEE (2009)
4. Cordasco, G., De Chiara, R., Fish, A.: Fast region computations for reducible Euler diagrams. Computation Geometry: Theory and Applications 44, 52–68 (2011)
5. Dau, F., Fish, A.: Conceptual Spider Diagrams. In: Eklund, P., Haemmerlé, O. (eds.) ICCS 2008. LNCS (LNAI), vol. 5113, pp. 104–118. Springer, Heidelberg (2008)
6. Eades, P., Lai, W., Misue, K., Sugiyama, K.: Layout adjustment and the mental map. JVLC 6, 183–210 (1995)
7. Fish, A.: Euler diagram transformations. ECEASST 18, 1–17 (2009)
8. Fish, A., Flower, J.: Abstractions of Euler diagrams. In: Proc. Euler 2004. ENTCS, vol. 134, pp. 77–101 (2005)
9. Fish, A., Flower, J.: Euler Diagram Decomposition. In: Stapleton, G., Howse, J., Lee, J. (eds.) Diagrams 2008. LNCS (LNAI), vol. 5223, pp. 28–44. Springer, Heidelberg (2008)
10. Fish, A., Flower, J., Howse, J.: The semantics of augmented constraint diagrams. JVLC 16, 541–573 (2005)
11. Flower, J., Fish, A., Howse, J.: Euler diagram generation. JVLC (2008)
12. Flower, J., Howse, J., Taylor, J.: Nesting in Euler diagrams: syntax, semantics and construction. Software and Systems Modelling 3, 55–67 (2004)
13. Howse, J., Stapleton, G., Taylor, J.: Spider diagrams. LMS Journal of Computation and Mathematics 8, 145–194 (2005)
14. Kauffman, L.: Knots and Physics. World Scientific (1991)
15. Kent, S.: Constraint diagrams: Visualizing invariants in object oriented modelling. In: Proc. OOPSLA 1997, pp. 327–341. ACM Press (October 1997)
16. Kestler, H.A., Müller, A., Kraus, J.M., Buchholz, M., Gress, T.M., Liu, H., Kane, D.W., Zeeberg, B.R., Weinstein, J.: Vennmaster: Area-proportional Euler diagrams for functional GO analysis of microarrays. BMC Bioinformatics 9, 67 (2008)
17. Riche, N.H., Dwyer, T.: Untangling Euler diagrams. IEEE VCG 16(6), 1090–1099 (2010)
18. Shin, S.-J.: The Logical Status of Diagrams. Cambridge University Press (1994)
19. Stapleton, G., Masthoff, J., Flower, J., Fish, A., Southern, J.: Automated theorem proving in Euler diagrams systems. Journal of Automated Reasoning (2007)
20. Swoboda, N., Allwein, G.: Using DAG transformations to verify Euler/Venn homogeneous and Euler/Venn FOL heterogeneous rules of inference. SoSyM 3(2), 136–149 (2004)
21. Thièvre, J., Viaud, M., Verroust-Blondet, A.: Using Euler diagrams in traditional library environments. In: Proc. Euler 2004. ENTCS, vol. 134, pp. 189–202. ENTCS (2005)
22. Weiler, K.: Polygon comparison using a graph representation. Computer Graphics (SIGGRAPH 1980 Proceedings) 14(3), 10–18 (1980)

# Speedith: A Diagrammatic Reasoner for Spider Diagrams

Matej Urbas[1], Mateja Jamnik[1], Gem Stapleton[2], and Jean Flower[3]

[1] Computer Laboratory, University of Cambridge, UK
{Matej.Urbas,Mateja.Jamnik}@cl.cam.ac.uk
[2] School of Computing, Engineering and Mathematics, University of Brighton, UK
g.e.stapleton@brighton.ac.uk
[3] Autodesk, UK

**Abstract.** In this paper, we introduce Speedith which is a diagrammatic theorem prover for the language of spider diagrams. Spider diagrams are a well-known logic for which there is a sound and complete set of inference rules. Speedith provides a way to input diagrams, transform them via the diagrammatic inference rules, and prove diagrammatic theorems. It is designed as a program that plugs into existing general purpose theorem provers. This allows for seamless formal verification of diagrammatic proof steps within established proof assistants such as Isabelle. We describe the general structure of Speedith, the diagrammatic language, the automatic mechanism that draws the diagrams when inference rules are applied on them, and how formal diagrammatic proofs are constructed.

## 1 Introduction

Diagrams have been used to prove theorems since ancient times. One can argue that diagrams often provide compelling and intuitive solutions to problems. Despite this, diagrams have rarely been formalised in proof tools to be used for reasoning. In this paper, we do just that: we present a new, formal diagrammatic theorem prover Speedith. Speedith's domain is the language of spider diagrams. It allows us to apply diagrammatic inference rules on conjectures about spider diagrams, and thus construct a proof. The entire proof construction process is carried out visually. The derived proof is certified to be (logically) correct. Here are the hypotheses we test and objectives we aim to achieve:

- We want to show that it is possible to design and implement a complete formal diagrammatic reasoner in the general domain of monadic first-order logic (MFOL) with equality, expressed using the language of spider diagrams.
- We aim to have the guarantee that the derived proofs are formally correct.
- We aim for our system to be standalone, yet also reasonably easily plugable into external proof tools, thus providing alternative problem representation and proof construction method for these tools.

P. Cox, B. Plimmer, and P. Rodgers (Eds.): Diagrams 2012, LNAI 7352, pp. 163–177, 2012.

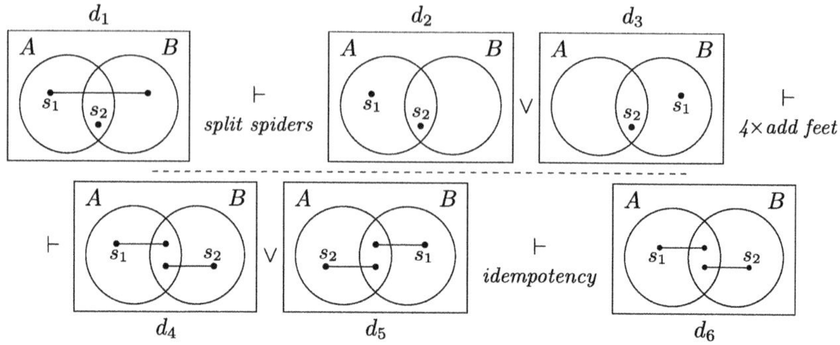


**Fig. 1.** A proof of a spider-diagrammatic statement. The proof establishes that given sets $A$ and $B$, if there are two elements $s_1$ and $s_2$ and one is in both of $A$ and $B$ and the other is either in only $A$ or only $B$, then we can deduce that one element is in $A$ and the other is in $B$. In this proof, we applied the split spiders, add feet, and idempotency inference rules. The rules are proved to be sound and their application in this proof is verified by Speedith to be correct. Hence, the proof is certified to be correct.

Whilst there exist other purely diagrammatic theorem provers, such as DIAMOND [1], Dr.Doodle [2], and Cinderella [3], they target different, more restricted domains (e.g., a small subset of natural number arithmetic, a subset of real arithmetic), and are hence able to prove only a limited class and number of theorems. They do not provide a provably sound and complete set of inference rules. They are also not designed to be readily integrated into external proof tools.

There are theorem provers that were developed for spider diagrams, but they worked only for fragments of the logic in this paper, and did not include any logical connectives, or only a limited number of them [4]. In Speedith we formalize the whole spider diagram logic, which is expressively equivalent to MFOL with equality. We also develop a set of sound inference rules, representing a conservative extension of the complete system in [5], which allow for more intuitive proof steps.

Speedith is an interactive proof assistant for the language of spider diagrams and allows its users to interactively apply diagrammatic (visual) inference rules on spider-diagrammatic statements. It checks whether the inference rules are used correctly and verifies that a spider-diagrammatic statement expresses a true fact – it is a theorem. Thus, Speedith's diagrammatic proofs are entirely formal and certified to be correct. Fig. 1 shows an example of Speedith's purely diagrammatic proof. Here, $d_1$ is a spider diagram which conveys some information about the relationships between two elements and two sets and proves that $d_6$ follows logically.

Speedith provides a graphical user interface through which all the diagrammatic proofs are constructed. It visually displays spider-diagrammatic statements; allows the user to specify which inference rules should be applied on what parts of the spider diagram; and displays the result of this visually.

Whilst Speedith is a standalone diagrammatic proof assistant, it is also designed to easily plug into external proof tools. This has the advantage that spider-diagrammatic proofs can be reconstructed in traditional logic, and thus certified with, for example, LCF-style general purpose theorem provers [6].

To confirm our hypotheses above and achieve our aims, we designed Speedith as a standalone system that incorporates the following components: full specification of Spider diagrams (Sec. 2) and their inference rules (Sec. 3), a reasoning kernel that manages the state of the proofs, controls how inference steps are applied, and manages the communication with external general purpose theorem provers (Sec. 4), and a visualisation component with input methods for constructing diagrammatic statements and interactively applying inference rules (Sec. 5). Lastly, we evaluate our prover (Sec. 6) and conclude with future directions and general observations (Sec. 7).

## 2 Spider Diagrams: Syntax and Semantics

We now introduce spider diagrams (see [5] for more details and examples). Spider diagrams use closed curves, called *contours*, to represent sets and assert relationships between those sets. For instance, the enclosure of one contour by another contour corresponds to a subset/superset relationship between the represented sets. Contours are named with labels (in $d_1$ in Fig. 1, the contour labels are $A$ and $B$). The set of contour labels used in a diagram $d$ is denoted by $L(d)$.

A *zone* is a region in a diagram that is inside some of the contours (possibly no contours) and not inside the rest of them. Formally, a zone is a pair of finite, disjoint sets of contour labels, $(in, out)$. Intuitively, $(in, out)$ is inside every contour of $in$, and outside every contour of $out$. So, in a diagram, the set of possible zones is formed by its contour labels (e.g., in $d_1$, the zones are $(\emptyset, \{A, B\})$, $(\{A\}, \{B\})$, $(\{B\}, \{A\})$, $(\{A, B\}, \emptyset)$). We denote the set of zones in a diagram $d$ by $Z(d)$. The zones from $Z(d)$ can be *shaded*, denoted $ShZ(d)$, and this places upper bounds on set cardinality: in a shaded zone, all elements are represented by spiders.

*Spiders* are trees used in spider diagrams to assert the existence of elements; they place lower bounds on set cardinalities (e.g., $d_1$ contains two spiders called $s_1$ and $s_2$ representing an element in only $A$ or only $B$, and an element in $A$ and $B$; there may be other elements too). The nodes of the trees are called *spider feet*, or simply *feet* (e.g., in $d_1$ the spider $s_1$ has two feet). The set of spiders in $d$ is denoted by $S(d)$ and the function $\eta\colon S(d) \to \mathbb{P}Z(d) - \{\emptyset\}$ (here $\mathbb{P}$ denotes the power set) returns the set of zones in which the spider is placed, called its *habitat* (e.g., in Fig. 1, $s_1$ is placed in the zones $(\{A\},\{B\})$ and $(\{B\},\{A\})$). To avoid ambiguity when talking about the habitats of spiders in more than one diagram, we write $\eta_d(s)$ to mean the habitat of spider $s$ in diagram $d$.

The diagrams considered so far are called *unitary* spider diagrams: they can be defined by a tuple $d = (L, Z, ShZ, S, \eta)$ as described above.[1] Spider diagrams can be negated and joined with binary connectives into *compound* diagrams: $\neg$

[1] We define unitary diagrams differently to [5] where no mapping function for spiders and their habitats was used. The mapping function provides a convenient translation of spiders into formulae where each spider is a variable (see Sec. 4.3).

to denote 'not', $\wedge$ to denote 'and', $\vee$ to denote 'or', $\Rightarrow$ to denote 'implies' (e.g., in Fig. 1, $d_2 \vee d_3$ forms a compound diagram), and $\Leftrightarrow$ to denote 'equivalent'.[2]

The semantics of spider diagrams are captured by *interpretations*, $I = (U, \Psi)$, where $U$ is a universal set, and $\Psi$ is an assignment of a subset of $U$ to each contour label. A zone, $(in, out)$, represents the set:

$$\Psi(in, out) = \bigcap_{l \in in} \Psi(l) \cap \bigcap_{l \in out} (U - \Psi(l))$$

where $\Psi(l)$ is the set assigned to contour label $l$. A set of zones represents the set which is the union of the sets represented by the individual zones.

In order to identify when an interpretation agrees with the meaning of a unitary diagram $d$ we define *missing zones*, $MZ(d)$ such that:

$$MZ(d) = \{(in, L(d) - in) : in \subseteq L(d) \wedge (in, L(d) - in) \notin Z(d)\}.$$

Intuitively, the missing zones are the zones that do not appear in the diagram due to its particular configuration, but could be specified using the labels from $L(d)$ (e.g., consider $d_1$ in Fig. 2 on page 167; nine zones can be specified, but do not appear in $d_1$, including $(\{A, C, D\}, \{B\})$ and $(\{A, B\}, \{C, D\})$). Briefly, we say that an interpretation, $I = (U, \Psi)$, is a *model* for unitary diagram $d$ if there exists a function $\psi\colon S(d) \to U$ (interpreting the spiders as elements) that ensures:

1. the missing zones represent the empty set, that is, $\Psi(MZ(d)) = \emptyset$;
2. each spider $s$ in $d$ maps to an element $\psi(s)$ of the set represented by the spider's habitat, that is, $\psi(s) \in \Psi(\eta(s))$;
3. no two spiders map to the same element, that is, $\psi(s_1) = \psi(s_2) \Rightarrow s_1 = s_2$;
4. the shaded zones contain only elements represented by spiders, that is, $\Psi(ShZ(d)) \subseteq \{\psi(s) : s \in S(d)\}$.

The definition of a model extends in the obvious way to compound diagrams.

## 3 Speedith's Inference Rules

We present some of Speedith's inference rules for spider diagrams. We introduce three new rules that conservatively extend the set of sound and complete rules from [5]. Speedith can use all of them, but we only present the ones needed for our examples. Our new rules are designed to allow making intuitive proof steps and to substantially reduce proof length. All inference rules are proved to be sound but, due to space restrictions, we omit the proofs in this paper.

Our first rule allows us to 'copy' a contour from one diagram into another diagram: we argue that this is a natural deduction step. To illustrate, consider the diagram $d_1 \wedge d_2$ in Fig. 2. In $d_2$ we can see that $C \subseteq E$ and $E \subseteq A$. We can copy the contour $E$ from $d_2$ to $d_1$ using this information and thus obtain $d_1'$. Here, $E$ is completely inside $A$ (since $E \subseteq A$) and $C$ is completely inside

[2] This extends the definition of compound diagrams in [5] which did not allow $\neg$ or $\Rightarrow$.

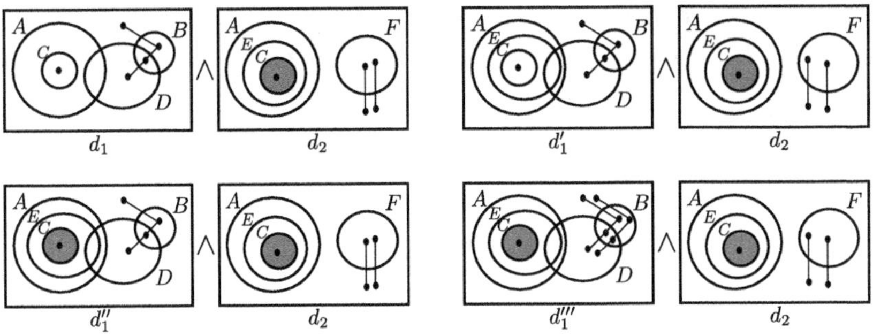


**Fig. 2.** Illustrating new inference rules: copying syntax

$E$ (since $C \subseteq E$). We do not know anything about the relationship between $E$ and $D$, thus we ensure that $E$ partially overlaps with $D$. The spider habitats are updated in line with how the zones have changed. In general, all shading is preserved in the same fashion.

To define the copy contour rule, we start by observing that each zone in the diagram, $d_1$, into which the contour is copied is either (a) completely outside the new contour, (b) completely inside the new contour, or (c) split into two zones by the new contour. In order to identify what happens to each zone, we need to inspect the contours of $d_2$.

**Definition 1.** *Let $d$ be a unitary diagram and let $\lambda$ and $\lambda'$ be in $L(d)$.*

1. *If (the contours labelled) $\lambda$ and $\lambda'$ have disjoint interiors then $\lambda$ and $\lambda'$ are* ***disjoint*** *in $d$, denoted $\lambda \cap_d \lambda' = \emptyset$.*
2. *If $\lambda$ is in the interior of $\lambda'$ then $\lambda$ is a* ***contained*** *by $\lambda'$ in $d$, denoted $\lambda \subseteq_d \lambda'$.*

For example, in Fig. 2, inspecting $d_2$ we have the following relations involving $E$: $E \cap_{d_2} F = \emptyset$, $E \subseteq_{d_2} A$ and $C \subseteq_{d_2} E$. Using these relations, we can determine the effect of copying $E$ from $d_2$ to $d_1$ on the zones. In particular, since $E \subseteq_{d_2} A$, all zones that are not inside $A$ will not be inside $E$; these zones are placed in a set called $Z_{OUT}$ ($OUT$ for 'outside').[3] Since $C \subseteq_{d_2} E$, all zones that are inside $C$ will be inside $E$; these zones are placed in a set called $Z_{IN}$ ($IN$ for 'inside'). The remaining zones will be split when $E$ is copied into $d_1$.

**Definition 2.** *Let $d_1$ and $d_2$ be unitary diagrams and let $\lambda$ be in $L(d_2) - L(d_1)$. We define three subsets of $Z(d_1)$ ($Z_{OUT}(\lambda, d_2)$, $Z_{IN}(\lambda, d_2)$, and $Z_{SPLIT}(\lambda, d_2)$) according to the following rules: let $(in, out) \in Z(d_1)$*

1. *$(in, out) \in Z_{OUT}(\lambda, d_2)$ provided there exists a contour label, $\lambda'$, in $L(d_2)$ such that either*
   *(a) $\lambda' \in in$ and $\lambda \cap_{d_2} \lambda' = \emptyset$, or*
   *(b) $\lambda' \in out$ and $\lambda \subseteq_{d_2} \lambda'$,*
2. *$(in, out) \in Z_{IN}(\lambda, d_2)$ provided there exists a contour label, $\lambda'$, in $L(d_2)$ such that $\lambda' \in in$ and $\lambda' \subseteq_{d_2} \lambda$,*
3. *finally, $Z_{SPLIT}(\lambda, d_2) = Z(d_1) - (Z_{IN}(\lambda, d_2) \cup Z_{OUT}(\lambda, d_2))$.*

[3] If $F$ occurred in $d_1$ then, since $E \cap_{d_2} F = \emptyset$ all zones inside $F$ would also be outside $E$.

**Rule 1** ***Copy a Contour.*** *Let $d_1$ and $d_2$ be unitary diagrams and let $\lambda$ be in $L(d_2) - L(d_1)$. Let $d_1'$ be the diagram whose components are defined as follows:*

1. *the contour labels are $L(d_1') = L(d_1) \cup \{\lambda\}$,*
2. *the zones are*

$$Z(d_1') = \{(in, out \cup \{\lambda\}) : (in, out) \in Z_{OUT}(\lambda, d_2) \cup Z_{SPLIT}(\lambda, d_2)\} \cup \{(in \cup \{\lambda\}, out) : (in, out) \in Z_{IN}(\lambda, d_2) \cup Z_{SPLIT}(\lambda, d_2)\}$$

3. *the shaded zones are*

$$ShZ(d_1') = \{(in, out \cup \{\lambda\}) : (in, out) \in (Z_{OUT}(\lambda, d_2) \cup Z_{SPLIT}(\lambda, d_2)) \cap ShZ(d_1)\} \cup \{(in \cup \{\lambda\}, out) : (in, out) \in (Z_{IN}(\lambda, d_2) \cup Z_{SPLIT}(\lambda, d_2)) \cap ShZ(d_1)\}$$

4. *the spiders are $S(d_1') = S(d_1)$, and*
5. *the habitat of each spider, $s' \in S(d_1')$, is*

$$\eta_{d_1'}(s') = \{(in, out \cup \{\lambda\}) : (in, out) \in (Z_{OUT}(\lambda, d_2) \cup Z_{SPLIT}(\lambda, d_2)) \cap \eta_{d_1}(s')\} \cup \{(in \cup \{\lambda\}, out) : (in, out) \in (Z_{IN}(\lambda, d_2) \cup Z_{SPLIT}(\lambda, d_2)) \cap \eta_{d_1}(s')\}.$$

*Then $d_1 \wedge d_2$ is logically equivalent to $d_1' \wedge d_2$.*

As well as enabling 'natural' proof steps to be made in a single inference step, our copy contour substantially reduces the number of proof steps required. In Fig. 2, if we used only the inference rules from [5], a proof establishing $d_1 \wedge d_2 \vdash d_1' \wedge d_2$ would require hundreds of proof steps (in part because, using the inference rules of [5], in proofs that $d_1 \wedge d_2 \vdash d_1' \wedge d_2$ the number of spider feet increases rapidly when contours are added and all of these spiders need to be split until they have single feet).

If we consider $d_1' \wedge d_2$ in Fig. 2, we can see that the shaded region that comprises the zone $(\{A, E, C\}, \{F\})$ in $d_2$ represents the set $C$ (this is the only zone inside $C$). There is a corresponding zone in $d_1'$, namely $(\{A, C\}, \{B, D\})$, that also represents the set $C$. Since these two zones contain the same spiders, we can copy the shading from $d_2$ over to $d_1'$, as shown in $d_1'' \wedge d_2$. In order to define this rule, we introduce some notation to denote the set of spiders whose habitat is a subset of a given region $r$ in a unitary diagram $d$: $S(r, d) = \{s \in S(d) : \eta(s) \subseteq r\}$. In addition, it is possible to syntactically identify when two distinct regions, $r_1$ and $r_2$, necessarily represent the same set [7]. Such regions are said to be *corresponding*. To illustrate the idea, in Fig. 2 the region inside $d_1$ that comprises the four zones outside of $A$ corresponds to the region that comprises the two zones outside of $A$ in $d_2$.

**Rule 2** ***Copy Shading.*** *Let $d_1$ and $d_2$ be unitary diagrams with corresponding regions, $r_1$ and $r_2$ respectively, such that:*

1. *$r_1$ contains at least one non-shaded zone in $d_1$,*
2. *$r_2$ is entirely shaded in $d_2$,*
3. *in $d_1$, all of the spiders that have a foot in $r_1$ are also in $S(r_1, d_1)$,*
4. *in $d_2$, all of the spiders that have a foot in $r_2$ are also in $S(r_2, d_2)$, and*

5. *there is a habitat preserving bijection,* $\sigma$*, from* $S(r_1, d_1)$ *to* $S(r_2, d_2)$ *(i.e.,* $\eta_{d_1}(s)$ *corresponds to* $\eta_{d_2}(\sigma(s))$*).*

*Let* $d_1'$ *be a copy of* $d_1$ *except that* $r_1$ *is entirely shaded. Then* $d_1 \wedge d_2$ *is logically equivalent to* $d_1' \wedge d_2$*.*

Our final new rule allows us to copy spiders from one diagram to another. This is illustrated in Fig. 2, where we can copy a spider from $d_2$ into $d_1''$, to give $d_1''' \wedge d_2$. The two spiders in $d_2$ that inhabit the region outside of $A$ tell us that there are at least two elements in $U - A$, where $U$ is the universal set. Since there is only one spider in the corresponding region of $d_1''$, that is, there is at least one element in $U - A$, we can copy across the second spider.

**Rule 3** ***Copy a Spider.*** *Let* $d_1$ *and* $d_2$ *be unitary diagrams with corresponding regions,* $r_1$ *and* $r_2$ *respectively, such that:*

1. $r_1$ *contains no shaded zones in* $d_1$*,*
2. *in* $d_1$*, all of the spiders that have a foot in* $r_1$ *are also in* $S(r_1, d_1)$*,*
3. *there exists a habitat preserving injective, but not surjective, map* $\sigma$ *from* $S(r_1, d_1)$ *to* $S(r_2, d_2)$ *(i.e.,* $\eta_{d_1}(s_1)$ *corresponds to* $\eta_{d_2}(\sigma_r(s_1))$*).*

*Choose a spider,* $s$*, that is in* $S(r_2, d_2)$ *but is not in the image of* $\sigma$ *such that there exists a region,* $r'$ *in* $d_1$ *that corresponds to* $\eta_{d_2}(s)$*. Let* $d_1'$ *be a copy of* $d_1$ *except* $d_1'$ *contains* $s$ *with habitat* $r'$*. Then* $d_1 \wedge d_2$ *is logically equivalent to* $d_1' \wedge d_2$*.*

Three other rules are used in our examples: *add feet* and *split spiders* (see both in Fig. 1) and *remove a contour* (see Fig. 7); their formal definitions are in [5].

**Rule 4** ***Add feet to a Spider.*** *Let* $d_1$ *be a unitary diagram that contains a spider,* $s$*, whose habitat does not include all of the zones in* $d_1$*. Let* $d_2$ *be a copy of* $d_1$ *except that* $s$ *contains additional feet. Then* $d_1$ *logically entails* $d_2$*.*

**Rule 5** ***Split Spiders.*** *Let* $d$ *be a unitary diagram containing a spider,* $s$*, with habitat such that* $|\eta(s)| \geq 2$*. Let* $\eta_1$ *and* $\eta_2$ *be a two-way partition of* $\eta(s)$*. Let* $d_1$ *(*$d_2$*) be the diagram obtained from* $d$ *by changing the habitat of* $s$ *to* $\eta_1$ *(*$\eta_2$*). Then* $d$ *and* $d_1 \vee d_2$ *are logically equivalent.*

**Rule 6** ***Remove a Contour.*** *Let* $d_1$ *be a unitary diagram and let* $\lambda \in L(d)$*. Let* $d_2$ *be the diagram obtained from* $d_1$ *by removing the contour,* $C$*, labelled* $\lambda$*, so* $L(d_2) = L(d_1) - \{\lambda\}$*. If, on the removal of* $C$*, two zones combine to form a single zone then spiders' habitats are updated in the same way. With regard to shading, if a shaded zone merges with a non-shaded zone then the shading is removed. Otherwise the shading remains. Then* $d_1$ *logically entails* $d_2$*.*

## 4 Architecture of Speedith

Speedith is the implementation of a diagrammatic theorem prover for spider diagrams described in Sec. 2, and the inference rules from Sec. 3 and [5]. It consists of four main components:

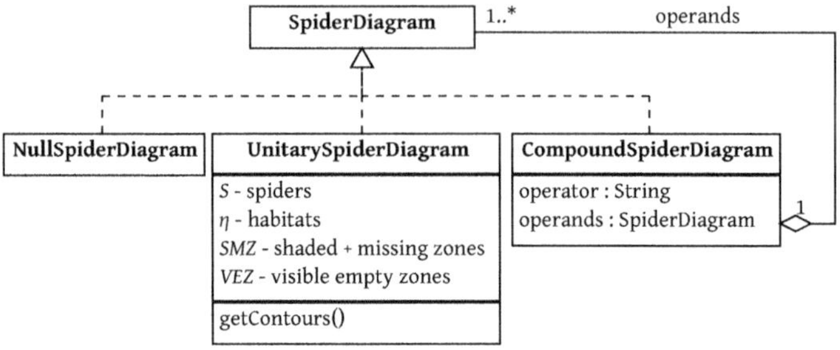


**Fig. 3.** A class diagram of the abstract representation of spider diagrams in Speedith

1. abstract representation of spider-diagrammatic statements (Sec. 4.1),
2. the reasoning kernel with proof infrastructure (Sec. 4.2),
3. verification of diagrammatic proofs, including input and output system for importing and exporting formulae in many different formats (Sec. 4.3), and
4. visualisation of spider-diagrammatic statements (Sec. 5).

### 4.1 Abstract Representation

Speedith uses an abstract spider diagram representation to express spider-diagrammatic formulae. This representation is captured by the class diagram in Fig. 3. The null spider diagram is the unitary diagram containing only the zone $(\emptyset, \emptyset)$, no spiders and no shading; it is used as the logical truth constant $\top$. Unitary spider diagrams contain the bulk of diagrammatic information. Finally, the compound spider diagrams build up more complex formulae by connecting spider diagrams through the usual logical connectives: conjunction, disjunction, implication, equivalence and negation. Thus, a compound spider diagram nests one or more other spider diagrams, as indicated with the diamond notation in Fig. 3.

To optimize performance, Speedith's abstract representation of spider diagrams removes some redundancies from the syntax in Sec. 2. In particular, Speedith does not explicitly store the sets $L(d)$ and $Z(d)$. Moreover, the sets $ShZ$ and $MZ$ are merged into $SMZ$, and the set $VEZ$ lists all the zones that are shaded but have no spider feet in them. This more closely matches the semantics of spider diagrams as the zones that convey no semantic information (i.e. zones with no spider feet and no shading) are not explicitly stored. However, Speedith does store the shaded zones, the missing zones, and the spiders with their habitats, which is needed when converting diagrams to sentential form for verification. Note that all sets from the tuple $d = (L, Z, ShZ, S, \eta)$ in Sec. 2, can still be computed. The set $L(d)$ is obtained via the method `getContours()`, which takes an arbitrary zone $(in, out)$ and computes $L(d) = in \cup out$. The set $MZ(d)$ is obtained through $SMZ - (VEZ \cup \text{habitats})$. Finally, $Z(d) = \{(in, L(d) - in) : in \subseteq L(d)\} - MZ(d)$.

An advantage of Speedith's abstract representation is the simplicity of converting spider-diagrammatic formulae to and from first-order logic formulae for the purposes of verification (see Sec.4.3). On the other hand, the user has no control on how the diagrams are drawn – Speedith lays them out automatically.

## 4.2 The Reasoning Kernel

One of the central components of Speedith is the reasoning kernel. It is responsible for correctly applying the inference rules on particular parts of spider-diagrammatic formulae. The kernel contains two types of inference rules: the ones for logical connectives based on the well-known logical equivalences, and purely spider-diagrammatic rules including the ones introduced in Sec. 3.

### 4.2.1 Proofs in Speedith

A proof in Speedith starts with a spider-diagrammatic formula $D$ (the *initial goal*, i.e., the theorem we aim to prove), proceeds by transforming $D$ with applications of inference rules, and ends with an empty set of goals. The specific rules and parts of the diagram on which they should be applied are chosen by the user. Rules can be applied in both backward and forward reasoning styles. Backward proof steps take a goal $D$ and transform it into a new goal $D'$, where $D' \vdash D$. In forward proofs, $D$ must be of the form $D_i \Rightarrow D_j$, and the rules transform $D_i$ into $D_i'$, where $D_i \vdash D_i'$, resulting in the new goal $D_i' \Rightarrow D_j$.[4] The user can switch between backward and forward proof styles due to the inference rule $(\top \Rightarrow D) \vdash D$. Either way, since the inference rules in both proof styles adhere to the entailment property for valid deductive steps, the proofs are of the standard structure:

$$\begin{array}{cl} \top & \\ \hline & \text{SD inference rule} \\ \vdots & \\ \hline & \text{SD inference rule} \\ \text{D'} & \\ \hline & \text{SD inference rule} \\ \text{D} & \end{array} \tag{1}$$

which together with the soundness of our rules justifies that $D$ is a theorem.

Speedith stores and manages the proof as a sequence of goals and inference rule applications. The diagram in Fig. 4 outlines the architecture for managing proof state and goals. Multiple simultaneous goals are supported, and the proof is finished only when all goals are converted to null diagrams.

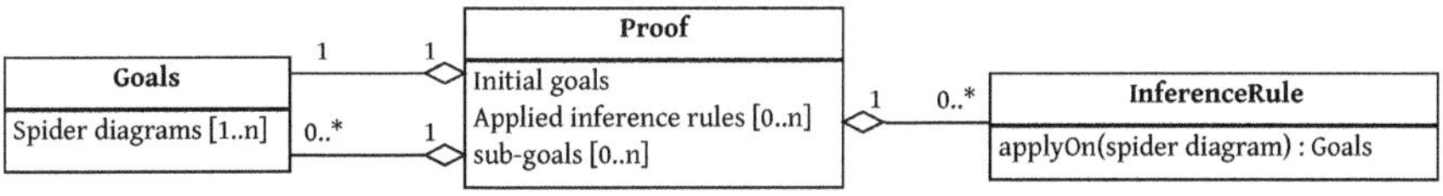


**Fig. 4.** A simplified class diagram of the part of the reasoning kernel responsible for tracking the proof state

Once a spider-diagrammatic theorem $D_t$ is proved, it is added to the database of theorems available for reuse in other proofs. This is possible in both backward and forward proofs through the *schematic inference rules* $\top \vdash D_t$ and $D_t \vdash \top$ respectively. In addition to theorems, Speedith allows the use of *axioms*: given an axiom $D_a$, it can be used in a proof through the inference rule $\top \vdash D_a$.

[4] Note that all our examples in Figs. 1, 6 and 7 use forward reasoning style.

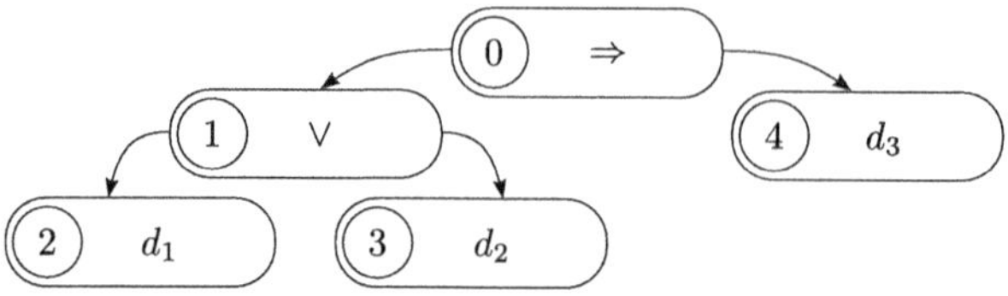


**Fig. 5.** The tree structure of the spider-diagrammatic formula $(d_1 \vee d_2) \Rightarrow d_3$, where $d_1$, $d_2$, and $d_3$ refer to unitary spider diagrams, and the logical connectives $\Rightarrow$ and $\vee$ refer to compound spider diagrams. The circled numbers are the indexes of sub-diagrams.

#### 4.2.2 Transforming Spider Diagrams

A compound spider diagram connects multiple sub-diagrams with logical connectives. In the abstract representation, sub-diagrams are children nodes of a compound spider diagram. This composition forms a tree structure. Every sub-diagram is assigned a sequential number – node indexes. The depth-first algorithm is used for assignment of these indexes. An example of the tree structure of a compound spider diagrams is shown in Fig. 5.

When an inference rule is to be applied, the tree structure of a spider diagram is traversed to a particular point where the relevant transformation takes place and thus returns new spider diagrams. This point is chosen by the user and supplied to the inference rule through the *rule application arguments*. The rule application arguments differ from rule to rule. Some rules, like the split spider rule, work on particular spiders in a unitary diagram. Thus, the split spider rule requires the sub-diagram index of the unitary diagram, the name of the spider, and also the region on which to split the spider's habitat. If a rule cannot be applied at a certain position, or if the sub-diagram does not satisfy all the requirements needed for a safe and valid rule application, the rule will not be applied and the user will be notified of this.

In the implementation of Speedith, the actual data structures of the abstract representation of a spider diagram do not change during the rule application. Rather, an entirely new spider diagram is constructed. For lower memory consumption and for efficiency purposes, the new spider diagram shares all unchanged sub-diagrams of the initial formula. This also improves the speed of syntactic equality comparisons in Speedith, as there cannot exist two different instances of syntactically identical spider diagrams.

### 4.3 Verification with External Tools

Proofs in Speedith rely purely on the soundness of the individual inference rules outlined in Sec. 3 and [5]. As proofs are derived by sequential application of these rules, they are guaranteed to be correct by construction (see Sec. 4.2.1). However, the implementation of Speedith cannot be guaranteed to be bug-free. Also, some users might trust Speedith more if its proofs are verified in other, established, theorem provers. Thus, we designed Speedith to easily plug into, and allow for seamless communication with external proof tools. This enables Speedith to verify particular proof steps externally: namely, spider-diagrammatic proofs can be reconstructed in traditional logic. To date, Speedith's proofs can be certified in the general purpose theorem prover Isabelle [8]. Other tools can be supported by supplying a plug-in which implements the communication with the new external tool. Conceptually, a proof step in Speedith can be proved correct by verifying that conjunctively connected sub-goals $D_i$ imply the initial goal $D$ (see Formula (1) in Sec. 4.2.1). Thus, in order to verify its proof steps, Speedith exports the following theorem which is to be proved by the external tool:

$$D_1 \wedge D_2 \wedge \cdots \wedge D_n \Rightarrow D.$$

Apart from exporting, Speedith can also import formulae. This enables the so-called heterogeneous reasoning, that is, constructing proofs that consist of diagrammatic inferences and traditional sentential logical inferences. We use it in our heterogeneous reasoning framework called Diabelli [8] that combines diagrammatic theorem proving in Speedith with sentential theorem proving in Isabelle [9].

#### 4.3.1 Input and Output Formats

Speedith supports different input and output formulae formats for importing and exporting. The standard input format of Speedith is the native textual representation outlined in Sec. 4.3.2 below. For export, on the other hand, Speedith uses formats that several other tools understand. For example, one supported export format translates abstract representations of spider diagrams into Isabelle/HOL formulae. Here is an example of how the diagram $d_1$ in Fig. 1 is translated to the Isabelle/HOL format:

$$\exists s1\ s2.\ distinct[s1, s2] \wedge s1 \in A \cap B \wedge s2 \in (A - B) \cup (B - A)$$

A new output or input format can be specified by providing a translation procedure that takes a set of spider diagrams in their abstract representation, a proof trace, or an inference rule application, and translates it to a specific textual format.

#### 4.3.2 Textual Representation

In addition to the data structures and object oriented model used in abstract representation, Speedith also provides a textual form of spider diagrams. This form is Speedith's default for exporting and importing diagrammatic statements to and from external tools. It is used when verifying particular steps in the diagrammatic proof, or verifying properties about spider-diagrammatic formulae in a sentential reasoner. Here is an example of the textual form of diagram $d_1$ in Fig. 1:

```
PrimarySD {
  spiders   = ["s1", "s2"],
  habitats  = [("s1", [(["A"], ["B"]), (["B"], ["A"])]),
               ("s2", [(["A", "B"], [])])],
  sm_zones  = [], ve_zones = []
}
```

## 5 Diagram Visualisation

The visualization component of Speedith, called iCircles, builds on the Euler diagram drawing software presented in [10]. We extended the drawing algorithm to spider diagrams, enabling the layout of spiders as well as including functionality to specify which zones are to be shaded.

The input to Speedith is a statement of the theorem. This can be entered through drawing commands, first-order logic formulae,[5] or the textual representation (as described in Sec. 4.3.2). A future direction would be to support additional input methods, such as free-hand drawing and shape recognition.

Next, the entered theorem is automatically drawn – Speedith uses the following drawing algorithm:

1. Draw the underlying Euler diagram, using the methods of [10]. This takes the set of zones to be present and determines how to draw the diagram with circles.
2. Next, shading is placed in the appropriate zones, using standard 'region shading' methods.
3. Finally, the spiders are laid out. Given a particular spider, $s$, with habitat $\eta(s)$, a point is found in each zone in $\eta(s)$. These points form the feet of the spider. The feet are then joined by edges: we prioritize drawing edges between feet in adjacent zones, until the feet form a tree. Care is taken to ensure that edges do not pass through feet belonging to other spiders. This is achieved by nudging the position of feet that lie on the path of a to-be-drawn edge in eight principal directions until a suitable position is found. In addition, when the initial points are found for spiders, we ensure that they do not lie on already drawn edges.

Fig. 6 is an example of Speedith using iCircles to display spider-diagrammatic formulae. The abstract representation of a spider-diagrammatic formula is converted into iCircles and composed to form arbitrary compound spider diagrams.

Speedith translates the abstract representation of unitary spider diagrams into a concrete description of the drawing. An important step of the translation is determining which shaded zones contain spider feet and must thus be present in the drawing. The decision on whether to display other shaded zones is left to the iCircles drawing algorithm.

After all unitary spider diagrams of a compound statement are drawn, they are laid out as operands of the logical connectives. Individual sub-diagrams are finally enclosed with a bounding box, which separates them spatially for clearer presentation and unambiguous nesting of operators.

Following the entry and drawing of the theorem, the user proceeds to apply inference rules on specific parts of the diagram. The exact target of the rule application is determined through a rule-specific sequence of clicking on select elements in the diagram. The inference rule to apply is chosen from a list of all available rules;

[5] Speedith currently supports only a specific form of first-order logic formulae in the Isabelle/HOL syntax.

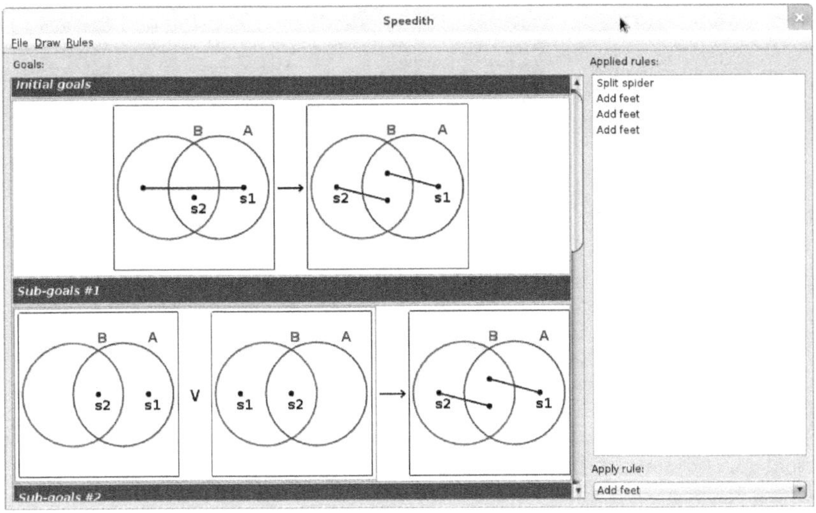


**Fig. 6.** Automatically drawing spider diagrams using iCircles

as we implement more inference rules in the future, we will introduce a filter that shows only the rules applicable to the part of the diagram that the user clicked on. Once the proof is completed, it is added to the suite of unit tests.

## 6 Results and Related Work

Speedith is implemented in Java and is currently under active development. Its sources are available from `https://gitorious.org/speedith`. With Speedith we are able to prove all theorems of MFOL with equality, expressed using spider diagrams – this is a significant range and depth of theorems.

We demonstrate the evaluation of Speedith's functionality with two diagrammatic proof examples. The first one was presented in Fig. 1. The proof makes use of diagrammatic rules that transform spiders and a disjunction equivalence rule. The second example, shown in Fig. 7, tests the inference rules which manipulate spiders, contours, shaded zones, and logical connectives. The proof in Fig. 7 essentially makes use of the information from the right conjunct to transform the diagram representing the left conjunct through a series of copying rules: first the contour $D$, then spider $s_1$, followed by shading of the zone $(\{C\}, \{D\})$, and finally eliminating the redundant right conjunct and removing contours $A$ and $C$ to deduce the theorem's conclusion.

One of Speedith's main contributions is its representation of formulae and proof steps. This differentiates it from interactive sentential theorem provers (such as Isabelle) in that it provides a domain-specific, visual, and thus perhaps more intuitive approach to proofs in MFOL with equality. Speedith's inference rules, which perform simple visual transformations of the diagrammatic statement are succinct and 'natural' – they capture the notion of truthfulness that humans find easy to understand. In contrast, proofs of the same theorems in

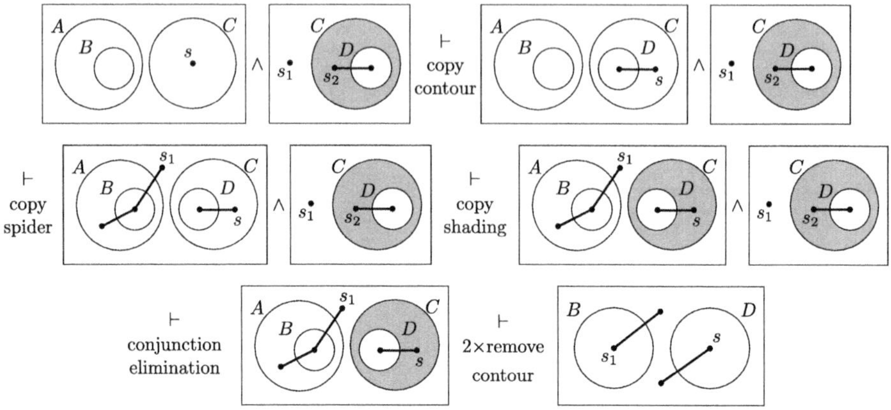


**Fig. 7.** A proof of a spider-diagrammatic statement using inference rules that work with spiders, contours, and shaded zones.

sentential theorem provers consist of lower-level, more fine-grained proof steps which make them longer and arguably harder to "see" the intuition behind the proof. Comparing Speedith's speed with other theorem proving tools remains work for the future.

Other diagrammatic theorem provers most related to Speedith are Edith, DIAMOND, and Cinderella. Whilst Edith is the closest to Speedith in terms of the domain it targets, it does not support spiders nor compound diagrams with logical connectives, and thus provides fewer inference rules. Edith also does not support external verification of its proof steps. DIAMOND, on the other hand, supports external verification, but the class of problems it tackles is different and narrower compared to Speedith. Cinderella targets the domain of geometry and uses a different approach to its diagrammatic proofs. The user gradually constructs the geometric model of the theorem, while in the background an automated theorem prover verifies that each construction step results in a valid geometric diagram. Thus, the steps in Cinderella are not guaranteed to be sound, and the proof process does not follow the standard inference rule application pattern (as described in Sec. 4.2.1).

Finally, Speedith was designed with language extensions in mind. Spider diagrams could be extended with non-monadic relations, functions, and universal quantification of spiders. Designing meaningful and complete diagrammatic inference rules for such extended language is hard and remains work for the future.

## 7 Future Work and Conclusion

By developing Speedith, we demonstrated the feasibility of diagrammatic reasoning systems that utilise a rule-based deductive proof approach. This is similar to the approach employed by general purpose proof assistants like Isabelle.

We also showed how to utilize existing state-of-the-art theorem provers to verify diagrammatic inference steps. Whilst we focused on spider diagrams, the approach can be used for other diagrammatic logics, such as existential graphs [11] or constraint diagrams [12].

Part of our future directions for Speedith includes extending the abstract representation to better control how diagrams are drawn. Moreover, the diagrams are currently laid out independently, and hence diagrams in consecutive proof steps can look radically different from each other. Thus, we aim to improve layout heuristics to take entire sequences of diagrammatic statements into account. In addition to better diagram visualisation, we also envision extensions to the language of spider diagrams, proof search automation, use of Speedith in practical settings [13,14], and a study of scalability of proofs and their visualisation in Speedith.

**Acknowledgements.** This work was supported by EPSRC Advanced Research Fellowship GR/R76783 (Mateja Jamnik), EPSRC Doctoral Training Grant and Computer Laboratory Premium Research Studentship (Matej Urbas).

## References

1. Jamnik, M., Bundy, A., Green, I.: On Automating Diagrammatic Proofs of Arithmetic Arguments. JOLLI 8(3), 297–321 (1999)
2. Winterstein, D., Bundy, A., Gurr, C.: Dr.Doodle: A Diagrammatic Theorem Prover. In: Basin, D., Rusinowitch, M. (eds.) IJCAR 2004. LNCS (LNAI), vol. 3097, pp. 331–335. Springer, Heidelberg (2004)
3. Kortenkamp, U., Richter-Gebert, J.: Using automatic theorem proving to improve the usability of geometry software. In: MUI (2004)
4. Stapleton, G., Masthoff, J., Flower, J., Fish, A., Southern, J.: Automated Theorem Proving in Euler Diagram Systems. JAR 39(4), 431–470 (2007)
5. Howse, J., Stapleton, G., Taylor, J.: Spider Diagrams. LMS JCM 8, 145–194 (2005)
6. Gordon, M.J., Milner, A.J., Wadsworth, C.P.: Edinburgh LCF. LNCS, vol. 78. Springer, Heidelberg (1979)
7. Howse, J., Stapleton, G., Flower, J., Taylor, J.: Corresponding Regions in Euler Diagrams. In: Hegarty, M., Meyer, B., Narayanan, N.H. (eds.) Diagrams 2002. LNCS (LNAI), vol. 2317, pp. 76–90. Springer, Heidelberg (2002)
8. Urbas, M., Jamnik, M.: Heterogeneous Proofs: Spider Diagrams Meet Higher-Order Provers. In: van Eekelen, M., Geuvers, H., Schmaltz, J., Wiedijk, F. (eds.) ITP 2011. LNCS, vol. 6898, pp. 376–382. Springer, Heidelberg (2011)
9. Wenzel, M., Paulson, L.C., Nipkow, T.: The Isabelle Framework. In: Mohamed, O.A., Muñoz, C., Tahar, S. (eds.) TPHOLs 2008. LNCS, vol. 5170, pp. 33–38. Springer, Heidelberg (2008)
10. Stapleton, G., Zhang, L., Howse, J., Rodgers, P.: Drawing Euler Diagrams with Circles. In: Goel, A.K., Jamnik, M., Narayanan, N.H. (eds.) Diagrams 2010. LNCS, vol. 6170, pp. 23–38. Springer, Heidelberg (2010)
11. Dau, F.: Constants and Functions in Peirce's Existential Graphs. In: Priss, U., Polovina, S., Hill, R. (eds.) ICCS 2007. LNCS (LNAI), vol. 4604, pp. 429–442. Springer, Heidelberg (2007)
12. Kent, S.: Constraint diagrams: Visualizing invariants in object oriented modelling. In: OOPSLA. SIGPLAN, vol. 32, pp. 327–341. ACM (1997)
13. Keslter, H., Muller, A., Kraus, J., Buchholz, M., Gress, T., Liu, H., Kane, D., Zeeberg, B., Weinstein, J.: Vennmaster: Area-proportional Euler diagrams for functional go analysis of microarrays. BMC Bioinformatics 9(67) (2008)
14. De Chiara, R., Hammar, M., Scarano, V.: A system for virtual directories using euler diagrams. ENTCS 134, 33–53 (2005)

# Algebra Diagrams: A HANDi Introduction

Peter C.-H. Cheng

Department of Informatics, University of Sussex, Brighton, BN1 9QH, UK
p.c.h.cheng@sussex.ac.uk

**Abstract.** A diagrammatic notation for algebra is presented – Hierarchical Algebra Network Diagrams, HANDi. The notation uses a 2D network notation with systematically designed icons to explicitly and coherently encode the fundamental concepts of algebra. The structure of the diagrams is described and the rules for making derivations are presented. The key design features of HANDi are discussed and compared with the conventional formula notation in order demonstrate that the new notation is a more logical codification of introductory algebra.

## 1 Introduction

This paper describes novel notational system for introductory (school level) algebra – Hierarchical Algebra Network Diagrams (HANDi). HANDi was invented as part of a programme of research that is developing the *Representational Epistemic* approach to the study of how notational systems encode knowledge and the potential cognitive benefits that novel codifications of knowledge may confer [1-3]. The core principle of the Representational Epistemic approach claims that notational systems designed to directly encode the fundamental conceptual structure of knowledge rich domains, using coherent notational schemes, will possess semantic transparent and thus enhance problem solving and conceptual learning (see [1-3]).

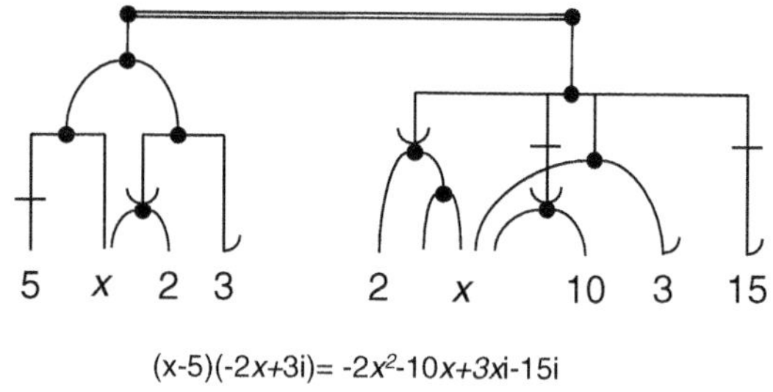


**Fig. 1.** A HANDi equation comprising two trees

This paper that has two purposes: (1) to provide a detailed description of HANDi, that includes a set rules for making HANDi derivations; (2) to highlight the main design features of the notation that attempt to more coherently encode the fundamental conceptual structure of algebra than the conventional formula notation. Thus, the paper has the 4 following main sections: section 2 present the basic graphical structure the HANDi notation; section 3 describes the rules for manipulating expression in the notation; section 4 provides

P. Cox, B. Plimmer, and P. Rodgers (Eds.): Diagrams 2012, LNAI 7352, pp. 178–192, 2012.

examples of derivation using the rules; section 5 discuses the how the HANDi notation encodes the key concepts of the domain.

# 2 HANDi Expressions

Fig. 1 shows an example of a Hierarchical Algebra Network Diagram for a quadratic equation in $x$ that involves a complex number. The equation in the conventional formula notation equation is shown below the diagram. This diagram consists of two *trees*, left and right, but other diagrams may have multiple trees. The full interpretation of the diagram will become clear as the component parts of HANDi equations are introduced.

## 2.1 Basic Operators

Fig. 2 shows the HANDi sub-tress for basic binary operators, which consist of a node, trunk and two branches. At the end of each branch is either an *argument* leaf that consists of a number, a variable, or another a node for an operator (as in Fig. 1). The *trunk* of a node is the line ascending from the top of the node. A *right-angle branch* represents addition of the arguments (or node) at the end of the branch. A right-angle branch with a horizontal bar represents subtraction of the argument at the end of the branch. An arc branch represents multiplication, in the same way. An arc branch with a diagonal bar represents division by the argument on the branch. The different styles of lines distinguish addition and subtraction from multiplication and division. The bars distinguish subtraction from addition and division from multiplication. All this conveys the notation that division is to multiplication as subtraction is to addition.

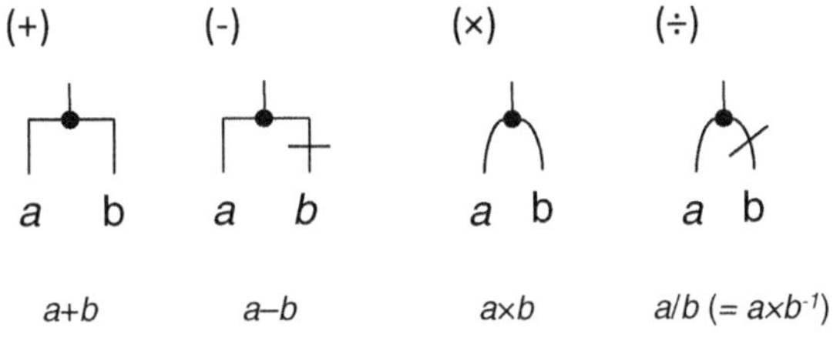


**Fig. 2.** Trees for the elementary algebraic operators

Different styles of branches may not be connected to the same node, and a diagonal bar may only be associated with an arc branch and a horizontal bar with a right-angle branch; Fig 3 shows three illegal diagrams.

## 2.2 Equation

Relations among numbers and variables are encoded as hierarchical networks, *trees*, composed of the basic operators. The numerical equality of trees is shown a double horizontal line connecting their topmost trunk; for example Fig 1. (Inequalities may be shown by a double horizontal line with a <, ≤, >, ≥ symbol written on the line.) A tree may only have a single occurrence of each variable; there is just

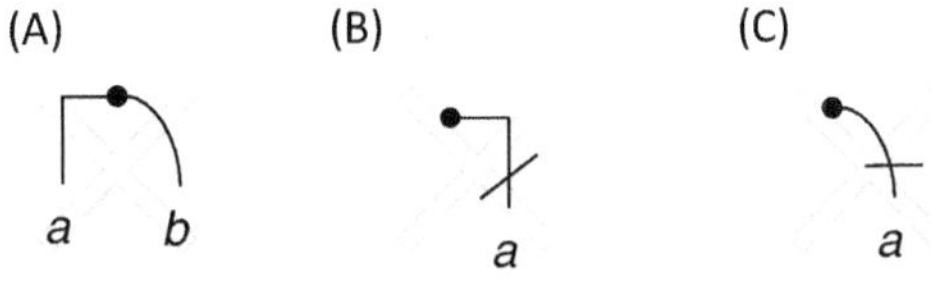


**Fig. 3.** Invalid tree and branch types

one letter '*x*' in each of the two trees in Fig. 1. This is a key feature of HANDi. However, there may be multiple instances of the same number in a tree, for two reasons: (a) the same numbers may stand for quantities that have different units, in which case they necessarily represent different things (e.g., 3 metres versus 3 seconds), so should have different symbols; (b) pragmatically, this provides a means to manage the complexity of the diagrams, particularly in order to avoid crossing branches.

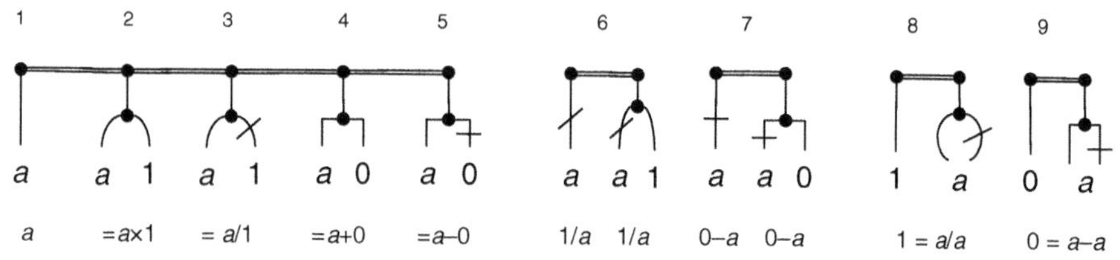


**Fig. 4.** Operations involving identify element

## 2.3 Identities and Inverse Operations

Fig. 4 shows how a sub-tree with a single branch may be drawn in place of a binary tree that has the same argument and a particular identity element. Fig 4.1 is a sub-tree with a single branch, and no bar, that is equivalent to Figs. 4.2-4.5. Figs. 4.6 and 4.7 show that single branches with a diagonal or horizontal bar are equivalent to trees in which the argument is the divisor of one or is subtracted from zero. Fig. 4.8 and 4.9 shows that an argument divided by or subtracted from itself are equal to one or zero, respectively.

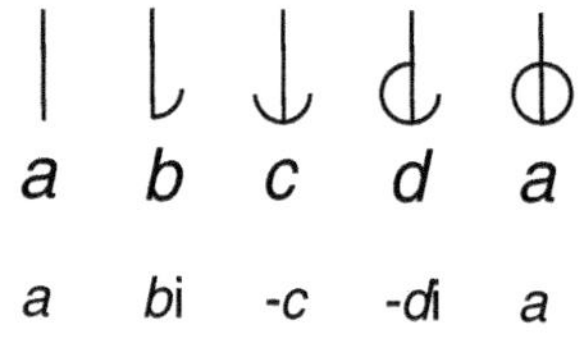


**Fig. 5.** Positive, imaginary, negative, negative imaginary and positive arguments

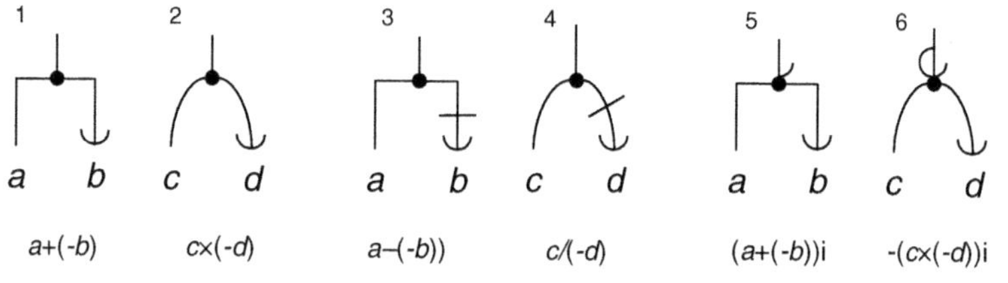


**Fig. 6.** Turns in trees

## 2.4 Negative and Imaginary Numbers – Unary Turn Operators

HANDi provides a notational scheme that coherently unifies positive, negative and imaginary numbers; see Fig. 5. A *turn* is a unary operator at the end of branch that indicates the extent of rotation of the argument in an Argand diagram. A quarter turn (1/4 circle) represents an imaginary argument, a half turn (semi-circle) represents a negative argument, a three quarter turn (3/4 circle) represents a negative imaginary argument, and positive argument is a branch with no turn or a complete (whole circle).

## 2.5 Combined Unary and Higher Order Operators

Fig. 6 shows unary turn operators in association with binary operators. Addition, subtraction, multiplication and division operators may act upon negative numbers; Figs. 6.1 to 6.4. HANDi makes an explicit distinction between negative numbers and subtraction in terms of the type of attachment to the branch, either a half turn or a bar, and in relation to the position of the attachment, either at the end or bisecting the branch. Fig. 6.5 and 6.6 indicate how unary turn operator may be applied to trees. Notice how in Fig. 6.6 the minus sign and 'i' occur at opposite ends of a conventional formula, but the same information is encoded as a single three quarter turn symbol in HANDi.

## 2.6 Repeated Operations – Recursive Addition and Powers

Figs. 7 and 8 show HANDi expressions that involve repeated applications of an operator to an argument. Fig. 7.1-7.4 shows multiple additions of a single argument, including recursive applications. Figs. 7.5-7.10 show how HANDi represents various power expressions for a single argument. In cases where multiple branches of a tree converge on a single sub-tree below (Figs. 7.4, 7.9 & 7.10 and Fig. 8.5), each branch represents a repeated operation on the sub-tree: Fig. 7.4 shows two additions of $3a$; Fig. 7.9 shows the product of three $a^3$; Fig. 8.5 shows the product of two $a \times b$. Notice that the top tree in Fig. 7.9 is equivalent to the top tree in Fig. 7.8, which both represent the cube of their respective arguments. Consider Fig. 7.10; raising an index of an argument (2 in $a^2$) to a power (3 in $a$^$2^3$) is represented by a number of the hierarchical repetitions of the sub-tree for the index (2) by the magnitude of the power (3): there is a stack of three pairs of branches. Fig. 8.4 shows that arguments raised to some power may be embedded as a sub-tree within a higher tree. If all the branches of the tree have diagonal bars, which indicate division by the argument, then the tree represents a negative power, as in Fig. 8.1.

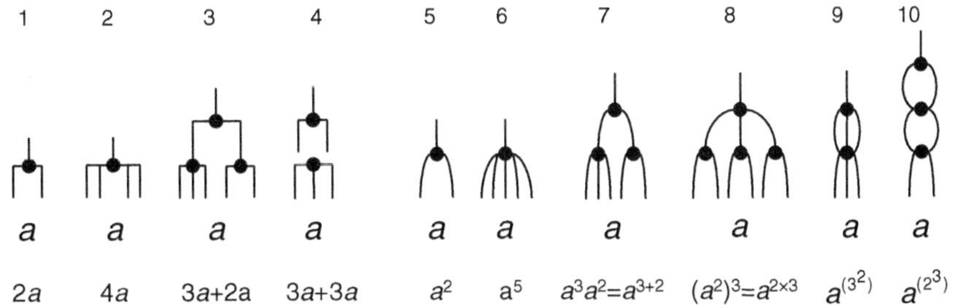


**Fig. 7.** Repeated addition and powers

To show non-integer fractional powers, branches that ascend as well as descend from the node are used. Fig. 8.2 has two ascending branches and one descending branch, so its power is 1/2 (square root). Fig 8.2 shows $a^{2/3}$, because two branches point down and three point up. By the same scheme, the two down and one up branches in Fig. 7.5 shows the argument raised to the "fractional" power of 2/1.

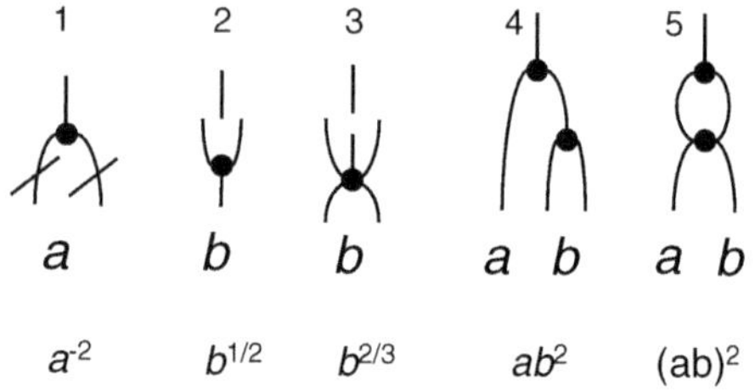


**Fig. 8.** Assorted power diagrams

That completes the description of the graphical structure of HANDi trees.

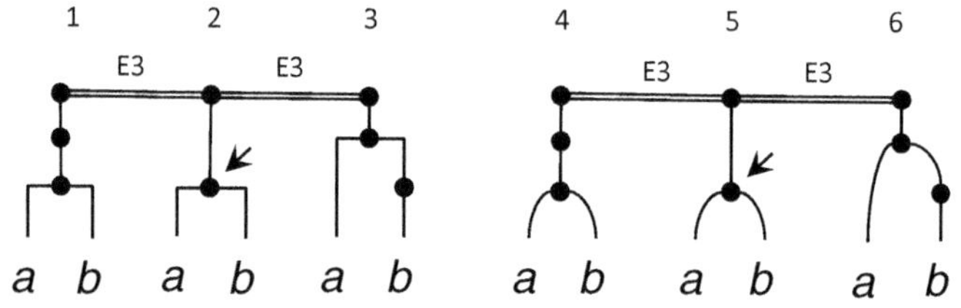


**Fig. 9.** Branch insertion or elimination

## 3 Elementary HANDi Transformation Rules

HANDi provides rules for the transformation of tree diagrams in to alternative forms. Elementary rules are the primitive transformations – considered in this section – and derived rules are more complex transformations composed by the successive applications of the elementary rules – considered in the next section.

HANDi has twelve elementary rules for transforming trees and equations.

**Rule E0** specifies that horizontal positioning of argument and nodes is arbitrary so their relative positions may be swapped at will; for instance, to improve the clarity of a diagram. The *E0* labels on the double equality line in Figs. 15 and 23 (below) show the application of this rule.

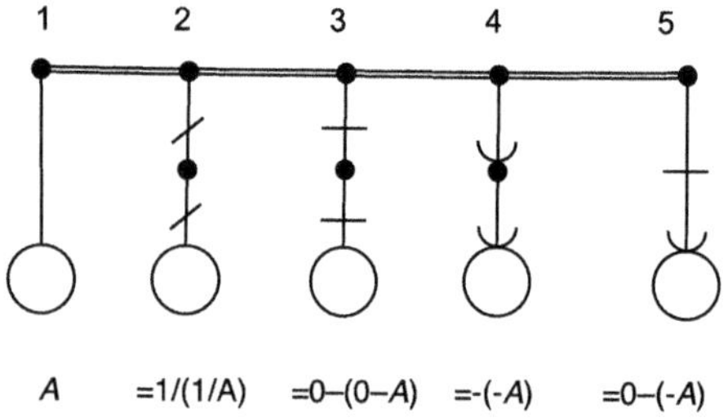


**Fig. 10.** Nest operations that are equivalent to a tree with a single plain branch

**Rule E1** allows binary trees for operations on an argument and an identity element to be replaced by a single branch for the argument, as described above and shown in Figs. 4.2-4.7. In the case of Rule E1.1 the single branches for the arguments have no bars (Fig 4.2-5), whereas in Rule E1.2 each branch has a bar (Figs. 4.6 & 4.7). The opposite also applies; a single branch may be transformed in to a binary tree (e.g., Fig 4.1 in to any of Figs. 4.2-4.6)

**Rule E2** also allows a binary tree to be replaced by a single branch, or vice versa, but in this case the argument on the branch is either one or zero, because the binary tree involves an argument divided by itself or subtract from itself, as described above and shown in Figs. 4.8 & 4.9.

**Rule E3** allows a single branch with no bar to be inserted (or removed) from a tree. Going from Fig. 9.2 to 9.1, or Fig. 9.5 to 9.4, an extra node and branch are inserted above the node in the tree indicated by the arrow. Going from Fig. 9.2 to 9.3, or Fig. 9.5 to 9.6, an extra branch and node are inserted below the indicated node of the tree. In the opposite direction (to Fig. 9.2 or to Fig. 9.5) the node and branch are eliminated.

**Rule E4** involves trees with single branches, Fig. 10. The circles and capital '*A*' stand for an argument or a sub-tree. Three trees have two nested branches for the same type the operator (Fig 10.2-4) and one has two different operators (Fig 10.5). In all cases the trees are equivalent to a single plain branch (Fig 10.1). For instance, Fig 10.4 states that taking the negative of a negative gives positive and Fig 10.5 shows that subtracting a negative is equivalent to a positive. Thus, the variants of E4 allow such nested branches to be replaced by a single plain branch, or vice versa.

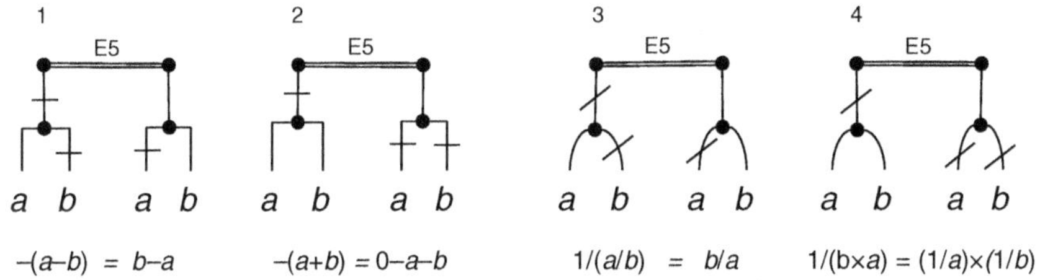


**Fig. 11.** Promotion and demotion of bars

**Rule E5** concerns the movement of bars up and down branches, as shown in Fig. 11. A bar on a trunk can be move down to any branch that has no bar, but if a bar already exists on the branch it is cancelled out by the moved bar. The rule applies equally to right-angle and arc trees. For example, Fig. 11.1 and 11.3 show that the bar on the *b* branch is eliminated and a bar added to the *a* branch, when the bar of the trunk is demoted.

**Rule E6** concerns the relation of right-angle trees to trees with arc branches. Fig. 12 is an example of how repeated additions of the same argument may be represented as a multiplication operation. A tree with multiple right-angle branches for a single argument may be transformed into a binary tree with arc branches, one for the argument and another with a number matching the quantity of right-angle branches.

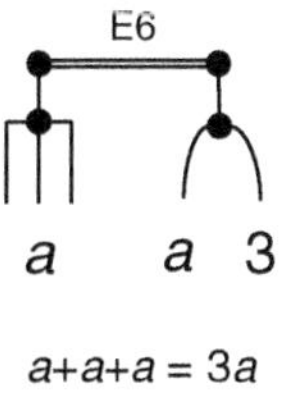


**Fig. 12.** Multiplication and repeated addition

**Rule E7**. Depending on the type of tree, transforms involving turns are governing by different conservation rules. In a tree with right-angle branches (with or without horizontal bars), the number of turns in successive right-angle branches must be preserved, with the proviso that one complete turn is equivalent to no turn. For example, for variable *a* in Fig. 13.1, there are no turns along the branches from the top of the tree down to the *a* leaf, so when a quarter, half and three-quarter turn are introduced on the trunk, Figs. 13.2–13.4, the *a* branch must be augmented with sufficient fractional turns that the total number of turns is the same, a complete turn (or none). Similarly, with branches *b*, *c* and *d*, which start with a quarter, half and three-quarter turn, respectively.

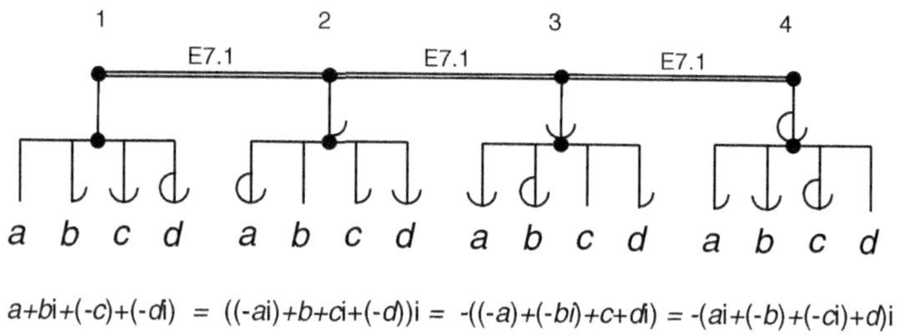


**Fig. 13.** Conservation of turns along successive right-angle branches in equivalent trees

The conservation of turns in trees with arc branches applies to the whole tree, so all the turns on all the branches are summed, but with the exception that turns on branches that also have diagonal bars must be deducted. In Fig. 14 all the trees have a total of one quarter turn. In Fig 14.1 and 14.2 the quarter turn is associated with one argument or the whole tree, respectively. In Figs 14.3-5 a quarter turn is associated with a branch that has a diagonal bar, variable *c*, which means that there must be a total of a half turn elsewhere in the rest of each of the trees, so that all the trees possess the same number of turns. In Fig 14.6 the half turn associated with that branch will cancel most of the three-quarter turn on the middle branch (variable *b*) to leave the requisite quarter turn.

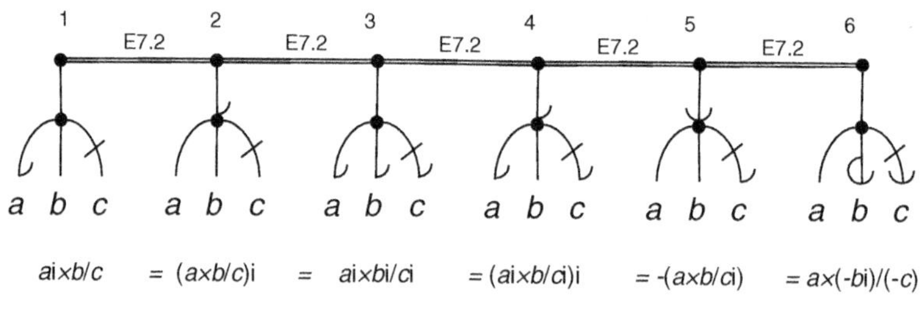


**Fig. 14.** Conservation of turns throughout equivalent trees with arc branches

**Rule E8**. This rule encodes the property of distribution of multiplication over addition. In Fig 15.1 the topmost tree has arc branches, one of which is for the variable '*a*' and the other attaches to a tree with right-angle branches (*b*+*c*). Now, as the arc branch attached to the right-angle tree has *no bar*, the right-angle tree may be promoted and its branches attached to two sub-tree with arc branches below, as shown in

Fig. 15.2. Note that the two new arc-branched trees now share the single variable ($a$) from the original arc-branched tree. Figs. 15.4 and 15.3 are variants of Figs. 15.1 and 15.2 that simply change the horizontal position of the variables according to rule E0. This rule applies to right-angle trees with more than two branches and may be derived by repeated application of Rule E8 to successive right-angle branches.

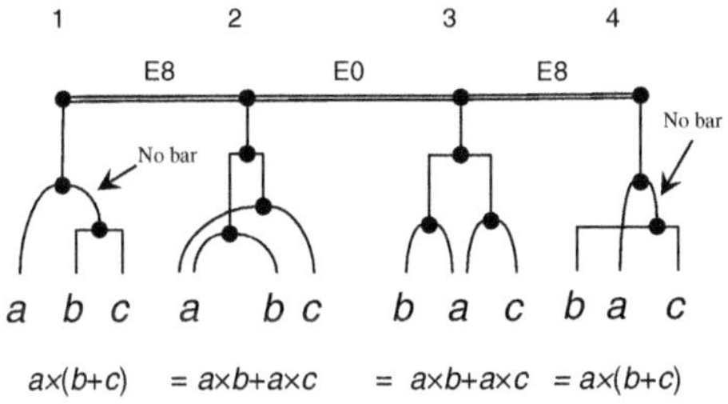


**Fig. 15.** Distributive property of multiplication over addition

Figs 16.1 and 16.2 show further two valid applications of rule E8 that also involve the presence of bars and turns. Again, note the absence of a bar on the critical arc branch, but the permitted presence of a turn on the branch in Fig. 16.2. In both cases the redistribution of argument $a$ does not directly affect the bars or the half turn, because they remain attached to their original branches. In contrast Fig 16.3 shows when E8 cannot be applied, because there is a bar on the critical arc branch, which may not be re-distributed through the right-angle tree when that branch disappears.

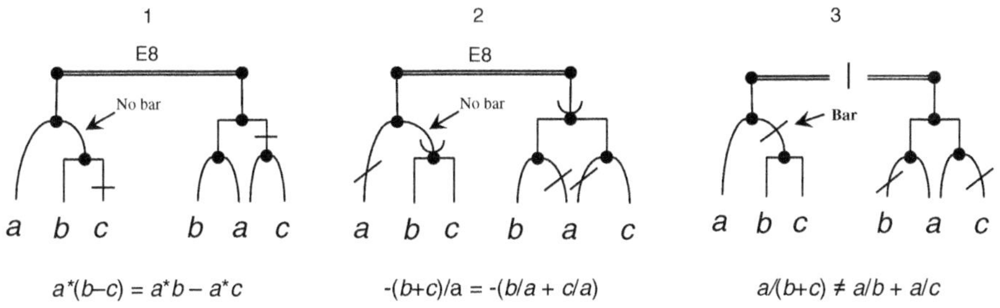


**Fig. 16.** Distributive property involving subtraction and division

**Rule E9** concerns the transformation of trees for a single argument that have multiple arc branches, which may be nested, such as those shown in Figs. 7.5-7.10 and Fig. 8. Transformations of such trees should simply maintain the total effective number of branches. For example, in Fig. 17.1 a single branch for the upper sub-tree connects to a sub-tree with three arcs and a sub-tree with two arcs, hence there is a total of five branches overall, as shown in Fig. 17.2. In Fig 17.4 there are three lots of three arcs, which may be redrawn as nine branches, Fig. 17.5, or as the recursive application of a three-branch sub-tree to a lower level tree with three branches, Fig. 17.3. In general, successive sub-trees with multiple branches at different levels are multiplied, so as trees with upward pointing arcs represent fractional powers, we simply multiply by the appropriate fraction. In Fig. 17.6 there is a sub-tree with a pair of arcs and above another with half an arc (two upside down arcs), thus the total number of branches is one, as shown in Fig. 17.7. Further, the transition from Fig. 17.1 to 17.2, and from Fig. 17.4 to 17.5 may also be interpreted as multiple applications of E3, which allows branches to be eliminated from trees.

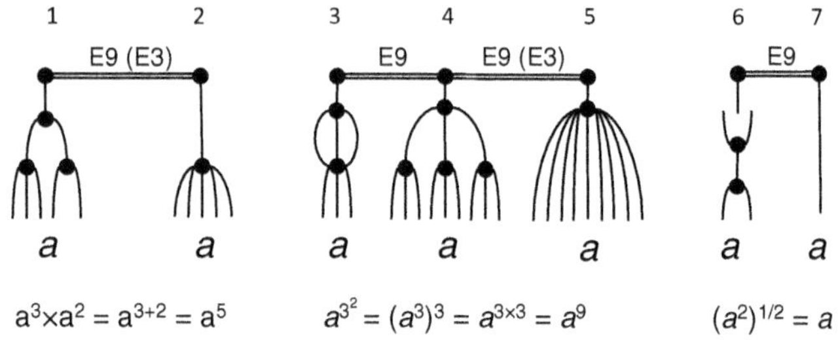


**Fig. 17.** Transformation of power expressions

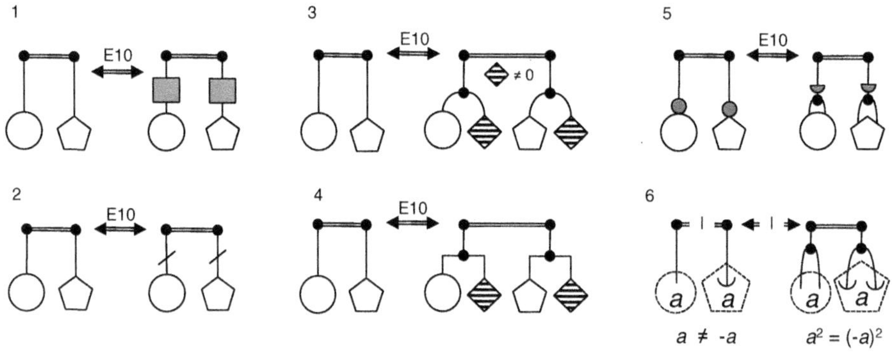


**Fig. 18.** Transformations to main trunks of equal trees

**Rule E10.** So far all the transformation rules have applied to individual trees. Consider now the transformation of multiple trees connected by the double horizontal line (for algebraic equality). This set of rules permits the same operation to be applied to all the trees attached to the double line, with some provisos. Rule E10.1 allows a horizontal or diagonal bar, or any turn, to be added to the top of both trees, or removed from both trees. Fig 17.1 shows this schematically, where the trees are represented by the circle and the pentagon and the grey squares stand for the same type of bar or number of turns that are to be added or removed: Fig. 17.2 is a particular example with diagonal bars. Rule E10.2 warrants the transformation of a tree with the incorporation (removal) of the same branch into both trees, as shown by the diamonds in Fig. 17.3 & 17.4. The one restriction is that the argument must not be a zero when the branches are curves, as this could yield trees are not equivalent (e.g., 2*0=3*0 but 2≠3). Further, Rule E10.3 states that the introduction of the operator must produce trees with equal overall numbers of turns, as illustrated schematically by the two grey circles and the two grey semi-circles on either side of Fig 17.5. Fig.

17.6 shows an invalid application of Rule E10.3, in which two trees with an equivalent number of turns (right) are operated on to produce two new trees that do not have equal numbers of turn (left).

**Rule E11**. To calculate magnitudes of trees whose leaves are numbers, one may simply replace nodes with the number that results from applying the operators to the given values. For multi-level trees the process is performed in a recursive manner starting with the leaves.

Table 1 summarizes the 12 elementary rules for transforming HANDi expressions.

**Table 1.** Elementary HANDi transformation rules

| Rule | Summary |
|---|---|
| E0 | Horizontal position of arguments/nodes is arbitrary. |
| E1 | Swapping a single branch with binary tree involving identities:<br>E1.1) Single branch with no bar, Figs. 4.1-4.5.<br>E1.2) Single branch with a bar, Figs. 4.6 & 4.7. |
| E2 | Introduce/eliminate an argument with a pair of inverse operators:<br>E2.1) Arc branches, Fig. 4.8 E2.2) Right-angle branches, Fig. 4.9 |
| E3 | Branch insertion/elimination at a node, Fig. 9. |
| E4 | Equivalence of nested operations on successive single branches:<br>E4.1) Two bars, Figs. 10.2 & 10.3; or two half turns, Fig. 10.4.<br>E4.2) A horizontal bar and half turn, Fig. 10.5. |
| E5 | Promotion/demotion of bars up/down equivalent branches, Fig. 11. |
| E6 | Multiplication is repeated addition, Fig. 12. |
| E7 | Conservation of number of turns with respect to:<br>E7.1) Successive right-angle branches within a tree, Fig. 13.<br>E7.2) Whole tree of arc branches (diagonal bars = subtraction), Fig 14. |
| E8 | Distribute arc branch over a right-angle tree, Figs. 15 & 16. |
| E9 | Equality of the total number arc branches for one argument, Fig. 17. |
| E10 | Operations applied to (removed from) top trunk of equal trees, Fig. 18. |
| E11 | Replace nodes with values calculated from number arguments. |

## 4 Derived Proofs and Composite Transformation Rules

The elementary transformations of HANDi diagrams presented in the previous section can be applied to derive proofs or generate composite transformation rules. This section considers a few examples.

Fig. 19 shows the two applications of elementary rules: rule E1.1 turns a single branch tree into a binary tree by introducing a second arc branch with the number 1 as its argument; rule E2.1 then converts the 1 into to a further binary tree that involves the simultaneous multiplication and division of a new variable. For the purpose of a further example below that uses this derivation, it will be called composite rule C1.

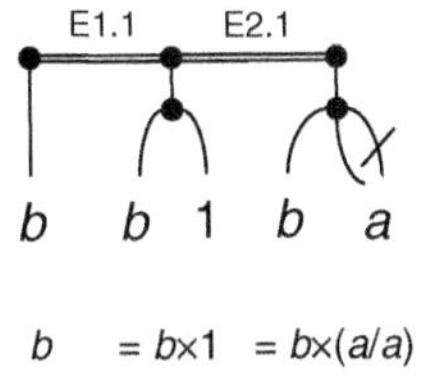


**Fig. 19.** Composite rule C1

Under the conventional approach the notion that addition and multiplication are associative is typically introduced in as fundamental property of algebraic formulas. In contrast, in HANDi the associative property may be treated as a composite rule that is derived from the elementary rule E3, for the insertion or elimination of a branch at a node. Fig. 20.2 may be redrawn as Fig. 20.1 or 20.3 by inserting a new right-angle branch to create a sub-tree. A diagram equivalent to Fig. 20.1-3 may also be drawn with arc braches for the associative property of multiplication. As E3 applies to any node, a branch may be inserted into a diagram with bars, for instance as shown in Fig. 20.4 and 20.5. However, branches with bar may not be inserted in this fashion, so Fig. 20.6 is not a valid transformation of Fig. 20.4 or 20.5: division is not associative. A diagram equivalent to Fig. 20.4-6 may also be drawn with right-angle branches showing the non-associative property of subtraction.

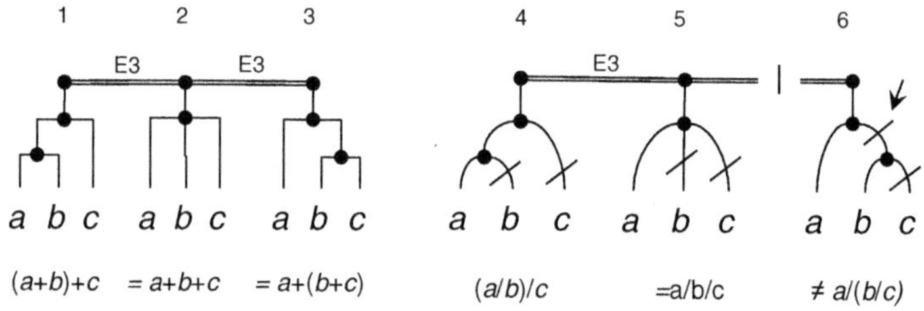


**Fig. 20.** Addition (& multiplication) is associative and division (& subtraction) is sometimes

Fig. 21 shows the proof that subtracting an argument is equivalent to applying the negation operator to the argument. Fig. 21.1 encodes elementary rule E4.2 that states that a positive argument is equal to the subtraction of the negative of the argument. A branch with a single horizontal bar may be inserted in both trees according to rule E10.2, so that left has one bar and the right has two, as shown in Fig. 21.2. Now, rule E4.1 permits the cancellation of the two bars on the right, so we obtain Fig. 21.3, which has just a half turn, the negation operator, to match the bar on the right, as required.

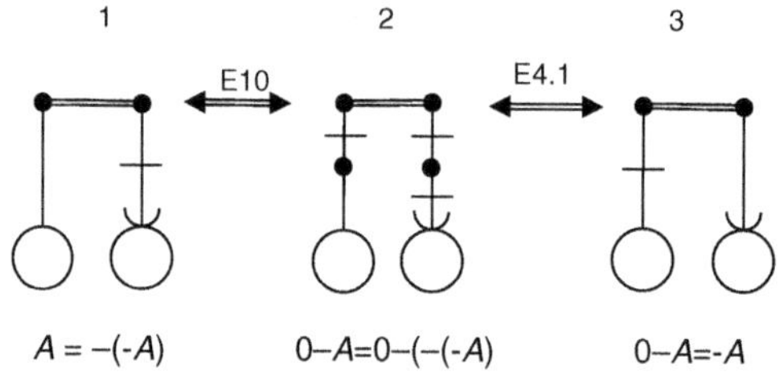


**Fig. 21.** Subtraction is equivalent to negation

The derivation of the product of two sums is shown in Fig. 22. This proceeds by applying the distributive transformation, rule E8, three times in succession. The parts of the diagram that are changed by each application of the rule are highlighted, which have the characteristic patterns found in Fig 15.1 and 15.2 (or their mirror image). In the last step of the proof, Fig 22.4 to 22.5, all the unnecessary right-angle branches are eliminated, by rule E3, to reveal the four products of the variables. Of course, one may treat Figs. 22.1 and 22.5 as a composite transformation rule that simply says draw arc sub-trees for each combination of variables in the two right-angle trees and joint them all together with top level right-angle tree.

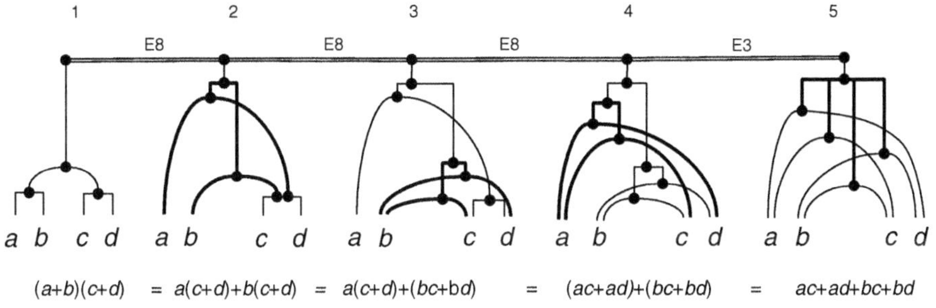


**Fig. 22.** Expansion of the product of two sums

Fig. 23 shows how the sum of reciprocals may be transformed by a diagrammatic procedure equivalent to cross multiplication. The first step is to form a sub-tree for each reciprocal in Fig. 23.1 with the variable of the other reciprocal by applying composite rule C1 (defined above, Fig. 19) to give Fig 23.2. The position of the pair of nodes may be swapped (applying rule E0) whilst the variables remain in place to yield Fig. 23.3, in which the interchanged nodes have been highlighted. Rule E3 may now be applied twice, as shown in Fig. 23.4, to introduce new branches to separate the arcs with bars from the ones without. As each sub-tree in Fig. 23.4 has branches with bars, the bars may be promoted to the next level by applying rule E5 (Fig. 23.5). The pairs of sub-trees in Fig. 23.5 are equivalent, so we have a situation like Fig 16.2 (right), where the sub-trees here map to '*a*' in that figure. Thus, rule E8 may be applied to coalesce the sub-trees to generate Fig. 23.6. With a little experience one learns to compose steps in similar proofs, for example jumping straight from Fig. 23.4 to Fig. 23.5 or even to Fig. 23.6. Again, the initial and final diagrams (Figs. 23.1 and 23.6) may be treated as a composite transformation rule.

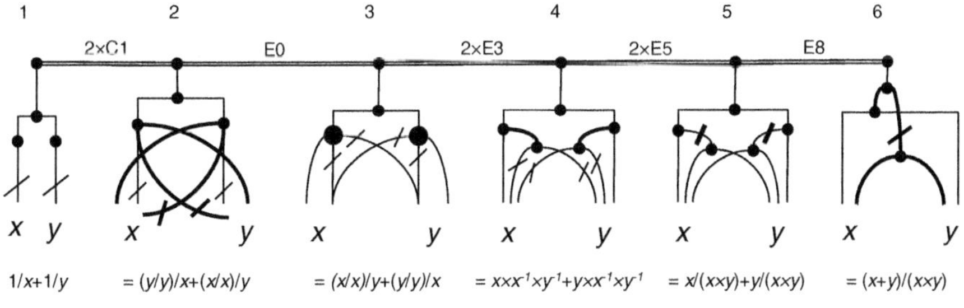


**Fig. 23.** Sum of two reciprocals

As a final example, consider how laws of indices may be derived. Fig. 24 shows a cubed variable divided by the square of the same variable. The proof simply involves: (i) demoting the diagonal bar on the trunk of the binary tree to its branches by applying E5 to give Fig. 24.2; (ii) eliminating branches of the upper tree using E3 twice, Fig. 24.3; (iii) eliminating pairs of arc branches with and without a bar to give branches ending in 1 by applying E2 twice, Fig. 24.4; and, (iv) deleting the branches ending in 1 by applying E1 twice, Fig 24.5.

That completes the introduction of HANDi and its transformation rules.

## 5 HANDi Design

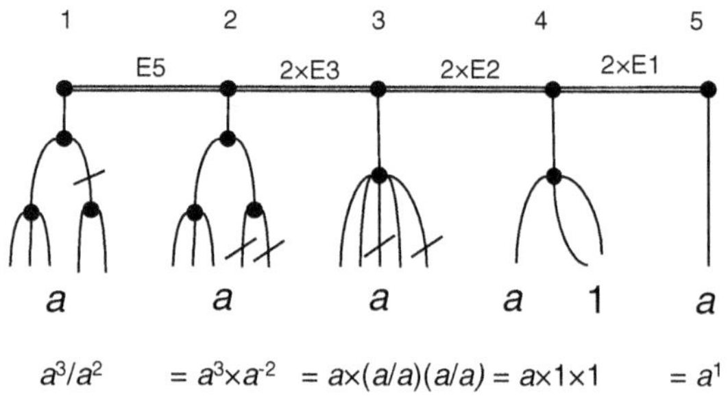


**Fig. 24.** Simplifying a power expression

The main motivation for the creation HANDi was to further test Representational Epistemic ideas about how to design notational systems to encode knowledge [1-3]. The core claim of this approach is that effective notational systems should directly encode the fundamental conceptual structure of their knowledge rich domains, within coherent notational schemes, and thereby possess semantic transparency that will enhance problem solving and conceptual learning [1-3]. Having presented HANDi in some detail above, this section makes explicit the design features of the new notation that attempt to achieve such a direct encoding of the knowledge of algebra.

HANDi attempts to provide a more rational codification than the conventional formula notation by coherently encoding fundamental concepts and properties of algebra. The notational schemes used to capture these concepts and properties will be considered in turn and contrasted with the conventional approach.

Representing variables and numbers, and relations among them, is obviously fundamental to algebra. HANDi uses individual sub-trees composed of a node, a trunk and branches to encode single relations among arguments. This is done for two reasons. First, a sub-tree is an explicit composite iconic symbol that stands for an operation among arguments, or other sub-trees (Fig. 5). Second, it is feasible to limit to just one instance the occurrence of letter standing for a variable in a tree. As the conventional notation (largely) relies upon the linear concatenation of symbols to capture relations among arguments, multiple occurrences of letters are often necessary, which means that the identity of arguments participating in a particular relations cannot always be read directly from the formula, so detailed examination of the arrangement of symbols in the formula is needed to determine the mathematical structure of the expression. In cognitive terms, HANDi exploits some of the well-understood benefits diagrams (e.g., [4]): information for each and every relation is co-located in the individual sub-trees of HANDi and the presence of just one letter for each variable in a tree reduces amount of deliberate search that is needed to match symbolic labels.

Relations in algebraic expression are hierarchical, so knowing the specific level of sub-expressions is essential to the correct interpretation and transformation of formulas. HANDi specifically uses the network structure of sub-trees, with nodes spatially distributed in the vertical dimension, to show the hierarchal structure. In contrast, the conventional notation makes the hierarchical structure of expressions rather opaque, because it relies upon nested parenthesis to define sub-expressions or requires the reader to mentally apply parsing rules (e.g., "BODMAS"). Both of these schemes in the conventional notational clearly demand more mental effort than the direct visual inspection of HANDi expressions (e.g., the levels of the nodes in Fig. 1 is more obvious than the levels of the expressions in the formulas below the diagram.)

The four elementary arithmetic operators possess important conceptual similarities and differences. In HANDi the respective shapes of branches, arc versus right-angle, captures the similarity between multiplication and division but distinguishes it from addition and subtraction. Bars on branches are not only used to distinguish subtraction and division from addition and multiplication, but also encode the asymmetric nature of the former two operations. The perceptive reader will have noted that no explicit rule relating to the commutative property was included among the elementary rules. The explanation for its absence is that it is built directly into HANDi by the particular design of branch shapes and bars. The benefits of this scheme reach further by providing a single definition of the circumstances in which the distribution rule holds when subtraction, division and negation operators are present, as shown in Fig. 16. Distribution is valid whenever the arc branch above the right-angle tree does not have a bar. For example, the asymmetry of trees determined by the location of the bar neatly encodes the validity of the right distribution of division but not the left distribution (i.e., $(b+c)/a$, Fig. 16.2, versus $a/(b+c)$, Fig. 16.3].

Powers and imaginary numbers extend the range algebraic operations. In the conventional formulas supplementary notation devices are built on top of the basic formulas: superscripts for powers and the 'i' symbol as the imaginary unit. Each device has it own particular set of rules. In contrast, the approach in HANDi is simpler. To encode power relations HANDi exploits the idea that powers are operations that repeat multiplication, so the hierarchical network of sub-trees in the diagrams naturally encodes repetitions of arc branches at the same level and by recursively spanning levels (Fig. 7). HANDi has a single unified scheme to deal with imaginary numbers and negative numbers, Fig. 5, 13 and 14, that builds the fundamental relations that exist among positive, negative and imaginary numbers into the design of the notation at a foundational level. In both the case of powers and imaginary numbers, HANDi does not require the introduction of unique sets of rules associated with supplementary notations.

A potential disadvantage is that HANDi expressions are more complex in terms of the sheer number of symbols, when considered at the level of nodes, branches and bars. However, with a little experience one quickly begins to read HANDi expressions at the level of sub-trees or higher, in which case its complexity is comparable to the conventional notation. Taking this notion further, one potentially significant difference between the two notations, in cognitive terms, is the extent to which HANDi may allow its users to exploit perceptual operators to recognize meaningful patterns and to make inferences (c.f. [4]). The primary notational scheme of the conventional approach is the linear concatenation of symbols, including parenthesizes, which tends to mask characteristic configurations of symbols. The network structure and principled design of the symbols (branches, bars, turns) of HANDi aims to provide distinctive patterns to associate with particular concepts. Thus, many of the HANDi transformation rules involve spotting patterns and drawing new configurations, such as the multiple application of the E8 distribution rule in Fig 15.

It has been argued that HANDi may be a more rational encoding of algebra than the conventional formula notation. However, reactions to a new notational system are sometimes negative, for at least two reasons. First, one may initially feel that HANDi expressions are arbitrary and its rules complex compared to the existing formula notation. However, such immediate judgments should be treated with caution, because one's relative expertise in the familiar notation masks the effort required to learn the notation in the first place, which is what one is experiencing during initial encounters with HANDi. Which notation better supports learning and problem solving is an empirical question, so studies to evaluate HANDi are planned. The second common negative reaction is to think that this is not what subject is "truly" about, because algebra *is* the writing and transformation of formulas. However, this falsely assumes that a topic and its notation are inseparable, perhaps because one has only experienced algebra in the one notation, and that there is a single valid codification of a topic. The scope of HANDi expressions and rules of derivations presented here provides an existence proof that an thoroughgoing rigorous alternative codification of algebra is feasible. HANDi is not a mere visualization of the formula notion, but a generative notation that re-codifies the content of this topic This, of course, opens up wider epistemic and pedagogic questions that must, unfortunately, be considered elsewhere.

**Acknowledgements.** My thanks go to members of the Representational Systems Lab in the Department of Informatics and to Alan Blackwell for their comments on early versions of this paper. Thanks also to the three anonymous reviewers for their handy comments.

## References

1. Cheng, P.C.-H.: Electrifying diagrams for learning: principles for effective representational systems. Cognitive Science 26(6), 685–736 (2002)
2. Cheng, P.C.-H.: Probably good diagrams for learning: Representational epistemic recodification of probability theory. Topics in Cognitive Science 3(3), 475–498 (2011)
3. Cheng, P.C.-H., Barone, R.: Representing complex problems: A representational epistemic approach. In: Jonassen, D.H. (ed.) Learning to Solve Complex Scientific Problems, pp. 97–130. Lawrence Erlbaum Associates, Mahmah (2007)
4. Larkin, J.H., Simon, H.A.: Why a diagram is (sometimes) worth ten thousand words. Cognitive Science 11, 65–99 (1987)

# Boolean Differences between Two Hexagonal Extensions of the Logical Square of Oppositions

Hans Smessaert*

Department of Linguistics
University of Leuven, Belgium
Hans.Smessaert@arts.kuleuven.be

**Abstract.** The classical Aristotelian Square characterizes four formulae in terms of four relations of Opposition: contradiction, contrariety, subcontrariety, and subalternation. This square has been extended into a hexagon by two different strategies of inserting intermediate formulae: (1) the horizontal SB-insertion of Sesmat-Blanché and (2) the vertical SC-insertion of Sherwood-Czeżowski. The resulting visual constellations of opposition relations are radically different, however. The central claim of this paper is that these differences are due to the fact that the SB hexagon is closed under the Boolean operations of meet, join and complement, whereas the SC hexagon is not. Therefore we define the Boolean closure of the SC hexagon by characterizing the remaining 8 (non-trivial) formulae, and demonstrate how the resulting 14 formulae generate 6 SB hexagons. These can be embedded into a much richer 3D Aristotelian structure, namely a rhombic dodecahedron, which also underlies the modal system S5 and the propositional connectives.

**Keywords:** square of oppositions, hexagon of oppositions, logical geometry, Boolean closure, 3D visualisation.

## 1 Introduction: The Aristotelian Square of Oppositions

The traditional relations of opposition are defined in terms of two formulae $\varphi$ and $\psi$ being *true together* ($\varphi \wedge \psi$) or being *false together* ($\neg\varphi \wedge \neg\psi$):

| | | | | |
|---|---|---|---|---|
| $\varphi$ and $\psi$ are *contradictory* | iff | $\mathsf{S} \models \neg(\varphi \wedge \psi)$ | and | $\mathsf{S} \models \neg(\neg\varphi \wedge \neg\psi)$, |
| $\varphi$ and $\psi$ are *contrary* | iff | $\mathsf{S} \models \neg(\varphi \wedge \psi)$ | and | $\mathsf{S} \not\models \neg(\neg\varphi \wedge \neg\psi)$, |
| $\varphi$ and $\psi$ are *subcontrary* | iff | $\mathsf{S} \not\models \neg(\varphi \wedge \psi)$ | and | $\mathsf{S} \models \neg(\neg\varphi \wedge \neg\psi)$, |
| $\varphi$ and $\psi$ are in *subalternation* | iff | $\mathsf{S} \models \varphi \rightarrow \psi$ | and | $\mathsf{S} \not\models \psi \rightarrow \varphi$. |

Two formulae are CONTRADICTORY when they *cannot be true together and cannot be false together*[1]. They are CONTRARY when they *cannot be true together*

* I wish to thank Fabien Schang for bringing the work of Czeżowski to my attention and Lorenz Demey for his Boolean advice. I am grateful to Fabien, Lorenz, Dany Jaspers and Alessio Moretti for their feedback on earlier versions of this paper.

[1] $\mathsf{S} \models \neg\alpha$ means that the formula $\neg\alpha$ is a tautology of the logical system $\mathsf{S}$. Hence $\alpha$ is a contradiction, i.e. $\alpha$ is false in all models of $\mathsf{S}$ (see [4]).

P. Cox, B. Plimmer, and P. Rodgers (Eds.): Diagrams 2012, LNAI 7352, pp. 193–199, 2012.

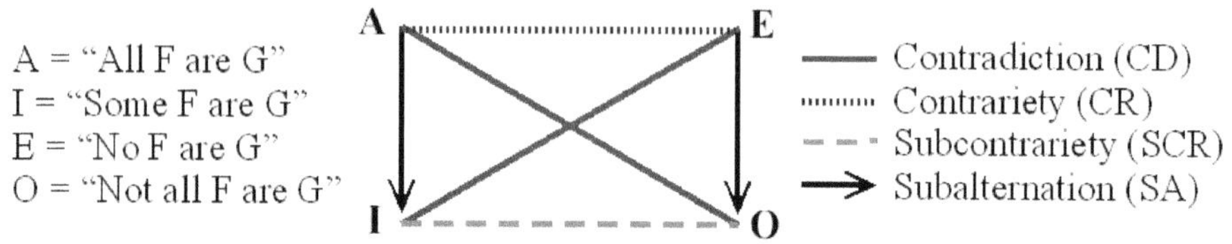


**Fig. 1.** The Aristotelian Square of Oppositions

*but may be false together* and SUBCONTRARY when they *cannot be false together but may be true together.* SUBALTERNATION holds between $\varphi$ and $\psi$ when $\varphi$ entails $\psi$ but not vice versa. Figure 1 then shows the classical constellation of four quantified formulae, together with their mnemonic conventions — *all* (A), *some* (I), *no* (E) and *not all* (O) — and the colour conventions for the four relations of Opposition. In Section 2 we consider the first hexagonal extension of the square by Sesmat and Blanché, whereas in Section 3 the second extension by Sherwood and Czeżowski is introduced. The crucial differences between the two lead to the definition of the Boolean closure of the SC-hexagon in terms of six different SB-hexagons in Section 4. The resulting structure is argued to be isomorphic to the modal system S5 and the system of propositional connectives in Section 5. This section also demonstrates how the six Aristotelian hexagons can be embedded into the 3D structure of a rhombic dodecahedron.

## 2 Sesmat-Blanché: From Square to SB-Hexagon

The Square of Oppositions in Figure 1 has been extended by Sesmat [12] and Blanché [2] to a logical hexagon by inserting a U vertex for *all or no* (A∨E) above the upper horizontal A-E connection and a Y vertex for *some but not all* (I∧O) below the lower horizontal I-O connection. In the resulting hexagon in Figure 2a the three red diagonals A-O, I-E and U-Y represent the relation of Contradiction (CD), whereas the blue relations of contrariety (CR) between A, Y and E in Figure 2b and the green ones of subcontrariety (SCR) between I, O, and U in Figure 2c yield two triangles interlocking into a star-like shape. The black arrows of Subalternation (SA) in Figure 2d then go from each vertex of the blue triangle to the two adjecent ones on the green triangle. The resulting constellation, henceforth referred to as the SESMAT-BLANCHÉ (SB) HEXAGON, not only contains the classical AIOE square SB1 in Figure 3a, but also the AYOU Square SB2 in Figure 3b and the IYEU Square SB3 in Figure 3c.

## 3 Sherwood-Czeżowski: From Square to SC-Hexagon

Czeżowski [3], on the other hand, proposes a hexagonal structure which inserts a third type of quantity on an intermediate layer in between the top layer of the universals *all* and *no* in (1-4) and the bottom layer of the particulars *some* and

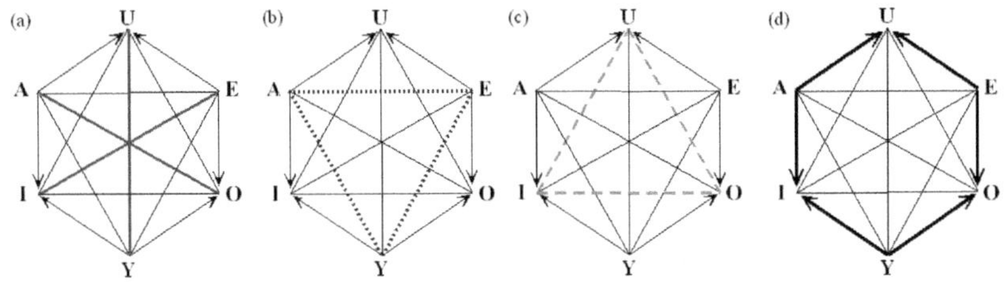


**Fig. 2.** Opposition relations in the SB-hexagon: (a) Contradiction (b) Contrariety (c) Subcontrariety (d) Subalternation

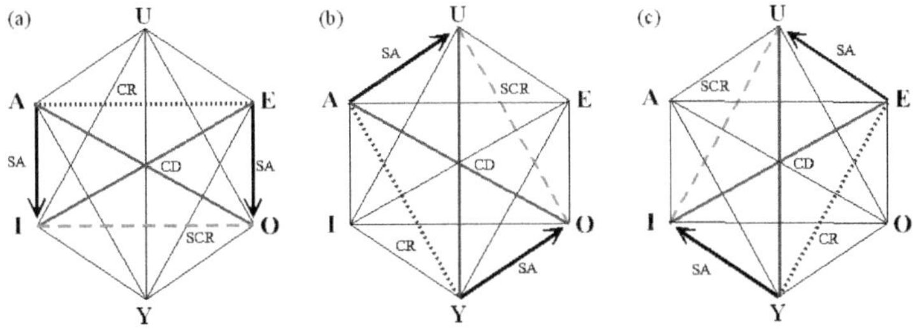


**Fig. 3.** Squares in SB-hexagon: (a) AIOE = SB1 (b) AYOU = SB2 (c) IYEU = SB3

*not all* in (3-6), namely the positive singular *This X is Y* in (2) in between **a** and **i** and the negative singular *This X is not Y* in (5) in between **e** and **o**[2].

| | | | | | |
|---|---|---|---|---|---|
| (1) | **a** *All men are asleep* | 0001 | (4) | **e** *No men are asleep* | 1000 |
| (2) | **u** *This man is asleep* | 0011 | (5) | **y** *This man is not asleep* | 1100 |
| (3) | **i** *Some men are asleep* | 0111 | (6) | **o** *Not all men are asleep* | 1110 |

The resulting hexagonal constellation in Figure 4 will henceforth be referred to as the SHERWOOD-CZEŻOWSKI (SC) HEXAGON[3]. Although the three diagonals in Figure 4a represent contradiction in much the same way as in Figure 2a, the SB-hexagon is fundamentally different from the SC-hexagon (cf. [6,5]): the horizontal SB-insertion of Sesmat-Blanché (A-**U**-E and I-**Y**-O) — with the vertical extra diagonal **UY** - is based on (sub)contrariety whereas the vertical SC-insertion of Sherwood-Czeżowski (a-**u**-i and e-**y**-o) — with the horizontal extra diagonal **uy** — is based on subalternation. As a consequence, the relations of contrariety in

[2] The expressions in (1-6) are of the general form Determiner(A,B) with A the noun *man*, B the predicate *be asleep* and **a** the unique element in A picked up by the deictic operator *this*. They are assigned a bit-string representation (see [13] for more details) where each bit-position corresponds to the truth value of one of the four disjuncts in the Disjunctive Normal Form: $[A \cap B = \emptyset] \vee [A \cap B \neq \emptyset \wedge a \notin B] \vee [A \not\subseteq B \wedge a \in B] \vee [A \subseteq B]$. We distinguish LEVEL 1 (L1), LEVEL 2 (L2) and LEVEL 3 (L3) expressions in terms of the number of values 1 in their bit-string.

[3] Khomskii [7] convincingly argues that this constellation was in fact discovered much earlier by William of Sherwood [8].

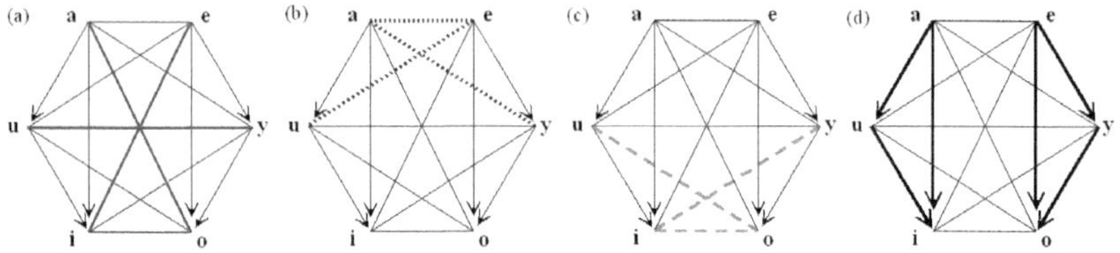


**Fig. 4.** Opposition relations in the SC-hexagon: (a) Contradiction (b) Contrariety (c) Subcontrariety (d) Subalternation

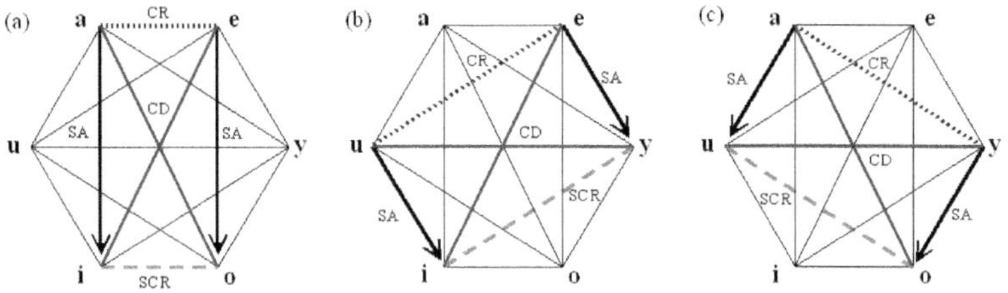


**Fig. 5.** Squares in SC-hexagon: (a) aioe = SC1 (b) uiye = SC2 (c) uoya = SC3

Figure 4b and subcontrariety in Figure 4c no longer constitute triangular shapes, and the arrows of subalternation in Figure 4d no longer constitute the outer edges of the hexagon. Furthermore, although the SC-hexagon also allows three different embeddings of a square of opposition, the arrows point downwards in Squares SC2 and SC3 in Figure 5b-c, whereas in Squares SB2 and SB3 in Figure 3b-c they point upwards[4].

## 4 Boolean Closure of the SC-Hexagon

The central claim of this paper is that the differences between the SB-hexagon and the SC-hexagon are due to the fact that the former is closed under the Boolean operations of meet (conjunction), join (disjunction) and complement (negation) whereas the latter is not. For any contingent formula $\varphi$ in the SB-hexagon, its negation $\neg\varphi$ is also contained in it, while for any two contingent formulae $\varphi$ and $\psi$ in the SB-hexagon, $\varphi \vee \psi$ and $\varphi \wedge \psi$ also belong to it[5]. The SC-hexagon is not closed under the Boolean operators in this way, however[6].

[4] Figures 3b-c are 120 degree rotations, counterclockwise and clockwise respectively, of Figure 3a, whereas in Figure 5 the corresponding rotations are only 30 degrees.

[5] In some cases conjunction yields a contradiction ($\mathbf{A} \wedge \mathbf{O}$) and disjunction yields a tautology ($\mathbf{A} \vee \mathbf{O}$). Although such trivial (non-contingent) formulae are not explicitly represented in the hexagon, they do belong to the Boolean closure. In a sense, both contradiction and tautology are "hidden" in the centre of the hexagon (see [4,13]).

[6] The disjunction $\mathbf{a} \vee \mathbf{o}$ (1001), e.g. does not belong to the SC-hexagon.

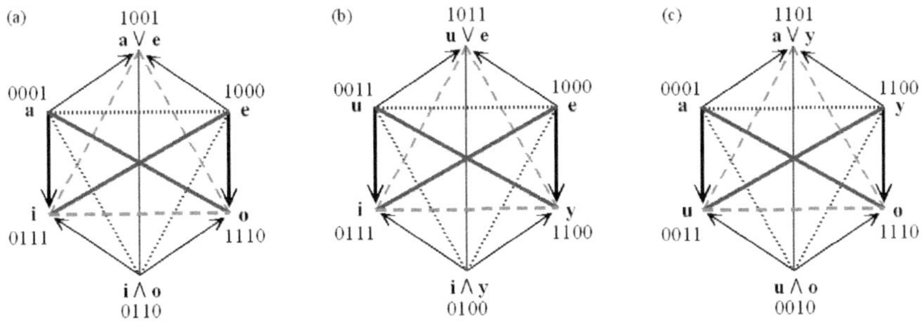


**Fig. 6.** (a) Hexagon SC1, (b) Hexagon SC2, (c) Hexagon SC3

### 4.1 Hexagonal Closure of the Three SC-squares

In a first step we now construct the Boolean closure of the three SC-squares in Figure 5 by systematically adding the disjunction of the square's two top nodes above their upper horizontal connection and the conjunction of the two bottom nodes below their lower horizontal connection[7]. Thus, the Boolean closure of the SC1 square **aioe** (1+3+4+6) in Figure 5a is obtained by adding the formulae (7-8) and yields the classical Sesmat-Blanché hexagon in Figure 6a. Similarly, the closure of the SC2 square **uiye** (2+3+4+5) in Figure 5b adds the formulae (9-10) yielding Figure 6b, whereas that of the SC3 square **uoya** (1+2+5+6) in Figure 5c incorporates the formulae (11-12) resulting in Figure 6c:

| | | | |
|---|---|---|---|
| (7) | $\mathbf{a} \vee \mathbf{e}$ | *No or all men are asleep* | 1001 |
| (8) | $\mathbf{i} \wedge \mathbf{o}$ | *Some but not all men are asleep* | 0110 |
| (9) | $\mathbf{u} \vee \mathbf{e}$ | *No men are asleep or this man is asleep* | 1011 |
| (10) | $\mathbf{i} \wedge \mathbf{y}$ | *This man is not asleep but some men are asleep* | 0100 |
| (11) | $\mathbf{a} \vee \mathbf{y}$ | *All men are asleep or this man is not asleep* | 1101 |
| (12) | $\mathbf{u} \wedge \mathbf{o}$ | *This man is asleep but not all men are asleep* | 0010 |

Notice that all hexagons in Figure 6 have two L1 expressions on the blue triangle, two L3 expressions on the green triangle, and two L2 expressions one on each triangle[8]. Thus the hexagons in Figure 6b and Figure 6c, which share the same L2 diagonal **uy**, differ as to which of **u** or **y** occurs on the blue triangle.

### 4.2 Three More Hexagonal Closures

A fourth hexagon can be obtained by taking the same L2 diagonal **a**∨**e**-**i**∧**o** as in Figure 6a, but this time putting **a**∨**e** on the blue triangle. The result in Figure 7a contains all six of the complex expressions in (7-8-9-10-11-12). The formulae

[7] Notice that conjunction and disjunction apply straightforwardly to the bit-strings: a conjunction only gets a 1 in positions where both conjuncts have value 1, a disjunction gets a 1 in positions where at least one disjunct has value 1 (see [13]).

[8] Furthermore, any formula $\varphi$ on the blue triangle can be defined as the conjunction of its two adjacent formulae on the outer hexagon, and any formula $\varphi$ on the green triangle can be defined as the disjunction of its two adjacent formulae.

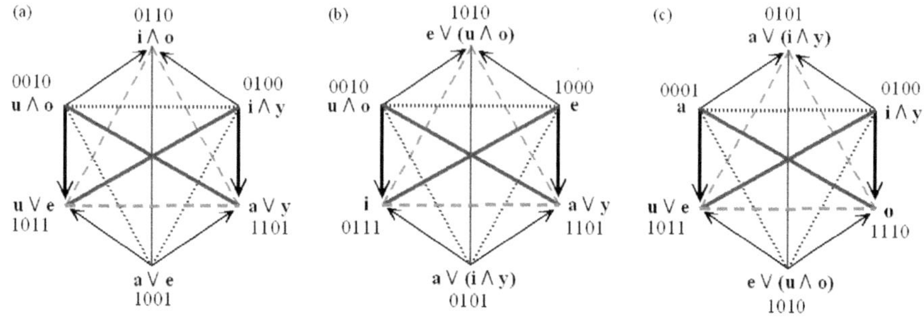


**Fig. 7.** (a) Hexagon SC4, (b) Hexagon SC5, (c) Hexagon SC6

in (1-6) and (7-12) contain all four possible L1 expressions and all four possible L3 expressions, but only four out of the six possible L2 expressions. The two remaining complex L2 expressions are given in (13-14).

| | | | |
|---|---|---|---|
| (13) | **e∨(u∧o)** | *No men sleep or this man does but not all men do* | 1010 |
| (14) | **a∨(i∧y)** | *All men sleep or this man does not but some men do* | 0101 |

These expressions constitute the third possible L2 diagonal: in combination with (3-4-11-12) this diagonal yields the fifth hexagon in Figure 7b, whereas upside down it combines with (1-6-9-10) into the sixth hexagon in Figure 7c.

## 5 Isomorphisms and the Rhombic Dodecahedron

The Boolean closure of the SC-hexagon in terms of 14 non-trivial formulae can easily be shown to be isomorphic to that of the modal system S5 ([1,5,9,13]) and that of the propositional connectives ([5,9,11]). The positive entailment sequence in (1-3) is isomorphic both to that of the modal system S5 [ $\Box\mathbf{p} \vdash \mathbf{p} \vdash \Diamond\mathbf{p}$ ], and to that of propositional logic, nl. [ $\mathbf{p}\wedge\mathbf{q} \vdash \mathbf{p} \vdash \mathbf{p}\vee\mathbf{q}$ ]. The negative sequence in (4-6) is isomorphic to both modal [ $\neg\Diamond\mathbf{p} \vdash \neg\mathbf{p} \vdash \neg\Box\mathbf{p}$ ] and propositional [ $\neg\mathbf{p}\wedge\neg\mathbf{q} \vdash \neg\mathbf{p} \vdash \neg\mathbf{p}\vee\neg\mathbf{q}$ ]. Hence, in the realm of nominal quantification the singular expression *This man is (not) asleep* fulfills a role similar to that of the null modalities **p** and ¬**p** in S5, and that of the unmodified single propositions **p** and ¬**p** with the propositional connectives[9]. In [4,13] the underlying algebraic structure has been visualised by means of the 3D polyhedral structure of a RHOMBIC DODECAHEDRON as in Figure 8a[10]. The six SB-hexagons in Figure 6 and Figure 7 then correspond to the 6 different ways in which the dodecahedron can be sliced in two equal parts, as in Figure 8b. Two hexagons sharing an L2 diagonal interlock as in Figure 8c.

---

[9] In other words, the middle area of "contingency" for *possibly but not necessarily* or *some but not all* is split into *possibly but not actually/some but not this one* and *actually but not necessarily/this one but not all.*

[10] In [9,10] it is called the $\beta 3$-structure and visualised as a so-called TETRA-ICOSAHEDRON, very much similar to the TETRA-HEXAHEDRON of [11]. In [4] the rhombic dodecahedron is applied to the realm of Public Announcement Logic.

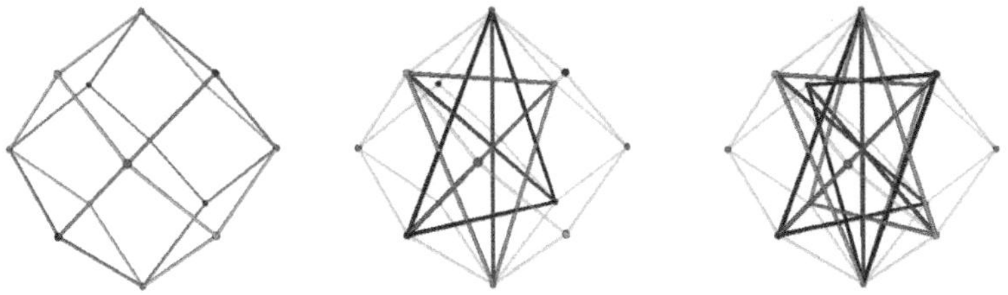

**Fig. 8.** (a) Rhombic Dodecahedron, with (b) Single Hexagon and (c) Double Hexagon

## 6 Conclusions

This paper has compared two hexagonal extensions of the Aristotelian Square: the SB-hexagon which is closed under the Boolean operators, and the SC-hexagon, which is not. The incomplete SC-hexagon gets a Boolean closure containing 8 more non-trivial formulae. The resulting 14 formulae generate 6 SB-hexagons embedded into a 3D rhombic dodecahedron of Oppositions.

## References


1. Béziau, J.Y.: New light on the square of oppositions and its nameless corner. Logical Investigations 10, 218–232 (2003)
2. Blanché, R.: Structures Intellectuelles. Essai sur l'organisation systématique des concepts. Librairie Philosophique J. Vrin, Paris (1969)
3. Czeżowski, T.: On certain peculiarities of singular propositions. Mind 64(255), 392–395 (1955)
4. Demey, L.: Structures of Oppositions in Public Announcement Logic. In: Béziau, J.Y., Jacquette, D. (eds.) Around and Beyond the Square of Opposition. Springer, Basel (2012)
5. Horn, L.R.: Hamburgers and truth: Why Gricean explanation is Gricean. In: Hall, K. (ed.) Proceedings of the Sixteenth Annual Meeting of the Berkeley Linguistics Society, pp. 454–471. Berkeley Linguistics Society, Berkeley (1990)
6. Humberstone, L.: Modality. In: Jackson, F., Smith, M. (eds.) The Oxford Handbook of Contemporary Philosophy, pp. 534–614. OUP, Oxford (2005)
7. Khomskii, Y.: William of Sherwood, singular propositions and the hexagon of opposition. In: Béziau, J.Y., Payette, G. (eds.) New Perspectives on the Square of Opposition. Peter Lang, Bern (2011)
8. Kretzmann, N.: William of Sherwood's Introduction to Logic. Minnesota Archive Editions, Minneapolis (1966)
9. Moretti, A.: The Geometry of Logical Opposition. Ph.D. thesis, University of Neuchâtel (2009)
10. Pellissier, R.: Setting n-opposition. Logica Universalis 2(2), 235–263 (2008)
11. Sauriol, P.: Remarques sur la théorie de l'hexagone logique de Blanché. Dialogue 7, 374–390 (1968)
12. Sesmat, A.: Logique II. Les Raisonnements. Hermann, Paris (1951)
13. Smessaert, H.: On the 3D visualisation of logical relations. Logica Universalis 3(2), 303–332 (2009)

# An Exploration of Visual Complexity

Helen C. Purchase[1], Euan Freeman[1], and John Hamer[2]

[1] School of Computing Science, University of Glasgow, University Avenue, Glasgow, G12 8QQ, United Kingdom
[2] Department of Computer Science, University of Auckland, 38 Princes Street, Auckland 1142, New Zealand

**Abstract.** Inspired by the contrast between 'classical' and 'expressive' visual aesthetic design, this paper explores the 'visual complexity' of images. We wished to investigate whether the visual complexity of an image could be quantified so that it matched participants' view of complexity. An empirical study was conducted to collect data on the human view of the complexity of a set of images. The results were then related to a set of computational metrics applied to these images, so as to identify which objective metrics best encapsulate the human subjective opinion. We conclude that the subjective notion of 'complexity' is consistent both to an individual and to a group, but that it does not easily relate to the most obvious computational metrics.

**Keywords:** Image complexity, visual aesthetic, image processing, empirical results.

## 1 Introduction

In assessing the usefulness of an interface, it is not only the functional interaction model and the provision of interactive elements that is important: non-functional aspects of the interface also have a part to play in how useful the interface is perceived to be, how much a user likes the interface, and how willing the user might be to engage with it for extended lengths of time. There is even research that suggests that the more 'aesthetically pleasing' an interface, the more likely a user is to perform their task correctly and efficiently [1].

In this paper, we consider an aspect of the aesthetics of visual design that has not been considered quantifiably before: that is, the 'visual complexity' of an image. If we were able to measure and combine computational features of an image so as to reliably match 'complexity' as judged by humans, then that would be the first step in being able to determine the effect (if any) of the visual complexity of images used on an interface and its effect on users' aesthetic judgments, preference, performance, or perceived usability.

This research therefore contributes to the growing area of investigating the effects of interface aesthetics, while also adding a new human perception focus to the field of image processing.

In this study, we created a set of nine computational metrics for image processing, applied them to 60 digital images, and removed those metrics which were highly

P. Cox, B. Plimmer, and P. Rodgers (Eds.): Diagrams 2012, LNAI 7352, pp. 200–213, 2012.

correlated with each other (while retaining one in each of the categories of colour, edge, intensity variation and file size). We then obtained both ranking and rating 'visual complexity' judgments from 54 participants, assessed their internal and external consistency, and then used regression analysis to devise a prediction formula. We then attempted to validate the formula on a further 12 images and 28 participants.

## 2 Background

### 2.1 Interface Aesthetics

There is an increasing recognition of the role of visual aesthetics in interface design, and several evaluation studies have been performed on assessing the aesthetics of interfaces (e.g. [2-4]). In addition, theoretical work has produced frameworks for the investigation of system aesthetics and on relevant and useful terminology (e.g. [5, 6]).

Recent research has demonstrated that the aesthetic appearance of an interface is important, not just with respect to users' preferences [7] and perception of usability [4] but with respect to their performance in visual search tasks [1].

Lavie and Tractinsky [6] provide an extensive list of aesthetic terminology to support subsequent experiments, and distinguish between "empirical studies of aesthetics" involving controlled studies with manipulation of visual variables, and an "exploratory approach" involving evaluation of existing stimuli. Most existing work falls into the latter category: while this has produced interesting results, without being able to relate the data back to quantitative or well-defined qualitative descriptions of the stimuli, the results remain descriptive rather than explanatory.

Ngo et al.[8] took a quantitative approach, and defined fourteen computational metrics for objectively quantifying different layout aspects of an interface, (e.g. symmetry, balance, equilibrium, sequence). They use the term "aesthetic" for these metrics, even though in effect they characterise simply the placement of objects on a 2D plane. These formulae produce values between 0 and 1, each an indicator of the presence of an aesthetic feature of the interface. Ngo validated these measures [9] and investigated whether they could be used to determine users' acceptance of data entry screens. He asked seven designers to rank 57 screens; from these results he proposed a regression formula and proved that this formula could predict (within a small range) the rankings of new participants on different screens.

The traditional definition of "aesthetics" with respect to the visual sense is much richer than simple object placement, encompassing the use of colour, texture and contrast. Lavie and Tractinsky [6] conducted studies using web sites which allowed them to classify perception as being of two dimensions; "classical aesthetics", featuring characteristics such as symmetry, clarity, order and organisation, and "expressive aesthetics", relating to creativity and originality.

Ngo's metrics addressed the quantification of the "classical" approach to aesthetic perception, based on the positioning of objects on the plane. The more nebulous notion of "expressive" aesthetics (being based on "originality and the ability to break design conventions") may be impossible to quantify objectively. There is, however, a perspective that falls between the simple positioning of objects (the "classical") and creativity (the "expressive"): it is the perception of the complexity of an image – colour, shapes, overlapping of visual objects, curvature etc. An image that is entirely blue is not complex; a photograph of a busy railway scene is highly complex – this complexity is not captured by either "classical" or "expressive" notions of aesthetic perception.

### 2.2 Visual Perception of Web Pages

In considering the visual complexity of interfaces, several studies have been performed on the perception of web pages. Such studies tend to focus on the overall perception of the entire collection of objects that make up a web page, rather than the perception of the individual objects themselves, and focus on preference judgements.

Knight and Pandir [7] asked participants to order printouts of twelve web pages according to their perception of "pleasingness", "complexity", and "interestingness" (with no further definition given to these terms). They found that the most pleasing web pages were neither the most interesting nor the most complex, and that when participants ranked web pages, complexity was not a predictor of aesthetic pleasure. Their data also showed that participants were generally in agreement on complexity judgements.

Michailidou et al. [10] also focussed on web pages, and asked participants to rank thirty web pages according to "visual complexity". Their correlation and regression analysis was based around a definition of complexity that included the structural and interactive features of the pages: the number of menus, number of images, number of links etc.

When considering users' perceptions of usability and "aesthetic appeal", Purchase et al. [11] asked participants to rank fifteen web sites, and related the ranking data to the values from Ngo's classical aesthetic layout metrics as applied to the positioning of the text, image and control objects on the web pages. They found that aesthetic appeal was strongly captured by Ngo's composite metric, and that colour was not a dominant factor in judging either aesthetic appeal or perceived usability. The relationship between aesthetic appeal and the classical metrics was strongest with respect to the placement of images (rather than text or control elements). They note that a limitation of their approach was that this classical method only took into account the *location* of the images, not their *content,* and that an image on the page that was wholly blue would have contributed to the aesthetic metric calculation as much as an image of a busy railway station of the same size and position.

### 2.3 Visual Complexity

The focus of this paper is the objective characterization of the visual complexity of a single image. We are therefore placing this research in between the "classical" and "expressive" definitions of Lavie and Tractinsky [6]. We are concerned with the aesthetic judgments of static images that may be used, for example, as the background for an interface, or as an item on an interface, or as a clickable image on a web page. By focusing on the static images themselves (rather than web pages), we are removing any factors that might be associated with interactive features. We aim to investigate whether we can devise objective, computational measures of visual complexity – comparable to those created by Ngo for the layout of objects.

The simplest objective indication of the visual complexity of a digital image is the compressed file size; JPEG compressed file size [12, 13] and GIF file size [14] have been found to correlate with human perceptions of visual complexity.

File compression metrics are, however, rather abstract, and are difficult to describe to both users and interface designers. Our research wished to investigate whether alternative, more visually concrete features of an image related to the perception of complexity – for example, colour, edges, extreme visual differences, etc.

Oliva et al. [15] found that people qualitatively characterised visual complexity using criteria such as “clutter”, “symmetry”, “open space” and “organisation”, and Knight and Pandir’s [7] participants used words like “overpowering”, “intense”, “daunting” and “unordered” when describing the most complex website, and “simple”, “unified” and “clean” when describing the least complex website. Our challenge was to see whether these visual terms could be captured computationally.

# 3 Method

## 3.1 Objective Measures of Complexity

In choosing our metrics for visual complexity, we considered definitions from Snodgrass and Vanderwart [16] as “the amount of detail or intricacy”, and as characterised by Oliva et al. [15] as being “principally represented by the perceptual dimensions of quantity of objects, clutter, openness, symmetry, organization, and variety of colors.” The work by Mario, et al. [17] suggests that analysis of the edges in an image could be used as a measure of complexity (where a sharp change in luminance in an image often represents the edge of an object).

**Table 1.** Computational metrics for the Visual Complexity of an image

| Name | Description | Values when applied to 60 experimental images | | |
|---|---|---|---|---|
| | | minimum | maximum | standard deviation |
| Colours | Number of unique RGB colours. | 10011 | 316630 | 55598 |
| RColours | Number of unique RGB colours, after colour reduction | 2070 | 114539 | 23460 |
| PColours | Number of unique RGB colours, after posterization. | 21 | 179 | 34.63 |
| SColours | Number of unique RGB colours, after pyramid segmentation. | 2 | 35 | 5.02 |
| EdgeArea | The area of the image occupied by edges. | 0.00029 | 0.1760 | 0.333 |
| GrayscaleSD | Standard deviation of pixel intensities in grayscale. | 14 | 80 | 16.55 |
| JPEG | JPEG file size when compressed. | 51243 | 382364 | 60202 |
| PNG | PNG file size. | 410820 | 2199954 | 410200 |
| GIF | GIF file size. | 403096 | 989309 | 131918 |

Based on this prior research, our own intuitions, pilot tests conducted as part of an associated research project [18] and considering the prior research that found the relationship between complexity and file size, we categorized our metrics into four types: colour, edges of objects, intensity variation, and file size. We implemented nine different computational metrics (Table 1): each metric takes as input a digital image file, and produces a metric value.

**Colours:** Digital images are typically represented as a grid of pixels, where each pixel has associated with it three values; the red, green and blue (RGB) components which make up the colour of that pixel. The most simple of the measures, Colours, is a count of the unique RGB colours which make up an image.

**RColours:** In an attempt to improve on the Colours metric, an algorithm was created which tried to normalize similar coloured regions by looking at adjacent pixels. This measure, RColours, is the number of unique colours after this algorithm has transformed an image. This algorithm compared each pixel with the pixel to the right, and below it. If either of these pairs of pixels were considered "similar" (within 10 units with respect to the CIE76 formula [19-21]) the adjacent pixel was coloured to match the original pixel. This had the effect of reducing the number of colours in similar coloured areas such as the sky.

**PColours:** The PColours metric is a count of unique RGB colours, after an image has been transformed using a posterization algorithm. The posterization process limits the red, green and blue components of an RGB colour to a specific number of areas. This has the effect of normalizing similar coloured regions of an image as these similar colours fall into the same area. In this study, a posterization parameter of 6 was used, allowing up to 63 colours.

**SColours:** In image processing, segmentation is the process of simplifying an image by organising pixels into a larger group of pixels, a "segment". Pyramid segmentation [22] constructs a pyramid, where pixels are compared with those on adjacent levels. The pyramid segmentation function is parametrised by two threshold values. The first of these defines the threshold below which two pixels on different levels are considered associated. The second threshold determines if two segments belong to the same cluster. Pyramid segmentation, with both thresholds set to 60, was used to transform an image to its segmented form. The SColour measure counts the unique RGB colours in this transformed image. The rationale for this metric is that the number of colours in a segmented image should roughly indicate the number of "objects" perceived to be in the image, as a segment (each a single colour) typically represents a single group of features in an image.

**EdgeArea:** The edge detection process searches for sharp changes in intensity [23] which typically represents the edges of objects. Using the Canny edge detection algorithm [24] implementation provided by OpenCV [22], a metric was created which calculates the EdgeArea; the ratio of edges in an image to non-edges. The two threshold values used for the Canny algorithm were 1 and 100, for the first and second thresholds respectively. To remove noise, images were first smoothed using an 11x11 Gaussian blur. The rationale for this metric is that images with a greater edge area may represent images containing a higher number of objects

**GrayscaleSD:** Unlike images which use the RGB colour system, grayscale images store the intensity of that pixel, typically in the range of 0 (black) to 255 (white). The GrayscaleSD metric takes the standard deviation in pixel intensities to give a measure of how much the image varies in intensity, thus representing high variations in intensity. Intensity variations can indicate presence of different objects in the image.

**JPEG, PNG and GIF:** These three metrics are the compressed filesize of the image. The JPEG representation of the image is compressed at 70% quality using ImageMagick [25].

### 3.2 Subjective Perception of Complexity

A within-subjects experiment was conducted to gather subjective rankings and ratings of visual complexity. The dependent variable in this experiment was participants' subjective view of "visual complexity".

Sixty images were used in this study, which were photographs taken by the second author. A wide range of image subjects were sought, including landscapes, domestic objects and city scenes. Images were resized to a resolution of 640x480 for the experiment and were shown in colour.

**Procedure:** Subjective data was gathered online, using a website created specifically for this study, allowing participants to take part at a time and place which suited them. After reading about the experiment and agreeing to volunteer to take part, participants began the first of two stages in the experiment.

In the first stage, participants were shown four images in a row and asked "*Please sort the images based on how visually complex you consider them to be.*" A drag-and-drop user interface allowed the participants to easily compare and sort images. Each image was shown twice during this stage, such that each participant completed thirty four-way comparisons. A four-way comparison was chosen over two or three based on pilot studies: we wished the task to be difficult enough that participants were required to think carefully about their considerations of visual complexity.

This first stage of the experiment allowed the participants to develop a good idea of what causes them to perceive an image as being complex. In the second stage of the complexity experiment, participants were shown each of the sixty images individually, and were asked to rate the visual complexity of each image using a five-point Likert scale.

Appropriate randomisation was used throughout, with different random orders used for each participant, both in the first stage (the selection of the four images to shown at any one time) and in the second stage (ordering of the images shown with the Likert scales).

At the end of the experiment, participants were thanked for their participation and given the chance to provide anonymous feedback. The mean time spent on the experiment per participant was 13 minutes. The experiment ran for a two week period, during which time 54 participants completed all stages of the experiment.

### 3.3 The Data

Each trial in the ranking stage of the experiment provided six pairwise orderings between the four images. As each of the 60 images was shown twice (a total of 30 trials), each participant provided 180 pairwise orderings. These orderings were represented in a directed acyclic graph of 60 nodes and 180 edges, one graph for each participant (called **G1**..**G54**). A composite graph (**GAll**) of 60 nodes was also created, representing all the pairwise orderings from all the participants.

In addition, each image was associated with the sum of two numbers, each representing the two ranking positions in which it was placed in its row-of-four (1 for least complex, 4 for most complex), summed over all participants (called **R1**..**R60**). The range was 154-408 (standard deviation = 63.9).

The third data measure collected was the mean Likert rating for each image over all participants (**L1**..**L60**). The range was 1.46-4.63 (standard deviation = 0.77).

All the metrics in Table 1 were implemented and applied to all 60 images.

## 4 Analysis

Several participants mentioned in the end-of-experiment questionnaire that they were confused as to what the term "Visual Complexity" meant. One participant remarked that "I found it hard to judge complexity of the scenes" and that he wasn't sure "how 'complex' they are." Another participant commented that it was "difficult to judge 'complexity' of an image due to the different concepts of complexity such as colour, shape and patterns." It was suggested by a third participant that "it'd help to explain in more detail what you expect from the term 'visually complex'." No description had been given on purpose, as previous work by Oliva et al. [15] found that perception of visual complexity is biased by how the term is described. Two groups in a between-subjects experiment were given different descriptions of complexity, and the reserchers found that these groups judged complexity differently according to the description they were given.

Before addressing the main research question of relating the complexity data to the computational metrics, we therefore needed to ensure that the ranking and rating data collected was robust enough to use. This analysis section therefore starts off with a consideration of consistency.

### 4.1 Consistency

Our consistency analysis considered two questions:

- Were participants consistent *within-themselves* as to their definition of 'visual complexity'? Each image was deliberately shown twice in the ranking task so that we could ensure that the participants were being consistent in their judgments (and were taking the task seriously).

- Were participants consistent *between-themselves* as to their definition of 'visual complexity'? We needed to be sure that our aggregate data over all participants was representative of a consistent view of complexity (rather than there being, for example, two opposing views, which would result in a meaningless aggregate).

In determining *within-participant* consistency, we laid out each of the 54 graphs (**G1**..**G54**) using a hierarchical graph layout algorithm which displays as many of the directed edges as possible in the same direction, from top-to-bottom. We then determined how many of the edges in each would need to be reversed for *all* the edges to flow downwards; that is: how many edges need to be reversed for the graph to contain no cycles; or, put another way, how many of the participant's pairwise judgments should be reversed for the graph to represent total consistency (Fig. 1).

22 of the 54 participants were wholly consistent, 13 were only inconsistent in one pairwise judgment, and only two made as many as six. When considering that each participant made three explicit pairwise judgments for each four-in-a-row ranking task,[1] and therefore a total of 90 placements, these are low numbers, with 6 judgments representing only 7% of all judgments. These numbers suggest that participants were internally consistent in their definition of visual complexity, and that they were all taking the ranking task seriously.

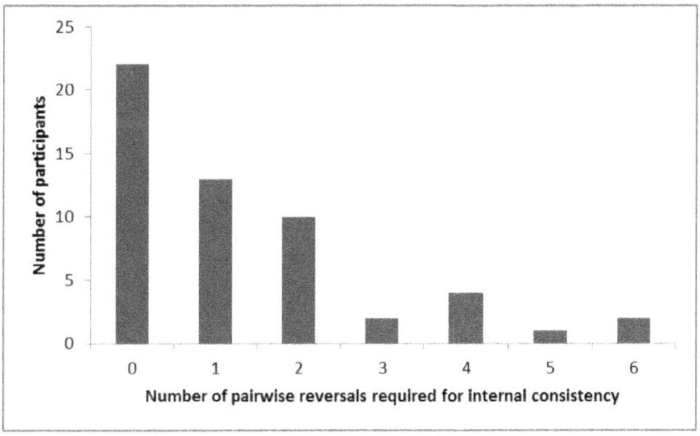


**Fig. 1.** Histogram showing the number of participants for whom consistency can be assured by reversing the given number of pairwise judgments

In determining *between-participant* consistency, we considered the **GAll** graph. This graph includes all 60 nodes (representing the images) and 9469 directed edges (representing 'is more complex than' judgments). All pairwise decisions made by all participants are included in this graph, excluding duplicate edges from the same participant: thus, if a participant had two opportunities to compare image A with

[1] An additional three pairwise rankings can be inferred in any four-way ranking; only three are explicit rank placements.

image B (and made the same decision as to which was more complex both times), this decision was represented only once in **GAll**. However, if two participants made the same complexity decision between image A and image B, or if a participants made two contradictory decisions about A and B, then both decisions are recorded.

For each pair of images, we computed an 'agreement index' – the proportion of edges going in the same direction; i.e. the proportion of those participants who had had an opportunity to make a pairwise judgment between the two images who agreed on their relative complexity. The range of this agreement index is [0.5-1.0], where 1.0 represents total agreement, 0.5 means a 50:50 split between the judgments made, and therefore no agreement.

Of the 1768[2] agreement indices, 572 (32%) represented total agreement with a value of 1.0, with 926 (52%) pairwise judgments representing agreement of at least 0.8 (Fig. 2). The mean was 0.79. Only 9% of the indices were 0.5, representing no majority agreement at all. We have no explanation for the unexpectedly higher number of 0.6-0.69 agreements than 0.7-0.79, but the general trend from right to left is clearly downwards. Our images were deliberately not chosen to represent extremes, but to cover a range of objects and scenes, and so the fact that there was not total agreement for all pairwise judgments is as expected. This data suggests that there was sufficient agreement amongst the participants as to what is meant by 'visual complexity' for the complexity ranking data to be worth analyzing further, and to be related to the computational metrics.

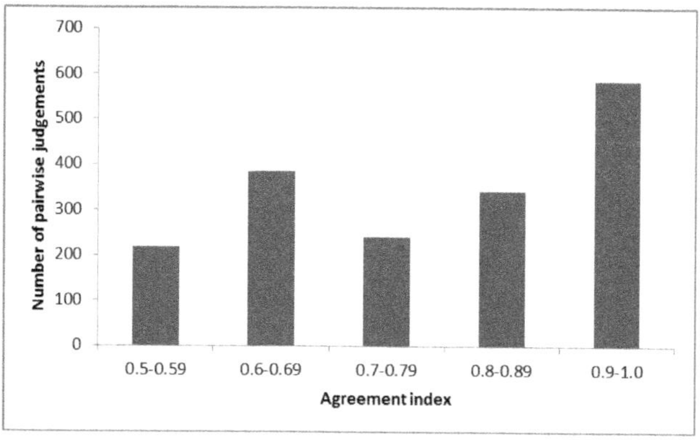


**Fig. 2.** Histogram showing frequency of agreement indices for 1768 pairwise complexity judgments: 1.0 represents total agreement; 0.5 represents no clear agreement.

Using these agreement indices, we identified those images for which there was most agreement: the top-ranked nine images have a mean agreement index of over 0.849 (Fig. 3).[3]

---

[2] 1768 is (60*59/2)-2. Two node pairs were never seen in the same group-of-four by any participant.

[3] 0.849 was chosen as the cut-off when considering the rank of agreement indices, as the difference between 0.849 and the next image's index (0.836) represented an apparent step change.

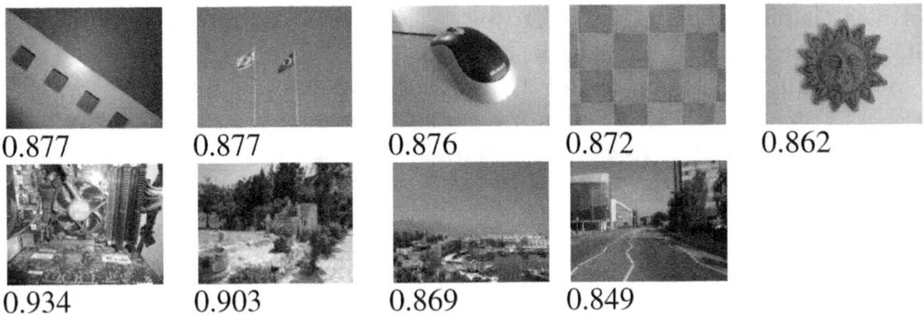


**Fig. 3.** (a) The nine images with the highest mean agreement amongst participants: the top row are 'the least complex', the bottom row are 'the most complex'

### 4.2 Computational Predictors of Visual Complexity

**Regression Analysis:** We have two measures of subjective relative complexity for each of the 60 images:

- the sum of the ranks (**R1..R60**), and
- the mean Likert value (**L1..L60**).

The correlation between R1..R60 and L1..L60 is 0.97 ($p<0.001$). We chose to use the Likert ratings in our analysis: the ranking values are relative (i.e. participants were forced to assign rank values to images in comparison with the others in each group-of-four), while the Likert ratings are absolute (i.e. participants were only looking at a single image when they gave it an individual Likert rating).

The predictor variables used in a multiple regression analysis should be independent where possible. While we had nine image metrics, we only needed one metric for each of the four visual categories: colour, intensity, edges and file size. By looking at the pairwise correlations between the values of all nine metrics when applied to the 60 stimuli, we eliminated those metrics for which there was a high correlation within the same type. This left us with four metrics: for colour (**SColour**), for edges (**EdgeArea**), for intensity (**GreyscaleSD**), and for file size (**GIF**). While these metrics were not all entirely independent (for example, there was a significant correlation between Edge Area and GIF), this was the best combination of metrics we could have chosen so as reduce the overall number of high correlations.

For a valid multiple regression analysis, it is generally accepted that the number of cases must substantially exceed the number of predictor variables: an absolute minimum ratio of 5:1 (and a preferably ratio of 10:1) [26]. In our case, we have 60 cases, and 4 predictor variables, a ratio of 15:1.

We used the SPSS multiple linear regressions tool, using the 'Enter' method first to identify the predictor variables, followed by the 'Backward' method which produced the statistically significant ($p<0.05$) co-efficients for these variables.

**The Regression Formula:** The analysis produces a formula $y = \varepsilon + \beta_1 x_1 + \beta_2 x_2 + ... + \beta_n x_n$ such that $y$ is the dependent variable (in this case, the visual complexity as represented by a Likert rating judgement between 1 and 5, called **VCL**), ε is the yintercept and $\beta_1$ to $\beta_n$ are coefficients for independent variables $x_1$ to $x_n$ (in this case, **SColour, GreyscaleSD**, **EdgeArea** and **GIF**).

The best-fitting model produced the following formula:

$$VCL = 1.945 + 0.013*GreyscaleSD + 0.053*SColour$$

The EdgeArea and GIF variables were not included in this model. The Durban-Watson statistic for this model is 1.992: being close to the generally agreed cut-off of 2.0, this value indicates, as we know, that the two predictor variables are sufficiently independent.

The goodness-of-fit of this model is represented by its $R^2$ value (0.248) which indicates that 25% of the variation in the dependent variable (VCL) is described by this model. The standard error of this model, 0.67, describes the standard deviation between observed dependent variables and the predicted dependent variables. To put this 0.67 value into context, the range for the dependent variable VCL is between 1 and 5.

## 4.3 Testing the Model

Although a statistically significant regression formula had been found based on two predictor variables, we had little confidence in it – this best fitting model only explains 25% ($R^2$=0.248) of variance in subjective ratings of visual complexity.

To see whether this model held any validity, we tested it against further, new experimental data. 28 participants underwent the same experimental process as before, with 12 new images provided by the first author. The model was used to predict the mean Likert rating for each image, and to rank the images in order of predicted visual complexity.

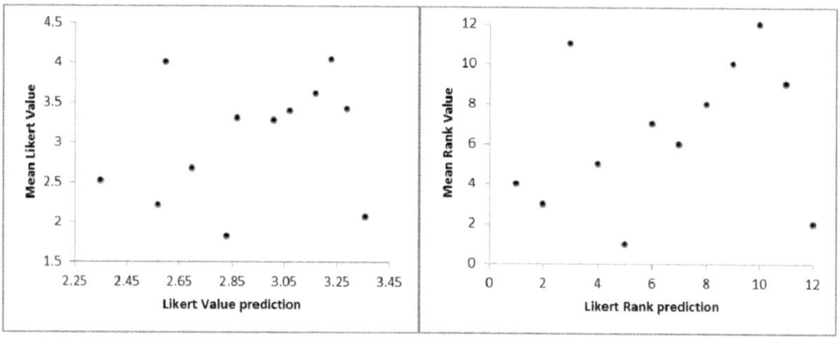


**Fig. 4.** The relationship between predicted and actual results. Note that both axes for the scatterplot on the left do not start at the origin.

Fig. 4 plots the mean Likert values for these validation images against the value predicted by the model (left) and the ranking of the validation images against their ranking predicted by the model (right).

The correlation co-efficient between the actual and predicted Likert values was 0.257; between the actual and predicted ranks it was 0.294. Neither of these values is significant. We removed image number 5 from the analysis; as a map, it was more of a schematic than an image, and we felt that, in retrospect, its inclusion had been inappropriate (and it was one of the obvious outliers from the data obtained: the lower-right data point in both plots in Fig. 4). Redoing the analysis without this map image produced revised correlation co-efficient of 0.450 and 0.473; again, neither of these is statistically significant.

## 5 Discussion

Several aspects of this study surprised us: even when they were not given a specific definition of an uncommon and abstract visual concept, participants are in broad agreement with each other as to what it means when applied to concrete instances. In addition, participants are internally consistent with their interpretation of a term that they are unlikely to have had to work with before. There seems to be no doubt as to what 'visually complex' means from the point of view of human perception – it may be difficult to precisely define, but people "know it when they see it."[4]

However, our exploratory study has shown that 'visually complex' is more difficult to define computationally than subjectively. That is, while it may be easy for us to devise computational metrics that measure various aspects of an image, finding 'the right' metrics that will adequately capture the human notion of 'visual complexity' is more challenging. Despite the fact that we used obvious visual variables of colour, intensity change, and extent of edges (in addition to the variable of compressed file size), it appears that there are other less obvious image features that need to be considered.

The participants' complexity judgments were based on two processes: the physical perception of the stimulus, as well as the semantic comprehension of the term 'visual complexity'. Our consistency analysis suggests that we need not concern ourselves overly with the latter issue, and that it is the former where further work should be focused. More complex image processing algorithms for feature extraction, pattern variation, intensity fragments, level of detail of edges (as in [17]) or spatial frequency analysis might provide more useful predictor variables – even if some of these features can only be completely defined computationally, are difficult to define qualitatively, and may be hard to describe to users or interface designers.

In our attempt to locate this study between "classical" and "expressive" aesthetics [6], we have made no attempt to include metrics representing the layout of objects in the image. This may have been an important omission, as it may be that 'complexity' also encompasses the notions of order, symmetry, clarity, balance (or the lack thereof). Image processing algorithms that can identify the main objects in image (and their location in the 2D plane), and then apply the Ngo metrics to derive a classical aesthetics measure may embody useful factors that are missing in our regression formula. Pointers to this idea can be found in a study on web site complexity [10], when it was found that "the more organised, clear, clean and beautiful a webpage is, the visually simpler the page was perceived by the participants" – it is possible that these classical features extend to non-interactive images as well.

---

[4] All the images (experimental and validation) are at www.dcs.gla.ac.uk/~hcp/Diagrams2012.

There are other higher-level perception processes that could also be integrated into studies of visual complexity, for example, the Gestalt laws and principles of pattern recognition [27]. Our approach has been bottom-up; additional studies could devise computational metrics that characterize images with respect to their overall visual pattern or structure and include these into the analysis.

Different results may be obtained by using stimuli that are diagrammatic or schematic in form, or that have abstract meaning, rather than concrete photographs. It may prove easier to quantify visual complexity with pictures that are less expressive than photographs.

Until a valid computational model encapsulating human perception of 'complexity' has been derived, experimental work on interface aesthetics that wishes to consider the complexity of images used will need to derive quantitative complexity rankings from supplementary human studies. This approach is similar to that used by Hassenzahl [3] who asked one set of participants to rank MP3 player skins according to 'beauty' and used the resultant 'most beautiful' and 'ugliest' skins as the independent variable in a subsequent experiment.

## 6 Conclusion

This paper has reported on an experiment that attempted to derive a computational formula for the human perception of 'visual complexity' of an image. Despite using metrics for typical visual features (i.e. colour, edges, intensity), and including a common image compression method, our regression formula gave poor predictions of the complexity of a validation set of images, as judged by human participants. It is clear that more subtle or advanced image processing algorithms will be needed to appropriately capture the nuances of the human perception of image complexity.

**Acknowledgments.** We are grateful to the many participants who took part in both parts of this experiment, to Mhairi McDonald for her statistics assistance, and to David Simmons and Paul Siebert for their useful image processing suggestions.

## References

1. Salimun, C., et al.: The effect of aesthetically pleasing composition on visual search performance. In: Nordic Human Computer Interaction Conference, pp. 422–431. ACM (2010)
2. Hartmann, J., Sutcliffe, A., De Angeli, A.: Investigating attractiveness in web user interfaces. In: Human Factors in Computing Systems (CHI) Conference, pp. 387–396 (2007)
3. Hassenzahl, M.: The Interplay of Beauty, Goodness, and Usability in Interactive Products. Human-Computer Interaction 19, 319–349 (2004)
4. Kurosu, M., Kashimura, K.: Apparent usability vs inherent usability: experimental analysis on the determinants of the apparent usability. In: Human Factors in Computing Systems (CHI) Conference (1995)

5. Hartmann, J., Sutcliffe, A., De Angeli, A.: Towards a Theory of User Judgment of Aesthetics and User Interface Quality. ACM Transactions on Computer-Human Interaction 15(4), 15 (2008)
6. Lavie, T., Tractinsky, N.: Assessing dimensions of perceived visual aesthetics of web sites. International Journal of Human-Computer Studies 60(3), 269–298 (2004)
7. Knight, J., Pandir, M.: Homepage aesthetics: The search for preference factors and the challenges of subjectivity. Interacting with Computers 18, 1351–1370 (2006)
8. Ngo, D., Teo, L., Byrne, J.G.: Modelling interface aesthetics. Information Sciences 152, 25–46 (2003)
9. Ngo, D., Byrne, J.: Application of an aesthetic evaluation model to data entry screens. Computers in Human Behavior 17(2), 149–185 (2001)
10. Michailidou, E., Harper, S., Bechhofer, S.: Visual Complexity and Aesthetic Perception of Web Pages. In: SIGDOC 2008 Conference, Lisbon, pp. 215–224 (2008)
11. Purchase, H.C., et al.: Investigating objective measures of web page aesthetics and usability. In: Lutteroth, C., Shen, H. (eds.) Australasian User Interface Conference, pp. 19–28. CPRIT, Perth (2011)
12. Donderi, D., McFadden, S.: Compressed file length predicts search time and errors on visual displays. Displays 26, 71–78 (2005)
13. Donderi, D.: An information theory analysis of visual complexity and dissimilarity. Perception 35, 823–835 (2006)
14. Forsythe, A., et al.: Predicting beauty: Fractal dimension and visual complexity in art. British Journal of Psychology 102, 49–70 (2001)
15. Oliva, A., et al.: Identifying the Perceptual Dimensions of Visual Complexity of Scenes. In: Cognitive Science Conference (2004)
16. Snodgrass, J.G., Vanderwart, M.: A Standardized Set of 260 Pictures. Norms for Name Agreement, Image Agreement, Familiarity and Visual Complexity. Journal of Experimental Psychology: Human Learning and Memory 6(2), 174–215 (1980)
17. Mario, I., et al.: Image complexity measure: a human criterion free approach. In: North American Fuzzy Information Processing Society, pp. 241–246 (2005)
18. Salimun, C., Purchase, H.C., Simmons, D.: Visual aesthetics in computer interface design: does it matter? In: 34th European Conference on Visual Perception, p. 220 (2011)
19. International Commission on Illumination: Colour Difference, `http://en.wikipedia.org/wiki/Color_difference#CIE76` (accessed February 28, 2012)
20. Robertson, A.: The CIE 1976 color-difference formulae. Colour Research and Application 2(1), 7–11 (1997)
21. Sharma, G.: Digital Color Imaging. IEEE Transactions on Image Processing 6(7), 901–932 (1997)
22. Willow Garage: OpenCV, `http://opencv.willowgarage.com/` (accessed February 28, 2012)
23. Ding, L., Goshtasby, A.: On the Canny edge detector. Pattern Recognition 34, 721–725 (2001)
24. Canny, J.: A Computational Approach to Edge Detection. IEEE Transactions on Pattern Analysis and Machine Intelligence 8(6), 679–698 (1986)
25. ImageMagick Studio LLC: ImageMagick, `http://www.imagemagick.org/` (accessed February 28, 2012)
26. Brace, N., Kemp, R., Snelgar, R.: SPSS for Psychologists, 2nd edn. Palgrave Macmillan (2003)
27. Ware, C.: Information Visualisation: Perception for Design. Elsevier (2004)

# Diagram Ecologies – Diagrams as Science and Game Board

Christoph Lueder

**Abstract.** This paper will examine two 'ecologies of thought', which encompass architectural theory, history, pedagogy, and practice.

A lineage of 'scientific' diagramming originates from scientific management and the Bauhaus-inspired curriculum introduced to Harvard by Walter Gropius; it incorporates diagrams into a problem-solving methodology, and is exemplified by the 'bubble diagram'. This scientific emphasis is extended by Christopher Alexander's urban analysis introducing mathematical set theory. In general, the scientific diagram emphasizes hierarchies and logical relations; it eschews visual resemblance to the subject of its analysis.

The second, post-war, trajectory privileges the semantic and syntactic potential of the diagram, and shifts emphasis from "solving a problem" to "learning a language"; it may be best understood through the 'Nine Square Grid' design exercise introduced by John Hejduk, resonating with positions articulated by Colin Rowe, Rudolf Wittkower, and Rudolf Arnheim.

The rendezvous of both trajectories with the digital screen sparks a new typology, diagrammatic controls.

**Keywords:** Diagram, annotation, architecture, design, heuristic, scientific management, syntactic, transparency, urbanism, grid, digital, screen, interface, diagrammatic control.

## 1 The Relational Identity of Diagrams

Unlike other forms of art, diagrams usually are inscribed and annotated with text; and diagrams can also be annotations to texts, ranging from technical manuals to theoretical manifestos. Diagrams are tools for explaining (Collins English Dictionary, 2003); they in turn are explained by their annotations. Indeed, diagrams have accompanied the earliest theoretical texts on architecture, such as Vitruvius Ten Books on Architecture, which refers to nine diagrammatic images (Carpo, 2001). Those diagrams are not included in the surviving (hand copied) exemplars; instead some copies, such as the Sélestat Codex, are annotated with new diagrams, revealing a dynamic relationship between text and diagram. In a more recent example of diagrammatic re-annotation, Bernhard Hoesli appends to his translation into German of Colin Rowe and Robert Slutzky's essay Transparency: Literal and Phenomenal exercises he had asked students to undertake, intending to complement their reading of the text through making diagrammatic models and drawings. Both the fabrication of diagrams, as well as their appendage to the translated text, augments the praxis of dynamic diagrammatic annotation.

P. Cox, B. Plimmer, and P. Rodgers (Eds.): Diagrams 2012, LNAI 7352, pp. 214–232, 2012.

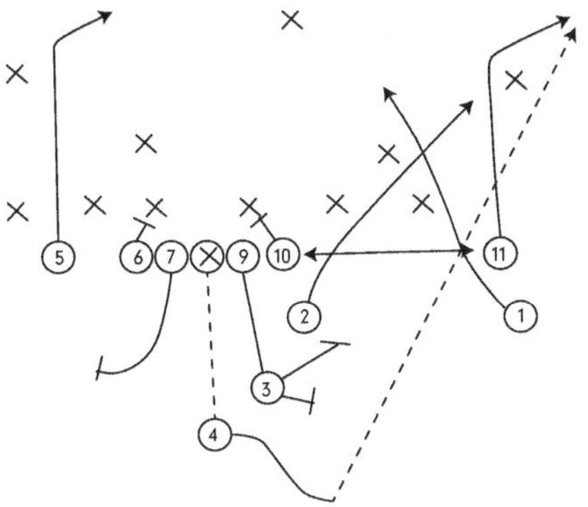

**Fig. 1.** American Football Coaches Association, Athletic Journal, 1939, Football offence diagram 11, Illinois High Schools

Diagrammatic charts can seamlessly integrate diagnosis and prognosis (or project and prognosis); other diagrams sustain parallel temporal and spatial readings, as witnessed by lines of longitude on world maps, which denote distance in space, but also mark hourly increments. Diagrams emanating from a particular discipline, such as biology, geography, dance, sport analysis (fig 1), architecture, or filmmaking can be appropriated by another discipline, transferring modes of operation and frames of reference. This often occurs in a semi-conscious, informal process.

The fundamentally relational identity of diagrams (Krauss, 1999, p. 3) places them in a supporting role, but also at pivotal, if not always acknowledged, positions within ecologies of thought. Hence, diagrams provide auspicious vantage points for describing and understanding such ecologies. Paraphrasing Anthony Vidler's statement that 'the diagram is both the instrument of thought and its mirror' (2006, p.20), it can be claimed that specific types of diagrams have become preferred instruments to specific ecologies of thought and act as their mirrors.

The science of ecology studies the relations between living organisms. The term Diagram Ecologies is appropriated to theory and used shift emphasis from away singular authors or schools toward dynamic and reciprocal links between praxis, education and theory. The 20th century has not produced a singular, unified diagram ecology; instead several approaches compete. The two most dominant diagram ecologies will be explored in detail, and at a set of scales ranging from human body, room, building, to city.

The first diagram ecology looks to science for reference; diagrams are employed as devices for solving problems, but not recognized as objects of inquiry in themselves, hence such heuristic diagrams could be described as transparent. These types of diagram historically have been associated with scientific management and Taylorism, however, their broader roots and wider reach merit attention. Bubble and flow diagrams are the most conspicuous manifestation in architecture; they remain indispensable to contemporary architecture and urbanism.

A second diagram ecology refers to the history of architecture itself, to art, and to Gestalt theory; it deems the diagram itself worth of interrogation. The nine square grid problem introduced by John Hejduk is its most conspicuous manifestation in education. Diagrams are exploited not only to solve, but also to invent and frame problems. Architecture becomes a language to be learned through diagrammatic exercises. This diagram typology resonates with characteristics of game boards. As practitioner and theorist, Peter Eisenman has long been its most vocal advocate, but its ethos persists in the work of OMA, MVRDV or SANAA, to name but a few.

This essay will argue that, while methodologies affiliated with these - potentially complementary - diagram ecologies are occasionally compounded by their users, their most prominent protagonists have conspicuously abstained from cross-references in their theoretical writing. I will attempt to assess the reach of both diagram ecologies into contemporary digital authoring environments, and the consequences of their newfound electronic proximity on the digital screen. However, a comprehensive report on the history of the diagram in the 20th century falls outside the scope of this article. The mental maps introduced by Kevin Lynch and the Deleuzian notion of diagrams acting 'as causes that are coextensive with the whole social field' (1988, p. 37), or the concomitant notion proposed by Robin Evans of the architectural plan as diagrammatic trace and description of the nature of human relationships (1997, p. 56), are fundamental to any discourse on diagrams, but will not be examined in particular.

## 2 Scientific / Heuristic Diagram and the Myth of Its Transparency

Hannes Meyer, successor to Walter Gropius as director of the Bauhaus, claimed that 'building is a biological and not an aesthetic process' (Meyer, 1930); thus also assigning a new role to the diagram within the modernist project. The implications occur at two scales: First, the human body becomes a subject of scientific diagramming, as the time and motion studies of Lillian and Frank Gilbreth and the flow diagrams of Christine Frederick and Alexander Klein most distinctly demonstrate. Second, the biological diagram of the body and its organs becomes a metaphor for the organization of buildings, exemplified by the early bubble diagrams of Le Corbusier, Percy Nobbs, Ernst Neufert, and the pronounced role of bubble diagrams at Harvard's Graduate School of Design under Walter Gropius.

Frank and Lillian Gilbreth's time and motion studies exploited photographic process and, unlike previous diagrams, no longer relied exclusively on hand and imagination of an author. Their chronocyclegraphs made routine movements visible by photographing over a long exposure the trace of an illuminated point attached to the body. The Gilbreth's advertized their research as a means to eliminate wasteful movement and thereby optimize production sequences; unlike architect's diagrams, their chronocyclegraphs and corresponding wireframe models of movements are not aimed at a spatial proposition. The architectural historian Siegfried Gideon, in his seminal book Mechanization Takes Command (1949, p. 17-30, 100-107), included the Gilbreth's chronocyclegraphs alongside chronophotographs by Etienne-Jules Marey; but even prior to that publication architects had taken note. Marey deemed his method of sequential photographic recording 'superior to all other modes of expression' as it is 'the language of the phenomena themselves' (1885, p. iv); his notion of the diagram affording an unbiased survey of its subject has resounded within the ecology of the scientific diagram. Scientific diagrams have accordingly been regarded by many authors to be transparent in the sense that typographers ascribe to a familiar font, which does not attract attention to itself, and therefore is transparent to the text it notates. Some skepticism appears justified as to whether chronocyclegraphs and chronophotographs can even be classified as diagrams at all, because they really are mechanical recordings of phenomena, whereas an intention to interpret and communicate should be at the basis of any diagram. On the other hand, one may equally argue that a large degree of interpretation takes place in editing such recordings, and presenting them in a particular context. However one may ultimately classify chronocyclegraphs and chronophotographs, their influence on a lineage of 'scientific' diagramming pursued by writers, scientists, and architects such as Gilbreth, Frederick, and Klein is crucial.

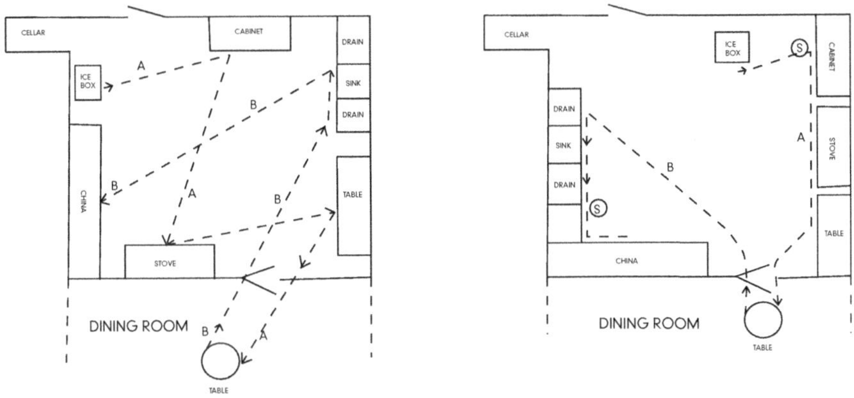


**Fig. 2.** Christine Frederick, 1913, The New Housekeeping

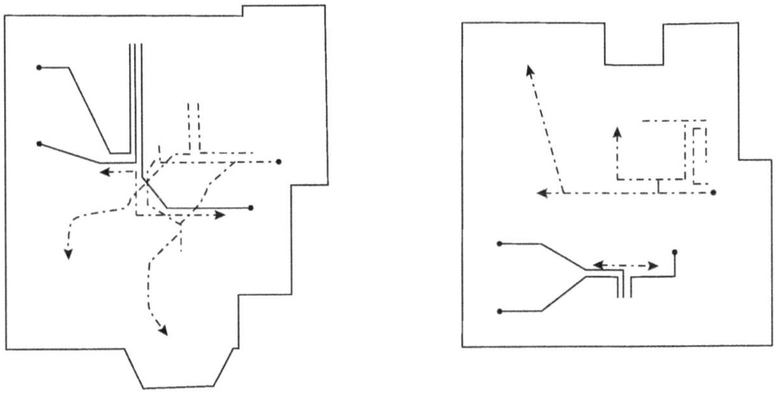

**Fig. 3.** Alexander Klein, 1927, Functional House for Frictionless Living

At the scale of the room, Christine Frederick's flow diagram of 1913 (fig 2) explains two alternative movement sequences in a kitchen while preparing an omelet, the first showing 'badly arranged equipment, which makes confused intersecting chains of steps', while the second demonstrates 'proper arrangement' and a 'simple chain of steps' (1913, p.53). The elimination of redundant steps in the process chart is of equal importance to spatial rearrangement.

At the scale of the apartment, Alexander Klein (1927, p.299) diagrams the flow of people; a 'bad example', which again features intersecting lines and numerous turnings of paths, is confronted with a 'good example', which cleanly separates daytime and nighttime paths (fig 3). Klein's traffic-way diagram was published under the title Functional Housing for Frictionless Living, reflecting his belief that diagrams provide an impartial basis for selecting the best proposal according to functional criteria. Christine Frederick's Routing diagrams comparing efficient and inefficient movement of the houseworker (1915, p. 74), (fig 4) examine the sequence in which tasks are performed; efficient as well as inefficient patterns of movement occur in a spatially identical setting.

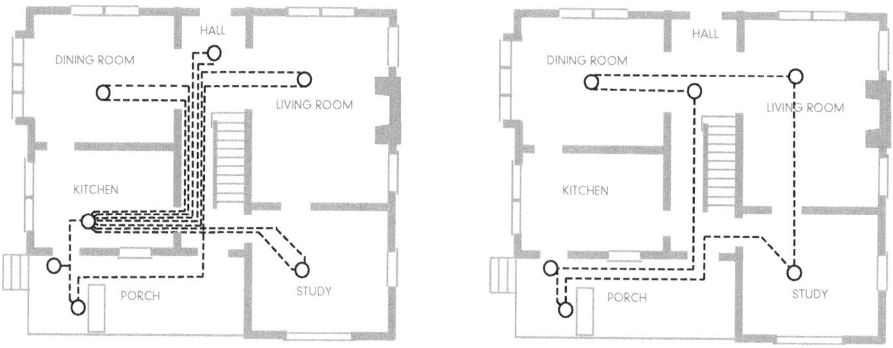


**Fig. 4.** Christine Frederick, 1919, Household Engineering

In the work of all three proponents of scientific optimization, the Gilbreth's, Frederick and Klein, it is hard to miss an obvious attention to diagrammatic beauty as well as graphic choices aspiring to elegance. Increased complexity of movement at the larger scale of the room or apartment, as opposed to the scale of the hand in the Gilbreth's chronocyclegraphs, necessitated a shift from mechanical or photographic recording to selective interpretation. Frederic omitted steps such as the washing of hands before preparation, and Klein eliminated the entrance sequence from his nighttime trajectory. Klein points out that 'general and objective valuation was until now difficult', and claims 'the graphical method of analysis differs from former methods of plan-valuation (…), by its means the qualities of the plan can be determined in an objective and clear manner' (1931, p. 166). Today's digital methods of analysis can rely on increased computational capability to handle large quantities of data in translating reality into diagrammatic representation. Design decisions are made by architects on the basis of such models. Taken to its logical conclusion, a genetic algorithm is trusted to reach those decisions in an evolutionary and quasi-biological simulation. Hence, Klein's argument is essentially reiterated, albeit at a higher level of complexity; but it needs to be remembered that any such simulation can only draw on edited and interpreted information and is never completely transparent to its content. Indeed, Robin Evans (1997, p. 85) has pointed out that the house for frictionless living articulates a cultural and social agenda far more than it presents an objective and clear evaluation.

Despite their declared scientific objective of optimization and rationalization, the graphic treatment of Frederick and Klein's diagrams, and a certain – unacknowledged - resemblance to dance notation, suggests another hidden agenda, aiming for prescriptive choreography of movement amalgamated with graphic celebration of efficiency.

Flow diagrams start with an analysis of movement at the scale of the human hand; at the scale of the apartment or house they point to a second biological analogy, to systems of circulation. Architectural bubble diagrams suggest a third biological reference, correlating programs and spaces with biological organs.

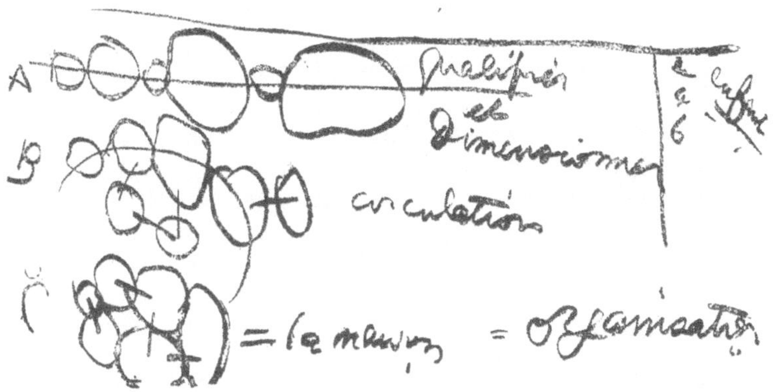


**Fig. 5.** Le Corbusier, 1929, Bubble diagram drawn during a lecture in Buenos Aires

An early bubble diagram (fig 5), by Le Corbusier (1930, p. 223) shows a sequence of evolutionary steps starting with a linear arrangement (A), which then is transformed into a more complex network labeled 'circulation (B)', and finally ends up as a densely packed arrangement of bubbles annotated with '= la maison (C)'. Bubble diagrams have been described earlier by other authors, such as Philip Sawyer (1923, p. 263), who maintained that 'once the problem is so stated it is hard to go wrong, (…) because the diagrams show the best solutions that can be evolved, and every move is toward that rather than being a mere attempt to install (…) a development of the existing arrangement.' Percy Nobbs (1937, p. 255), recommends laying out the components of the program, and then to 'draw in lines and loops among the circles to show the connexions', in a next step 'the circles should be regrouped to simplify the diagram' (a topological transformation), resulting in an arrangement where closely related programs become adjacent, and certain spaces, such as the hall and pantry reveal themselves as nodes in the network (fig 6). Nobbs observes that 'the plan at this stage is like a medical student's dissected animal, with all its organs neatly spilled out and some of the connections stretched, but none broken.'

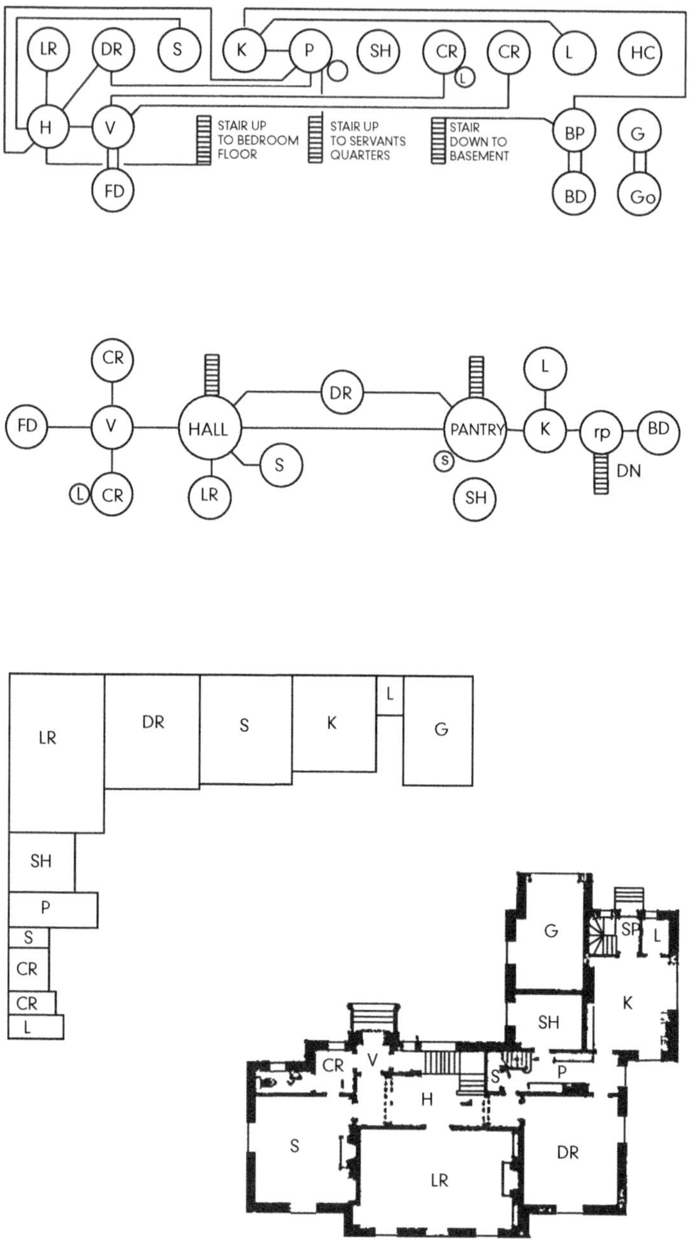


**Fig. 6.** Percy Nobbs, 1937, Successive steps in solving a problem of planning

Percy Nobbs suggests the use of bubble diagrams in a design process which is not only ripe with biological connotations, but also implies an understanding of the diagram as a 'transparent' device through which the best solution, the evolution of the 'fittest organism' can be simulated, apparently independent of the architectural language to be employed in the materialization of the project. Notwithstanding the obvious differences between flow and bubble diagrams, between Alexander Klein's agenda and that of Percy Nobbs, their approaches are comparable inasmuch as they both view the diagram as a quasi-scientific tool, a neutral device independent of architectural preferences and style.

The propagation of the bubble diagram by Gropius during his tenure at the Harvard Graduate School of Design helped to incorporate it into dominant design methodology; and turned it into a conspicuous symbol of the functionalist agenda. Thus, the bubble diagram was often indicted alongside the Bauhaus methodology when attacked as 'too dogmatic (...) of for having inhibited individuality' (Herdeg, 1983, p. 3). Corbusier's labeling of his sketched bubble diagram at its final stage with '= la maison' (1930, p. 223) may at first sight indeed suggest a simplistic approach; however, elsewhere in his South American lectures, he lists five distinct acts, ('to classify, to dimension, to circulate, to compose, to proportion') which are illustrated by a much wider variety of diagrams (1930, pp. 124-135), and associated with a much more complex and layered design process. Ernst Neufert, describing his very systematic design process, took care to point out that the initial, diagrammatic phase needs to be followed by a creative process (1936, p. 34).

The attacks against Bauhaus methodology reverberate in the critique against early attempts to introduce the computer to the architectural design process. Anthony Vidler (2000, p. 16) is representative of a widely held view when he comments on 'the simplistic and often rigid models based on functional analysis proposed by design-methods theorists like Christopher Alexander in the early decades of computerization'. In this context it is worth recalling Christopher Alexander's comment that the 'machine is distinctly complementary to and not a substitute for man's creative talent' (1963, p. 166).

The cross-section of early proponents of the bubble diagram exposes an eclectic group; Sawyer, Corbusier, Nobbs, Gropius, and Neufert could hardly be any more diverse in their architectural approach and style. None of their diagrams bear direct visual resemblance to their buildings, and this, along with references to biology or other sciences, may be the defining criteria of scientific diagramming.

In that sense, aspects of the work of Christopher Alexander, and in particular his essay 'A City is not a Tree' (1965) can be classified as an augmentation of the scientific project of diagramming. Alexander draws on mathematical set theory to diagram selected cities, identifying tree-like networks (fig 7) with 'artificial' cities, planned in the 20th century, as opposed to semi-lattice structures (fig. 7) observed in the social structure of existing cities. Alexander argues that planners find it easier to conceptualize hierarchical, tree-like arrangements, and thus, in a 'self-conscious' process of urban design, such organizations come to dominate, as opposed to a natural process of design, in which urban patterns are incrementally extended following received knowledge, which can result in semi-lattice type structures.

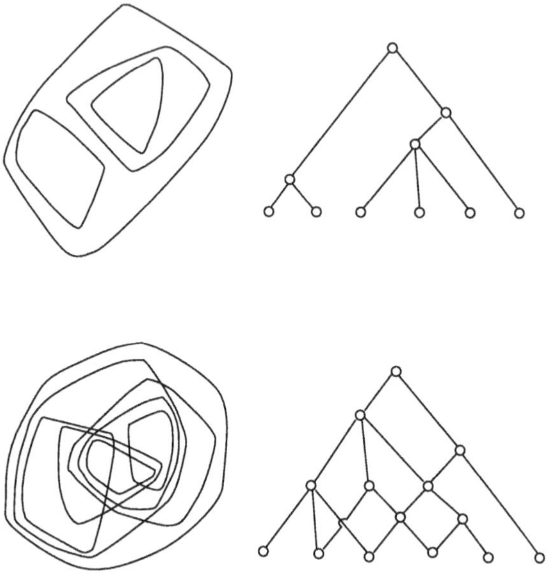

**Fig. 7.** Christopher Alexander, 1965, Tree and Venn diagram of tree
Christopher Alexander, 1965, Semi-lattice and Venn diagram of semi-lattice

Here, Alexander verges on assigning agency to the diagram and to its visual representation. However, this is not his argument. Alexander's criticism is not directed at the diagram, but rather at habits and cognitive limitations of planners. The diagram merely serves as a device to diagnose and explain these constraints. Alexander's diagrams do not attempt to visually resemble, but rather to translate urban structure into the language of mathematics. His Venn diagrams adapt the typology of the bubble diagram to an urban argument, broadening the repertoire of the scientific diagram. However, by including a reproduction of a painting by Simon Nicholson, which is used to exemplify spatial overlap and ambiguity, he paraphrases in terms of urban structure (but not acknowledges), the argument on architectural space Colin Rowe and Robert Slutzky had declared two years earlier in Transparency, Literal and Phenomenal (1963). Rowe and Slutzky's assertion will be explored in the context of the second diagram ecology, the ecology of the syntactic diagram.

## 3 Syntactic Diagram and Game Board

In a pointed remark, Peter Eisenman states that 'many see the diagram's initial emergence in Rudolf Wittkower's use of the nine square grid in the late 1940's to describe Palladian villas' (Eisenman, 1999), thereby purposefully negating the legacy of the functionalist, scientific diagram.

Wittkower's diagrams (1952, p. 27), (fig 8) eliminate any indication of function or links between rooms, and, beyond staircases, do not give any indication of movement.

Unlike flow or bubble diagrams, they are not concerned with assemblage of program and activities to form an efficient organism. To the 11 diagrams of Palladian villas, Wittkower appends a 12th diagram, showing a rectangle subdivided into 12 fields by a symmetrical grid. Wittkower reveals the Palladian floor plans to be variations originating from a shared system of geometrical rules, suggesting that collectively they may form a single conceptual project. While he does not refer to the concept of the game board, his diagrammatic catalogue bears irrefutable visual resemblances, and suggest a latent reading of the plan diagrams as topological permutations, or positions occurring as a game proceeds. Time is no longer a function of movement through a building; instead the diagrams capture snapshots taken during a permutational game played between 1547 and 1567.

Eisenman (1999, p. 27) recounts that the academic pedigree of Wittkower's diagram 'continued to develop in the form of the nine square problem (which) was seen as an antidote to the bubble diagramming of the Bauhaus functionalism rampant at Harvard (…) and to the parti of the French academy.' John Hejduk and others developed the nine square grid problem at the University of Austin in 1954 (Caragonne, 1991, p. 190-195). Students were given a grid onto which architectural elements and fragments could be placed. The exercise was underpinned by Colin Rowe's The Mathematics of the Ideal Villa (1947), which extends the premises of Wittkower's diagrammatic comparison to encompass Le Corbusier's Villa Stein at Garches, asserting dialectic of ideal diagram (paradigm) against its elaboration as a specific plan (program). This process of negotiating ideal against specific is re-enacted by the student on the nine square board of Hejduk's exercise.

The terms of the exercise were further refined by Transparency, Literal and Phenomenal (Rowe and Slutzky, 1963), written in 1955-56, while both Rowe and Slutzky were teaching at Austin. Rowe and Slutzky argue that the material definition of transparency, i.e. seeing one object through another, overlapping object, needs to be complemented by a spatial definition, manifest in a person inhabiting two overlapping spaces simultaneously. Their architectural examples work with a variety of elements, which have the ability to imply spatial boundaries, such as walls, parapets, facades, and floor surfaces. These elements usually are fragmented and operate in conjunction with imaginary planes to partially enclose overlapping spaces; they are bound by and organized on an underlying grid. Thus, the nine square grid problem can act as a laboratory table on which experiments are conducted, compositional strategies evaluated and spaces elaborated. However, the diagram ceases to function as a transparent device through which to study alternatives, and then identify the best proposal according to functional criteria; instead it provides an idealized starting point for an educational design process. Architecture is defined through syntactical relationships between elements, and perceived as a language to be learned, no longer as a functional problem to be solved. Indeed, the nine square grid problem prepares the ground for a continuing lineage of diagrammatic narratives, which are invented in order to motivate idiosyncratic architectural vocabularies and languages, as evidenced in the work of OMA or MRDV.

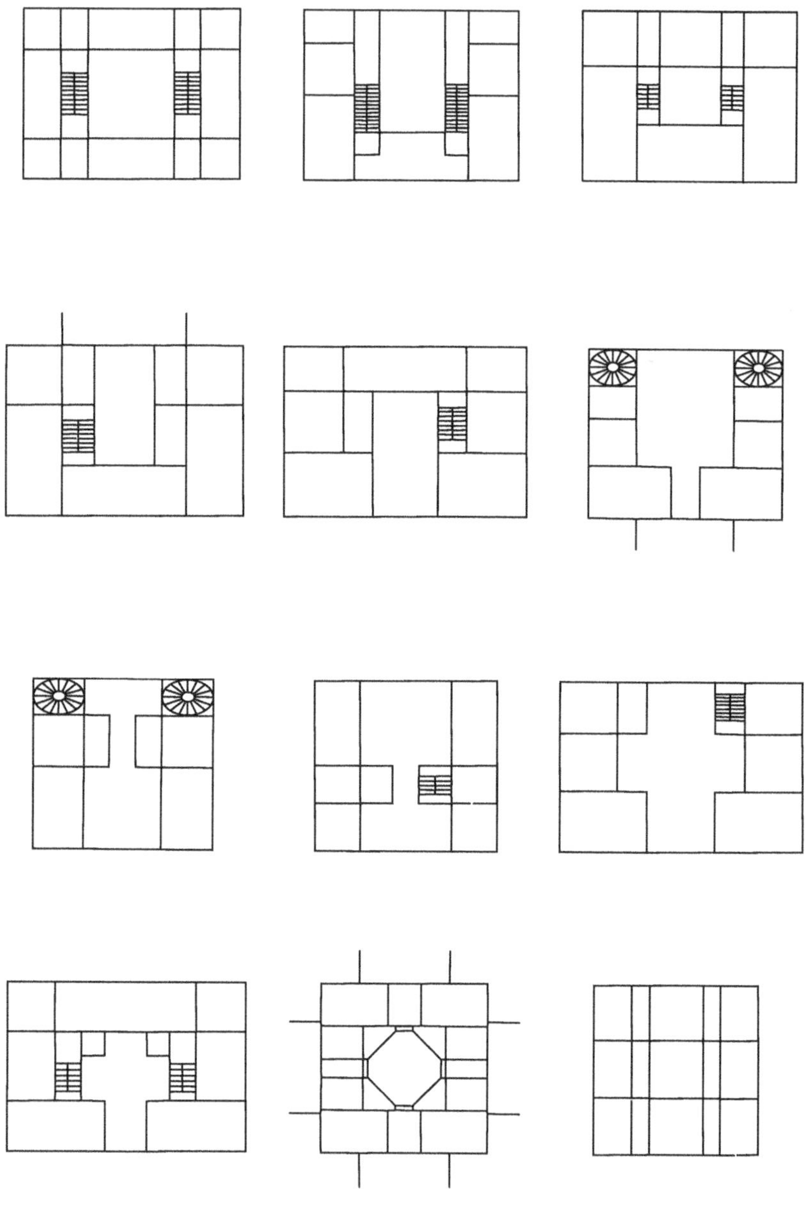

**Fig. 8.** Rudolf Wittkower, 1952, Palladian Villa Types

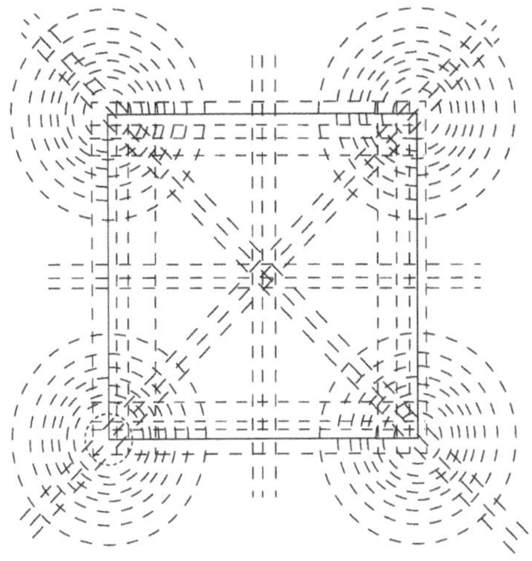

**Fig. 9.** Rudolf Arnheim, 1954, Structural map of the perception of a square

The agenda of the nine square grid extends beyond syntactical analysis of experienced space through diagrams; in fact it applies the same analytical methodology to the diagram itself. This second level of analysis, turning the formerly transparent diagram into an object of experimental and theoretical inquiry, is driven by the publication of Art and Visual Perception by Rudolf Arnheim (1954). Arnheim interrogates drawings and diagrams (fig 9) much as Rowe examines buildings, investigating what he defines as 'perceptual forces' acting on diagrams, which determine the 'percepts', which we form when we see.

Alexander Caragonne (1991, p. xvii - xviii) offers a student perspective on Rowe and Hejduk's teaching methodology: 'As I continued to sketch, tracing overlay after overlay gradually I became aware of a curious phenomenon. If I could somehow manage to align certain key walls, even columns, (…) in place of a courtyard flanked by two galleries, another, grander space, richer and more complex, yet allowing each to retain its identity, would miraculously appear. (…) For vocabulary or no, I had begun to "see" architectural space for the first time.' Caragonne's sketches and diagrams become prerequisite and enhancement to seeing, and cease to be transparent devices dedicated to impartial scientific analysis. The diagram, which Alexander Klein trusted to guard against ambiguity, and to impartially identify the single best solution, has, under Rowe and Hejduk inverted its ambition and now celebrates spatial ambiguity.

The close biographical links between Wittkower, Rowe, Hejduk and Eisenman, and the rather specific nature of the method of diagrammatic analysis that developed from unique interaction between writing and teaching, at first suggests that their approach might be more aptly described as a school of thought. However, the impact of their conceptions has sparked an active ecology of ideas sustained by the students who graduated from their programs at Austin, Cornell, the Cooper Union and many other universities.

This claim is substantiated by the argument Albert Pope, in his book Ladders (1997), elicits from his urban analysis of the postwar American city. Pope compares what he describes as a centrifugal grid emanating from city to surrounding field of landscape (beautifully illustrated with an image of a street sign announcing '339 Avenue' in the middle of nowhere), to a centripetal grid looking inwards, which can be found in fragments of grids cut off from their surroundings, e.g. in malls or some inner city districts (fig 10).

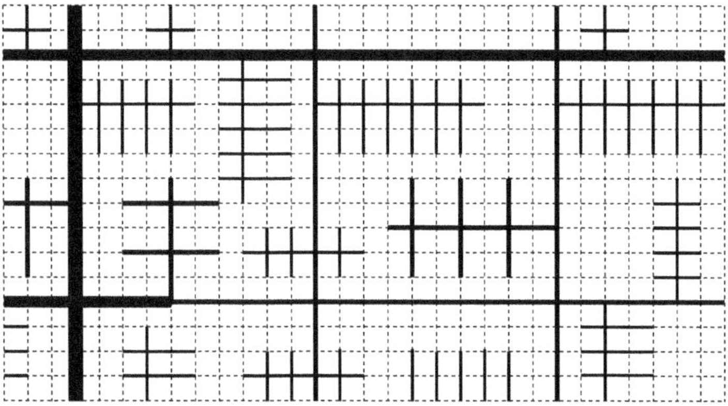

**Fig. 10.** Albert Pope, 1997, The Superblock

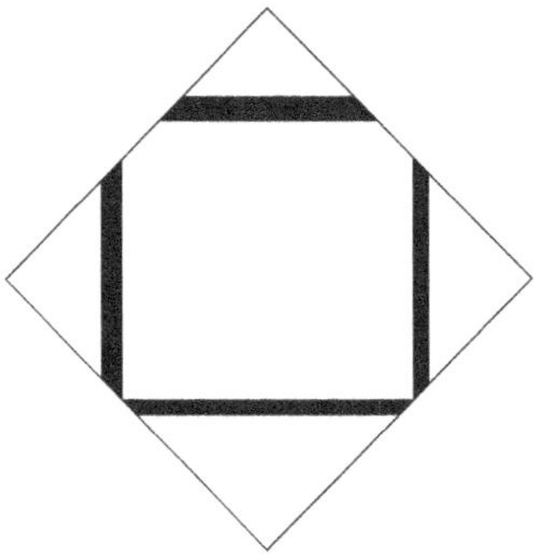

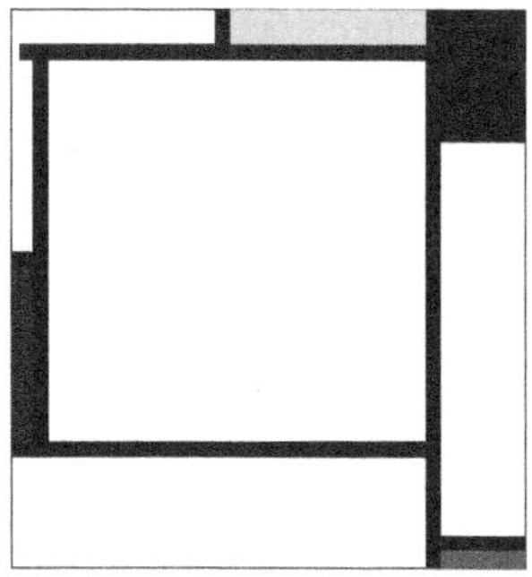

**Fig. 11** and **Fig. 12.** Piet Mondrian, Composition IA, 1930 and Composition 2, 1922

Pope introduces his discourse on urban space with a reference to Rosalind Krauss' essay Grids (1979, p. 9-22) which explores a subtle difference in how Piet Mondrian interacts with the edges of his canvas (fig 11, 12). In one set of Mondrian paintings, black lines seem to extend beyond the canvas, suggesting an infinite grid. In some cases Mondrian has reinforced this centrifugal perception of the grid by rotating

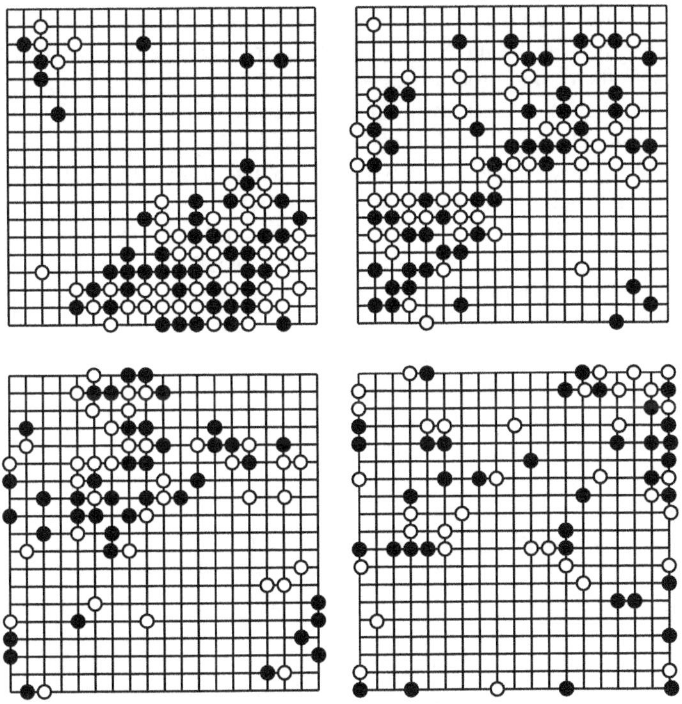

**Fig. 13.** Hoshino Toshi and Yanabe Toshiro, 1950, longest recorded game of Go

canvas boundaries by 45°. Conversely, in another set of Mondrian Paintings, a centripetal reading is suggested by grid lines stopping just shy of the edge of the canvas. The simultaneous presence of both readings is described by Krauss as 'cheerfully schizophrenic', and finds its equivalent in a similarly ambiguous state of the urban grid. By introducing his diagnosis of an urban condition with a reference to perceptual forces acting on Mondrian's paintings, Pope expands the scope of syntactic analysis to the scale of the city, while simultaneously extrapolating the terms of the nine square grid exercise, which are predicated on a bounded surface, to an exploration of the grid itself and to an examination of its boundaries. The diagram has indeed become 'the matter of architecture' (Somol, 1999, p. 7). Gridded cities figure prominently in both Alexander's and Pope's urban diagnosis, yet their conclusions could not deviate more, due to their opposing approaches to diagramming, if nothing else. By including an illustration of the game of Go (fig 13), Ladders explicates the references to the game board, which are latent, but remain uncommented, in Wittkower's Palladian diagrams and Hejduk's nine square grid problem.

## 4 Scientific and Syntactic Diagram in Digital Context

The relationships between the scientific and syntactic diagram ecologies, and also those of their most articulate protagonists, are complex. The nine square grid exercise, and the curriculum at the University of Austin it was situated in, 'was seen as an antidote to the bubble diagramming of the Bauhaus functionalism rampant at Harvard' (Eisenman, 1999, p. 27). Nevertheless, bubble diagrams are used by Bernard Hoesli at Austin to explain the design process (Caragonne, 1991, p. 85), Hoesli suggests their use in a – rather functionalist - 'pre-drawing phase of design' (Caragonne, 1991, p. 102), and both two-dimensional and three-dimensional bubble diagrams made by students are documented (Caragonne, 1991, p. 93, 104), suggesting a duplicity of polemical critique directed against the bubble diagram's dominance, which is confuted by its pragmatic usage. In respect to Christopher Alexander's bubble and tree diagrams, the position is inverted to an equally schizophrenic stance of polemical embrace versus negation of discourse. R. E. Somol bookends his seminal introduction to Diagram Diaries with demonstrative acknowledgment of Alexander's contributions, but sidesteps any meaningful engagement, echoing the similarly double-edged attitude of his mentor Peter Eisenman. The remarkable lack of a genuine discourse between both diagrammatic approaches again manifests itself during the legendary debate between Christopher Alexander and Peter Eisenman at Harvard University (Alexander and Eisenman, 1983), when Alexander repeatedly and ostentatiously pleads ignorance to references brought up by Eisenman.

At first, this divide appears reiterated in the digital realm. On the one hand, flow and bubble diagrams find an obvious corollary in equivalent diagram typologies used to optimize computer code; their biological connotation is mirrored by the use of genetic algorithms for scripting and programming. On the other hand, the bounded surface of the nine square grid, set up as a field of syntactic manipulation, visually resembles and operationally corresponds to the digital screen, which becomes a scene of operations, which sometimes reference game board metaphors. However, a closer look suggests that such an apparent divide between code and screen is too simplistic; it reveals that bubble and flow diagrams have migrated to the digital screen; indeed they are the sole interface metaphor used by the audio editing program MaxMSP (fig 14).

As diagrams establish themselves in computing, their most significant impact may be invested in the emergence of diagrammatic controls on the digital screen, further compounded by the convergence of authoring environments, which have originated in a wide range of creative disciplines, filmmaking, animation, architecture, graphic design, and even navigation and game design, towards a singular language of shared diagrammatic controls. Within the extensive inventory of those controls, tree-like hierarchical representations (Windows explorer, scene explorers, and digital effect libraries) sit alongside graphs mapping the parameters of motion, undo-redo history palettes, and the ubiquitous digital timeline, which stands out amongst the multitude of diagrammatic conceptualizations mapping time onto space. On the digital screen, the dual legacies of the scientific and syntactic diagram are only separated by a single mouse click. Hierarchical parent-child relations represented in a tree-like folder structure (Cinema 4D, fig 15), procedural controls storing geometric information as an elaborate narrative of editable transformation (Houdini) or the aforementioned rhizomatic networks of objects (MaxMSP) coexist within a single digital authoring environment or even a single software package. In fact, within this heterogeneous set, controls can act as diagrammatic annotations to each other; changes are updated electronically and instantly, in vastly accelerated and automated reenactment of the process of annotation and re-annotation known since Vitruvius Ten Books on Architecture, albeit written text no longer partakes.

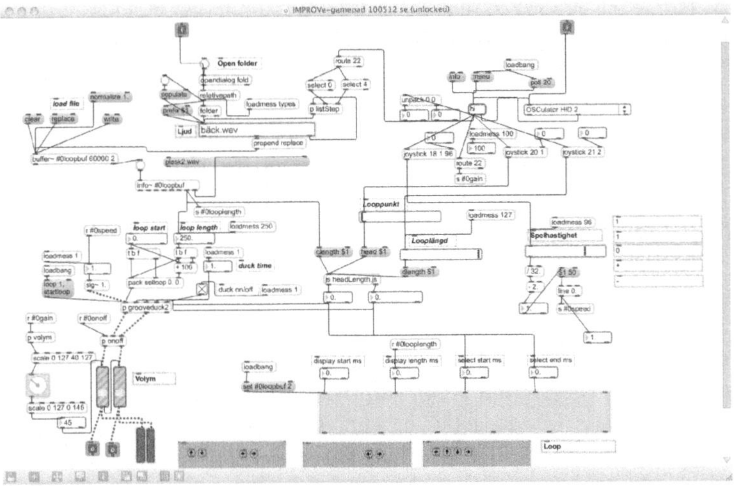


**Fig. 14.** MaxMSP audio editing software, 2011, screenshot

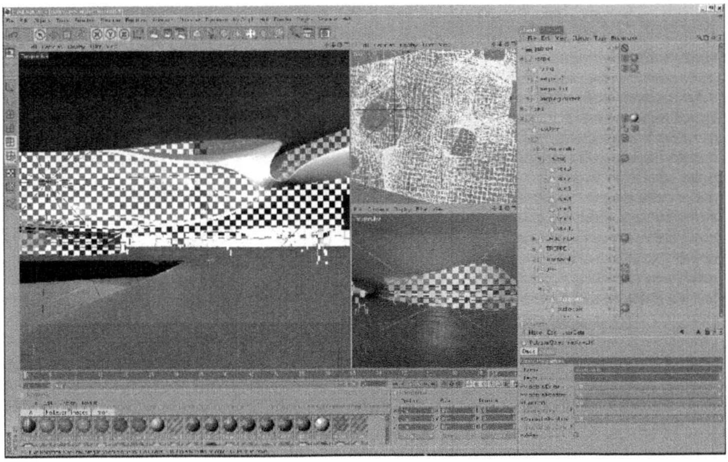

**Fig. 15.** Cinema 4d Animation Software, 2011, screenshot

Despite the omnipresence of diagrammatic interfaces controlling and thereby conceptualizing time and space on screen, and the highly specialized and original options of manipulation they afford their users, the myth of the transparent diagram reappears in a new guise; the diagrammatic metaphors that travel between disciplines alongside these controls, changing the way designers and architects think about time and space in their work, rarely are registered and interrogated. A shift in the spatiality of architect to drawing, from the horizontal surface of the drawing board to a chasm between vertical screen of output and horizontal keyboard and trackpad of input, is similarly underappreciated; it has been examined elsewhere (Lueder, 2011).

Sherry Turkle has remarked that 'the culture of simulation includes a new emphasis on visualization and the development of intuition through the manipulation of virtual objects' (1996, p. 52). In that sense, diagrammatic controls on screen are 'objects-to-play-with' and increasingly their significance is grasped through manipulation, rather than instructed by material experience, technical manual or theoretical text. Alexander Caragonne's account of his education to 'see architectural space' through sketching (and reading) at the University of Austin comes to mind. Caragonne's contemporary revenant may now learn to 'see architectural space and time' through digital authoring environments and the metaphors carried by their diagrammatic controls. Learning to 'see architectural space' under the guidance of carefully calibrated studio exercises rooted in an elaborate body of theoretical writing is an extended process, and only its motivated student will be rewarded with augmented spatial perception. Therefore it is not surprising that, while long a dominant voice in theoretical discourse, the syntactic diagram has not been able to extend its reach beyond that discourse to a wider audience; perhaps its protagonists never intended for it to do so. By contrast, digital diagrammatic controls, such as the ubiquitous timeline, have been appropriated by a vast public as objects to play with, and thus inadvertently are becoming objects to think with for many of their users.

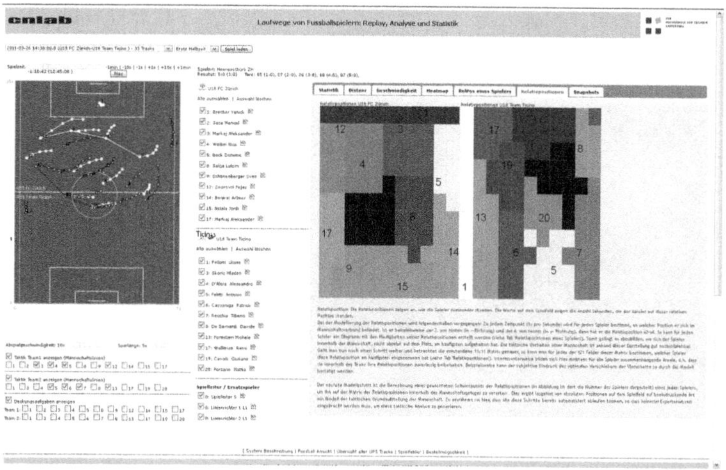


**Fig. 16.** CNLab, 2011, Laufwege von Fußballspielern, Replay, Analyse und Statistik

The avid football fan, for example, is assisted in his match analysis by digital interfaces such as the one furnished by CNLab (CNLab, 2011), (fig 16), which displays, and dynamically links to each other, no less than eleven diagrammatic representations of the match, ranging from animated flow diagram, timeline, statistical table, distance-time graph for each individual player, heatmap, positional matrix, relative positioning, to diagrammatic snapshots. Such interfaces may still be atypical, but they portend modes of conceptualizing dynamic links between time and space which architecture – and theory - cannot afford to ignore.

## References


Alexander, C., Chermayeff, S.: Community and Privacy. Toward a new architecture of humanism. Doubleday & Co., Garden City (1963)

Alexander, C.: A City is not a Tree. Architectural Forum 122(1&2) (1965); reprinted in: Thackara, J. (ed.) Design After Modernism: Beyond the Object, pp. 67–84. Thames and Hudson, London (1988)

Alexander, C., Eisenman, P.: Contrasting Concepts of Harmony in Architecture. Lotus International 40, 60–68 (1983)

Arnheim, R.: Art and Visual Perception: A Psychology of the Creative Eye. University of California Press, Berkeley and Los Angeles (1954)

Caragonne, A.: The Texas Rangers: Notes from the Architectural Underground. MIT Press, Cambridge (1991)

Carpo, M.: Architecture in the Age of Printing. MIT Press, Cambridge (2001); for a list and discussion of the Vitruvian diagrams

Collins English Dictionary, A diagram is defined as 'a sketch, outline, or plan demonstrating the form or workings of something'. HarperCollins Publishers (2003)

Deleuze, G.: Foucault. Athlone, London (1988)

Eisenman, P.: Diagram: An Original Scene of Writing. In: Eisenman, P. (ed.) Diagram Diaries. Thames and Hudson, London (1999)

Frederick, C.: The new housekeeping: efficiency studies in home management. Doubleday, Pape & Co., Garden City (1913)

Frederick, C.: Household engineering. American School of Home Economics, Chicago (1913)

Evans, R.: Figures, Doors, and Passages. In: Evans, R. (ed.) Translations from Drawing to Building and Other Essays, p. 56. Architectural Association, London (1997)

Giedion, S.: Mechanization Takes Command. Oxford University Press, Oxford (1948)

Herdeg, K.: The Decorated Diagram: Harvard Architecture and the Failure of the Bauhaus Legacy. MIT Press, Cambridge (1983)

HSR Hochschule für Technik Rapperswil, CNLAB (2011), `http://www.cnlab.ch/fussball/` (accessed August 20, 2011)

Klein, A.: Die Baugilde (November 1927); Reprinted as: Illustrations of German Efficiency Studies. Architectural Record, 299 (March 1929)

Klein, A.: Judging the Small House. Architectural Forum 55, 166–172 (1931)

Krauss, J.: Information at a Glance, On the History of the Diagram, OASE 48, Diagrams (1999)

Krauss, R.E.: Grids, You Say. In: Grids: Format and Image in 20th Century Art. Pace Gallery, New York (1978)

Le Corbusier: Precisions on the present state of architecture and city planning (1930); translated by Edith Schreiber Aujame. MIT Press, Cambridge (1991)

Lueder, C.: Thinking between diagram and image: the ergonomics of abstraction and imitation. Architectural Research Quarterly 15, 57–67 (2011)

Marey, E.-J.: La méthode graphique dans les sciences expérimentales et principalement en physiologie et en medicine, 3rd edn. G. Masson, Paris (1885)

Meyer, H.: My ejection from the Bauhaus. Open letter in the newspaper: Das Tagebuch (August 16, 1930); Reprinted in: Schnaidt, C., Meyer, H.: Bauten, Projekte und Schriften. Buildings, projects and writings, transl. D. Q. Stephenson, p. 103. Architectural Book Publishing Co., New York (1965)

Neufert, E.: Design Method, Architect's Data. Crosby Lookwood, London (1936); translated by Herz, R., et al. (1970)

Neufert, E.: Bau-Entwurf Arbeitsvorgang, Bauentwurfslehre, 3rd edn. Bauwelt Verlag, Berlin (1936)

Nobbs, P.: Design: A Treatise on the Discovery of Form. Oxford University Press, Oxford (1937)

Pai, H.: The Portfolio and the Diagram. MIT Press, Cambridge (2002)

Pope, A.: Ladders. Princeton Architectural Press, Princeton (1997)

Rowe, C.: The Mathematics of the Ideal Villa. Architectural Review (01), 101–104 (1947)

Rowe, C., Slutzky, R.: Transparency: Literal and Phenomenal. Perspecta 8, 45–54 (1963)

Sawyer, P.: The Planning of Banks. Architectural Forum 38, 263–272 (1923)

Somol, R.E.: Dummy Text, or the Diagrammatic Basis of Contemporary Architecture. In: Eisenman, P. (ed.) Diagram Diaries. Thames and Hudson, London (1999)

Turkle, S.: Life on the Screen. Weidenfeld and Nicholson, London (1996)

Vidler, A.: Diagrams of Diagrams: Architectural Abstraction and Modern Representation. Representations 72 (2000)

Vidler, A.: What is a Diagram Anyway. In: Cassara, S., Eisenman, P., Vidler, A., Kipnis, J. (eds.) Peter Eisenman: Feints. Skira, Milan (2006)

Wittkower, R.: Architectural Principles in the Age of Humanism. Warburg Institute, University of London, London (1949)

# 5 Illustrations

Fig 1: American Football Coaches Association, Athletic Journal, 1939, Football offence diagram 11, Illinois High Schools
Fig 2: Christine Frederick, 1913, The New Housekeeping
Fig 3: Alexander Klein, 1927, Functional House for Frictionless Living
Fig 4: Christine Frederick, 1919, Household Engineering
Fig 5: Le Corbusier, 1929, Bubble diagram drawn during a lecture in Buenos Aires
Fig 6: Percy Nobbs, 1937, Successive steps in solving a problem of planning
Fig 7: Christopher Alexander, 1965, Tree and Venn diagram of tree, Semi-lattice and Venn diagram of semi-lattice
Fig 8: Rudolf Wittkower, 1952, Palladian Villa Types
Fig 9: Rudolf Arnheim, 1954, Structural map of the perception of a square
Fig 10: Albert Pope, 1997, The Superblock
Fig 11: Piet Mondrian, Composition IA, 1930
Fig 12: Piet Mondrian, Composition 2, 1922
Fig 13: Hoshino Toshi and Yanabe Toshiro, 1950, Longest recorded Go game
Fig 14: MaxMSP audio editing software, 2011, screenshot
Fig 15: Cinema 4d Animation Software, 2011, screenshot
Fig 16: CNLab, 2011, Laufwege von Fußballspielern, Replay, Analyse und Statistik, screenshot

Fig 1–4, 6-13 are redrawn by the author.

# Dynamic Diagrams: A Composition Alternative

Richard Lowe[1] and Jean-Michel Boucheix[2]

[1] Curtin University, Australia
r.k.lowe@curtin.edu.au
[2] University of Burgundy, France
Jean-Michel.Boucheix@u-bourgogne.fr

**Abstract.** A major problem that learners face in comprehending animated diagrams is in decomposing the presented information into a form that furnishes appropriate raw material for building high quality mental models. This paper proposes an alternative to existing design approaches that shifts the prime focus from the nature of the external representation to the internal composition activity learners engage in during mental model construction.

**Keywords:** dynamic diagrams, decomposition, composition, comprehension.

## 1 Introduction

Current standard practice with explanatory animations is to portray the subject matter as a coherent whole, that is, as a dynamic external representation of a fully operational system. However, learners may run into problems relatively early in their processing of this conventional type of animation [1]. In particular, learners can fail to internalize key high relevance aspects from the continuous flux of information that confronts them in these rich depictions. As a consequence, the quality of the mental model that learners ultimately construct on the basis of their animation processing may be compromised. The efforts of those who design animations (typically a subject matter expert and a graphic designer) are primarily concerned with the attributes of the animation as an *external representation*, rather than with the learner's task of constructing an *internal representation* of the depicted content. Further, the prevailing idea in animation design that the external representation should at all costs be faithful to its referent is sometimes erroneously equated with a need to depict the subject matter in a realistic way [2,3]. However, realism can actually make information harder for learners to process, particularly where the depiction portrays complex unfamiliar content. Under such circumstances, there is a fundamental mismatch between the characteristics of the representation and the constraints of the human visual processing system.

Recent efforts to improve the effectiveness of explanatory animations typically involve approaches such as adding various types of cues, dividing the animation into segments, making the animation user controllable, adding accompanying learning activities (e.g. self-explanation or prediction), and giving strategy training [4,5,6,7,8,].

P. Cox, B. Plimmer, and P. Rodgers (Eds.): Diagrams 2012, LNAI 7352, pp. 233–240, 2012.

However, these interventions are ancillary to the animation itself in that they do not alter the core assumptions of its design.

It is our contention that in order to make further progress on improving the effectiveness of animated explanations, a re-conceptualization of how animations present information to learners is needed. There is a fundamental incompatibility between the activities of (a) following an animation's ever-changing information stream, and (b) simultaneously processing that information effectively. Rather than requiring learners to adapt to the presentation, perhaps the way that animations are designed needs to better fit the way learners actually process dynamic information. The Animation Processing Model (APM) [9] offers a basis for developing such an alternative. It was devised with the aim of providing foundations for a more principled and coherent approach to the design of animations. This five phase model targets the psychological processes learners use to deal with conventional animations that present complex, unfamiliar dynamic subject matter (Figure 1). It incorporates perceptual aspects of animation processing that have hitherto been neglected by other more cognitively-oriented accounts [e.g., 10]. Further, it places special emphasis on the psychological effects of the visuospatial and dynamic character of animations and how they affect the learner's construction of a mental model [11,12].

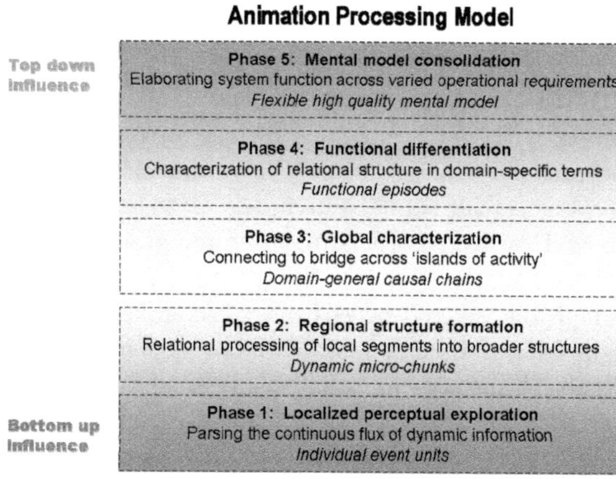


**Fig. 1.** Phases of the Animation Processing Model indicating the role that both bottom-up and top-down influences have in the construction of a mental model from an animated presentation

The Animation Processing Model has its origins in characteristics of the fundamental perceptual and cognitive processing capacities that govern operation of the human visual system when it deals with animations (see [9] for a detailed account). Limitations in the capacity of this system mean that we process all but the simplest of animations in an incremental and cumulative fashion rather than all at once. Eye tracking research shows that learners gather information from an animation discontinuously via a series of individual fixations separated by saccades, not through smooth pursuit of moving entities in the display [13, 14].

According to the APM, phase 1 processing involves the learner decomposing the animation's unbroken information flow into discrete event units (i.e., entities plus their associated behaviours). During this decomposition, capacity limits on human perception and competition for attention mean that learners extract subsets of the information available in the display whilst neglecting other aspects. Unfortunately, the information that they attend to then extract under these circumstances is not necessarily what is most relevant to the task at hand. Instead, learners may by-pass crucial information in favour of material that is more conspicuous but less relevant. As a result of this sub-optimal breakdown of the animation, the mental models they build from the extracted material may not be of high quality. The question then arises as to whether the considerable problems learners have in obtaining appropriate raw material for building high quality mental models could be circumvented by changing the way information is presented.

## 2 Decomposition versus Composition

Humans readily decompose familiar everyday experience by partitioning it into coherent objects, events and scenes that are the basis for our understanding of the world around us [15]. However, with an animation that depicts complex unfamiliar content, decomposition not only tends to be extremely demanding for the learner, but can also fail to produce the required content understandings. Given the constraints on human information processing capacity, it would make sense to do everything possible to help learners devote their maximum processing capacity to composition. This cannot happen if learners must also allocate their scarce perceptual and cognitive resources to decomposition, a process that can be regarded as peripheral rather than central to mental model construction. Learner implemented decomposition is not only inefficient, but it is also likely to produce unsatisfactory outcomes.

We propose that learners be relieved of the burden of having to decompose animations in order to supply themselves with the components from which to compose their mental models. This would involve abandoning the current reliance on conventional ('comprehensive') animations. Our proposed alternative approach is to feed learners far smaller and more constrained pre-selected subsets of the entire information corpus that they need to internalize. We will use the term *relation sets* to describe these selections, indicating that they will be connected by means of various relationships in order to build the mental model. Relation sets have three fundamental constituents; (i) entities that are the actors in the depicted phenomenon, (ii) events in which those entities are engaged, and (iii) relations that bind the entities and events together into a coherent, stand-alone assemblage. In effect, relation sets are composed of two or more interlinked event units. In the next section, we consider how relation sets might be generated and illustrate this approach with an example.

## 3 Composition Components

The approach we are suggesting is directly derived from the way learning from animation is characterized in the APM. It particularly targets phase 1 processing activity in seeking to minimize the extent to which learners need to carry out their own parsing. To illustrate how the proposed relation sets might be determined, we consider the case of an animated hydraulic circuit diagram (Figure 2). This diagram portrays the circuit of a hydraulically operated system for clamping then drilling a work-piece (apply clamp, apply drill, retract drill, retract clamp). The animated version of this diagram provides a dynamic depiction of how the various components of the hydraulic circuit contribute to this overall functioning. When depicted as an entire functioning system, multiple events occur simultaneously or in close succession at separate locations. These temporal and spatial features of the system's dynamics present a challenging set of processing demands to learners if they are all depicted as they actually occur [16].

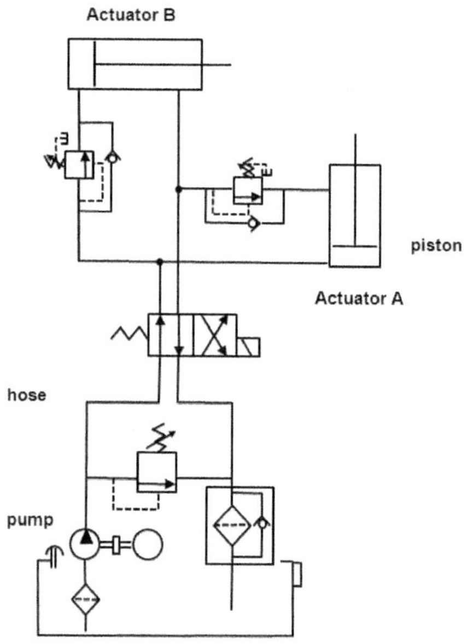


**Fig. 2.** Hydraulic circuit diagram for a system that clamps then drills work-pieces. An animated version of this diagram would show operation of dynamic components (pump, actuators, etc.)

Instead of presenting a conventional animation of the full system, the composition approach we are suggesting would present information already broken down into individual relation sets. An example will make clear the need for relation sets. The basic reason for Actuator A's piston being extended is pressurization of the hydraulic fluid by the pump. However, this cause and its effect are widely separated in the original display. Further, the symbol for the pump is far less conspicuous than that for the actuator, due its size, position and dynamics. A conventional animation would present these two aspects submerged in the rich context of the whole system's entire corpus of dynamic changes. Differences in the conspicuity of the cause (pump) and effect (actuator) together with competition for attention from the rest of the dynamic display prejudice the learner's chances of singling out this important relationship from the animation. As a result, learner decomposition is not likely to be an efficient and effective way of establishing this key aspect of the system's functionality.

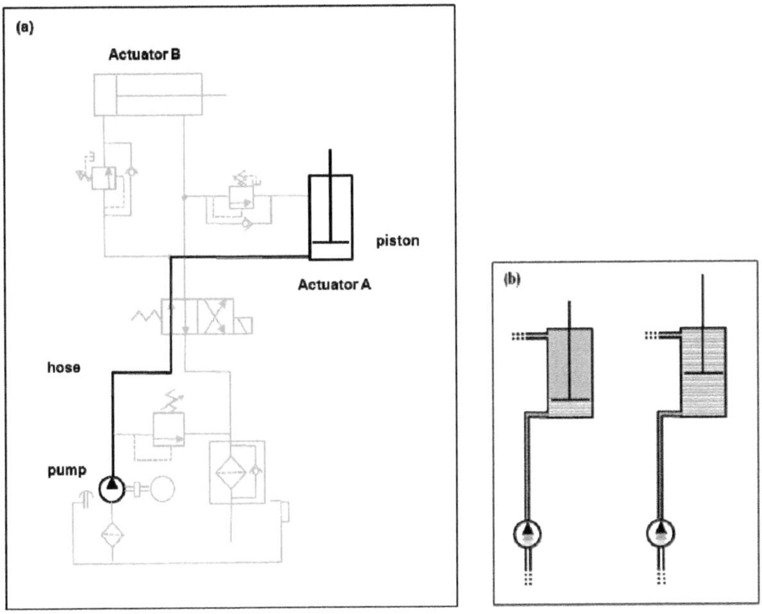


**Fig. 3.** (a) Pump-actuator causal chain in context of whole diagram (b) relation set for the same causal chain. Note pump/actuator re-alignment and simplification/shortening of hydraulic hose

By contrast, a composition approach would present only those aspects that are fundamental to this cause-effect pairing. Omitting other components of the display would both reduce the apparent visual complexity of the display and remove undesirable competition for attention. In addition, simplifying the route of the hydraulic hose that provides the explicit cause-effect connection between the pump and the actuator could allow the linkage to be more readily perceived. For example, the pump and actuator could be more directly aligned and the hydraulic hose shortened as shown in Figure 3b. This would facilitate perceptual routines such as the use of visual scanning to make the comparisons of event units required to establish the presence of a relationship. Further, it is consistent with Johnson-Laird's view that effective mental models use tokens to represent external subject matter in order to facilitate internal manipulation of that content within the constraints of our limited cognitive capacity. The simplifications applied to produce the example relation set shown in Figure 3b is a step on the path to abstraction and should facilitate the formation of such tokens.

## 4 Sequencing Composition

Appropriate timing is an important consideration in external visualizations [17]. Task relevant information needs to be presented when the viewer is ready to benefit from it. The sequence in which raw material is made available for composing a mental model is therefore likely to affect its construction. The APM characterizes learner construction of mental models from an animation as a cumulative, iterative process. A single pass through a complex, unfamiliar animation is simply insufficient to allow the learner to extract all the necessary information [5]. The sequencing used to present information via relation sets should therefore be one that facilitates their accumulation towards the desired mental model. It would also need to accommodate repeated cycles of learner activity.

If the animation is not being presented in its entirety but rather in the piecemeal fashion being suggested in the proposed compositional approach, effects such as initial competition for attention from other aspects of the display are reduced or eliminated. By using a series of relation sets, the animation designer can have many more opportunities for guiding how learners carry out mental model construction activities. These opportunities arise once it is accepted that the sequence in which information is presented does not have to correspond to its order of occurrence in a behaviorally realistic conventional animation. Rather, the order in which relation sets are presented can be manipulated so that it is as consistent as possible with efficient building of the desired mental model.

Once freed from the restriction of having to ensure behavioural realism, animation designers can introduce treatments of the content that both relieve the learner of non-core processing demands and foster cohesive knowledge structures. For example, dynamics that take place simultaneously in different parts of the display or in rapid cascades can be teased out so that split attention effects are eliminated; similar relation sets that apply to various parts of the display but at different times can be grouped spatially and temporally during presentation to indicate their commonality.

## 5 Conclusion

This paper argues that conventional animated diagrams which present a dynamic depiction of an entire functioning system subject learners to unnecessary information processing burdens. The composition approach being proposed here may seem to bear a superficial resemblance to existing 'build' animations in which graphic items depicting different aspects of the content are progressively added to the display. However, the building up of a conventional animation in this way by accretion of its parts is the antithesis of the composition approach. This is because the fundamental goal of using relation sets is to support the construction of an *internal* representation, not an *external* one.

A fundamental feature of our alternative approach is that animation design should begin with a careful consideration of the nature of the mental model that instruction is seeking to produce, not simply the characteristics of an external representation of the to-be-learned subject matter. The composition approach opens up new opportunities for using cues in more targeted and powerful ways. Instead of being used to support decomposition, cues could be used to aid the composition of information subsets into higher order structures by signaling the relations that are the basis for combining their constituent event units. The possibility of such relational cueing has previously been canvassed by Lowe and Boucheix [18].

Moving from the theoretical ideas presented in this paper to their practical implementation will require an extensive program of research and development. Fundamental issues that need to be explored in the first instance include principled approaches to designing effective relation sets, facilitating learners' linking of those sets, and helping learners to structure the internal representations they compose according to the referent content's functionality.

## References


1. Lowe, R.K., Schnotz, W., Rasch, T.: Aligning affordances of graphics with learning task requirements. Applied Cognitive Psychology, 452–459 (2010)
2. Hegarty, M.: The cognitive science of visual-spatial displays: Implications for design. Topics in Cognitive Science, 1–29 (2011)
3. Schnotz, W., Lowe, R.K.: A unified view of learning from animated and static graphics. In: Lowe, R.K., Schnotz, W. (eds.) Learning with Animation: Research Implications for Design, pp. 304–356. Cambridge University Press, New York (2008)
4. de Koning, B.B., Tabbers, H.K., Rikers, R.M.J.P., Paas, F.: Towards a framework for attention cueing in instructional animations: Guidelines for research and design. Educational Psychology Review 21, 113–140 (2009)
5. Lowe, R.K., Boucheix, J.-M.: Cueing complex animations: Does direction of attention foster learning processes? Learning and Instruction 21, 650–663 (2011)
6. Kriz, S., Hegarty, M.: Top-down and bottom-up influences on learning from animations. International Journal of Human Computer Studies 65, 911–930 (2007)
7. Amadieu, F., Mariné, C., Laimay, C.: The attention-guiding effect and cognitive load in the comprehension of animations. Computers in Human Behavior 27, 36–40 (2011)
8. Mayer, R.E., Mathais, A., Wetzell, K.: Fostering understanding of multimedia messages through pre-training: Evidence for a two-stage theory of mental model construction. Journal of Experimental Psychology: Applied 8, 147–154 (2002)
9. Lowe, R., Boucheix, J.-M.: Learning from Animated Diagrams: How Are Mental Models Built? In: Stapleton, G., Howse, J., Lee, J. (eds.) Diagrams 2008. LNCS (LNAI), vol. 5223, pp. 266–281. Springer, Heidelberg (2008)
10. Mayer, R.E.: Research-Based Principles for Learning with Animation. In: Lowe, R.K., Schnotz, W. (eds.) Learning with Animation. Research Implications for Design, pp. 30–48. Cambridge University Press, New York (2008)
11. Johnson-Laird, P.N.: How We Reason. Oxford University Press, Oxford (2006)
12. Jahn, G., Knauff, M., Johnson-Laird, P.N.: Preferred mental models in reasoning about spatial relations. Memory & Cognition 35, 2075–2086 (2007)
13. Hyönä, J.: The use of eye movements in the study of multimedia learning. Learning and Instruction 20, 172–176 (2010)
14. Jarodzka, H., Scheiter, K., Gerjets, P., Van Gog, T.: In the eyes of the beholder: How experts and novices interpret dynamic stimuli. Learning and Instruction 20, 146–154 (2010)
15. Kurby, A., Zacks, J.: Segmentation in the perception and memory of events. Trends in Cognitive Science 12, 72–79 (2007)

16. Lowe, R.K., Boucheix, J.M.: Unfair Competition: Static. Cues in Animated Graphics. Poster presented at the Comprehension of Text and Graphic, SIG 2 Meeting, Tubingen, Germany (2010)
17. Khooshabeh, P., Hegarty, M.: Inferring Cross-Sections: When Internal Visualizations Are More Important Than Properties of External Visualizations. Human Computer Interaction 25(2), 119–147 (2010)
18. Lowe, R., Boucheix, J.-M.: Supporting Relational Processing in Complex Animated Diagrams. In: Stapleton, G., Howse, J., Lee, J. (eds.) Diagrams 2008. LNCS (LNAI), vol. 5223, pp. 391–394. Springer, Heidelberg (2008)

# Diagrammatically-Driven Formal Verification of Web-Services Composition

Petros Papapanagiotou, Jacques Fleuriot, and Sean Wilson

School of Informatics
University of Edinburgh
United Kingdom
p.papapanagiotou@sms.ed.ac.uk, jdf@inf.ed.ac.uk, sean.wilson@ed.ac.uk

**Abstract.** This paper describes a diagrammatic approach to the formal verification of web-services composition. We present a set of graphical composition rules that map to proof steps in Classical Linear Logic (CLL) and can be used to drive the proof assistant HOL Light purely through interactive, diagrammatic reasoning. The end result is a verified, workflow-like diagram that provides a visual account of the composition process and of the information flow between the services making up the composite service. Our approach thus removes the need to interact directly with HOL Light and provides a mean of visualising and carrying out the whole verification process at an intuitive, yet fully rigorous, level.

## 1 Introduction

In recent work [7] using the HOL Light theorem prover [5], we showed how Classical Linear Logic (CLL) can be used to verify the correctness of web services composition interactively. In short, by specifying each service as a CLL statement, the composition process corresponds to finding a proof for a requested service, with the available services stated as assumptions. If a proof is found, this means a valid composition exists, and then a process calculus realisation of the composite service can be extracted automatically. This particular approach, provides a verified result where it is guaranteed that the inputs and outputs of each web service are matched appropriately and that all exceptions are handled systematically. This property is particularly useful in systems where a certain level of trust is required in the interaction among the involved parties and information provenance can be explicitly tracked.

Although our approach can successfully deal with complex compositions involving numerous web-services, it asks for some familiarity with CLL and, more importantly, it also requires the user to have a decent level of expertise in the use of HOL Light. While these requirements are not intractable, it does make the methodology quite demanding for those whose main interest lies in the design of web-services and their composition rather than theorem-proving. However, during the development of our framework, we realised that we often used diagrams in order to explore the various ways of composing services and illustrate

P. Cox, B. Plimmer, and P. Rodgers (Eds.): Diagrams 2012, LNAI 7352, pp. 241–255, 2012.

the information flow between them. This observation provided us with the motivation for developing a systematic diagrammatic approach that can abstract from the low-level text-based theorem proving. We first developed pen-and-paper versions of a set of diagram-transformation rules – we call these *actions* – that could be applied intuitively during the interactive composition of web services (see Section 3). The actions were then implemented as part of a Java-based GUI that drives the theorem prover HOL Light in the background during a purely graphical composition process. The overall result is a new composition approach embodied into a tool that does not require any knowledge of CLL or theorem proving on the part of the user and yet provides a formally-verified composition process.

The organisation of the paper is as follows: We describe the representation of web services in our system, both in its internal logical form and its corresponding diagrammatic notation in Section 2. We then analyze the defined composition actions through various examples and provide actual screenshots from our implemented tool as well as the corresponding formal proof trees in Section 3. A use case from a real estate domain is presented in Section 4, followed by some details of the implementation of our interface in Section 5. Finally, we briefly cover some related work in Section 6 and mention plans for future work in Section 7, before summarizing our results in Section 8.

## 2 Representation

The main focus in designing our diagrammatic representation was to provide enough expressivity to allow the specification of any web service, while hiding the complicated syntax and attached semantics of the underlying logic. In what follows, we briefly explain how web services are specified internally in our system in Section 2.1 and then describe the diagrammatic notation for these specifications in Section 2.2.

### 2.1 Classical Linear Logic Specifications

The theorem-proving part of our implementation uses the multiplicative additive fragment of propositional CLL (MALL) to describe web services. This is a relatively expressive logic that allows us to incorporate all the necessary information, including exceptions, when composing services.

We will not delve into all the details of the various operators of MALL and their formal semantics here. We instead refer the interested reader to other papers where these have been given [7,8]. For the purpose of this article, we only briefly explain how various relevant operators can be used to express web services functional properties. Note also that whenever we refer to CLL in the rest of this paper, we are implicitly referring to MALL.

We focus on representing the types of the inputs, preconditions, outputs, postconditions, and exceptions of each service. From here on in this paper, unless otherwise stated, every reference to inputs corresponds to the types of both inputs and preconditions, whereas any reference to outputs corresponds to the types of both outputs and postconditions, as these are treated similarly. In general, negated terms $(.^{\perp})$ correspond to inputs, whereas positive literals correspond to outputs or exceptions. A number of basic operations are allowed on inputs and outputs and can be expressed as follows in CLL:

- $A^{\perp} \parr B^{\perp}$ indicates the simultaneous input of $A$ and $B$ (composite input)[1].
- $A \otimes B$ indicates the simultaneous output of $A$ and $B$ (composite output).
- $A^{\perp} \& B^{\perp}$ indicates that $A$ or $B$ but not both will be given as input (optional input).
- $A \oplus B$ indicates that $A$ or $B$ but not both will be given as output (optional or exceptional output).

Based on the above we can specify any web service using logical consequence in one-sided sequent calculus CLL statements.

### 2.2 Diagrammatic Notation

We can now introduce our diagrammatic notation for capturing these CLL specifications. Each service is represented as a node with dangling edges standing for its inputs and outputs. Solid edges correspond to composite inputs/outputs whereas dashed edges correspond to optional inputs/outputs. Moreover, lighter (grey) edges correspond to information that is not explicitly handled by the receiving service but merely being buffered through it.

For example, we can represent a service $Pa$ with inputs $A$ and $B$ and outputs $X$ and $Y$ in CLL and our graphical notation as shown in Fig. 1.

$$\vdash A^{\perp}, B^{\perp}, X \otimes Y$$

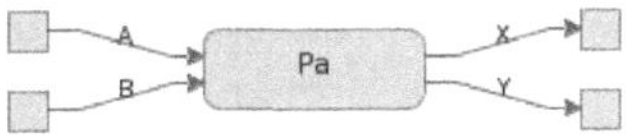


**Fig. 1.**

Similarly, a service $Pb$ with inputs $A$ and $B$ and either an output $X$ or an exception $E$ is represented in Fig. 2.

$$\vdash A^{\perp}, B^{\perp}, X \oplus E$$

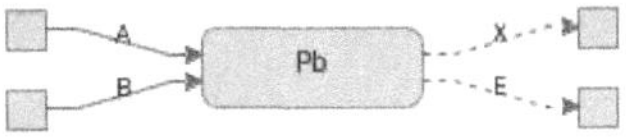


**Fig. 2.**

More complicated cases can be represented in the same way. For example, a service $Pc$ with inputs $A$ and $B$, and outputs $X$ and $Y$ that may throw an exception $E$ can be represented as shown in Fig. 3.

[1] Note that, in CLL, $\vdash A^{\perp} \parr B^{\perp}$ and $\vdash A^{\perp}, B^{\perp}$ are logically equivalent.

$\vdash A^{\perp},\ B^{\perp}\ (X \otimes Y) \oplus E$

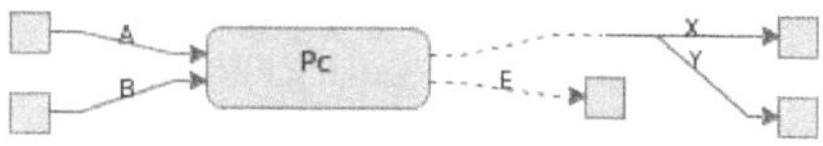


**Fig. 3.**

# 3 Proof-Based Diagrammatic Composition

As already mentioned, the core part of our composition methodology involves finding a CLL proof that the requested composite service is achievable, using the available services as assumptions. Thus, a diagrammatic interface should allow us to drive the theorem prover to find similar mechanical proofs, thereby guaranteeing the correctness of the graphical composition.

Although CLL allows for a wide range of formal statements, only a restricted subset of these actually correspond to sensible web service specifications. This makes our task more tractable (than if we had to deal with fully diagrammatic CLL proofs) as we only need to capture a small set of generic actions that a user needs when interactively composing web services in our diagrammatic interface. Each of these actions corresponds to a custom, fully-automated, HOL Light proof tactic that applies a number of CLL inference steps in order to generate the verified result at the logical level, thereby ensuring the action's correctness. By using these actions the user is relieved from the tedious process of applying a large number of primitive CLL rules since these are now subsumed by each action.

We next present three generic actions — JOIN, TENSOR, and WITH — that are sufficient to accomplish non-trivial compositions. These correspond to three generic types of compositions between two services, namely sequential, parallel, and optional, respectively. Note that our system can be extended to accommodate other custom actions, provided a corresponding HOL Light proof tactic can be constructed for each one.

## 3.1 The JOIN Action

The JOIN action is perhaps the most intuitive one when composing web services, but also the most complicated one to handle in CLL. It reflects the connection of two services in sequence, ie. where (some of) the outputs of a service are connected to the corresponding inputs of another.

We illustrate it by means of an example: Let us assume the user wants to connect the output of service $P$ to service $Q$. At the diagrammatic level, the user can accomplish the JOIN action by clicking on an edge that corresponds to one of the outputs of $P$ and then right-clicking on an edge that corresponds to a matching input of $Q$.

In order to verify the result of this action in CLL, we have to consider different cases of outputs for $P$, namely single (atomic) output, multiple (composite) outputs, and optional outputs. Each of these cases is handled by a different set of proof steps. The diagrammatic representation before and after the application

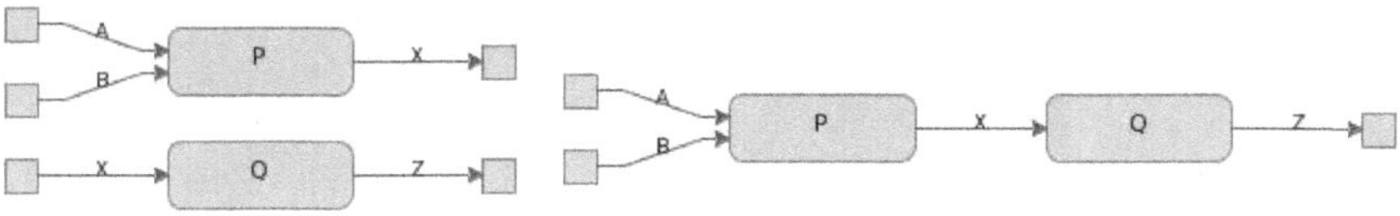


(a) Before (b) After

$$\frac{\vdash \mathbf{A}^{\perp}, \mathbf{B}^{\perp}, \mathbf{X} \quad \vdash \mathbf{X}^{\perp}, \mathbf{Z}}{\vdash \mathbf{A}^{\perp}, \mathbf{B}^{\perp}, \mathbf{Z}}\ Cut$$

(c) Proof tree

**Fig. 4.** The JOIN action between service $P$ with an atomic output and service $Q$ with a matching input

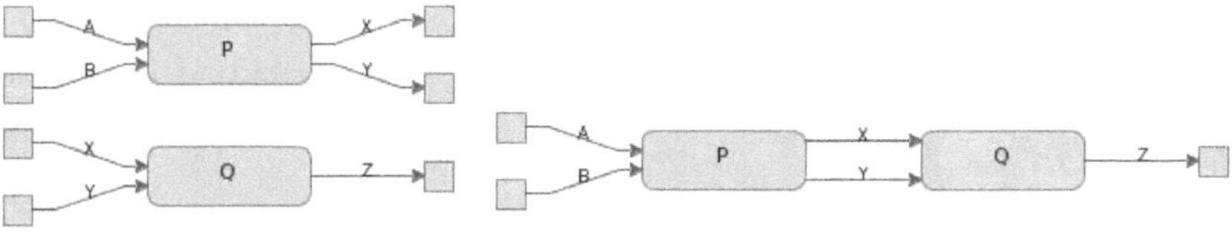


(a) Before (b) After

$$\frac{\vdash \mathbf{A}^{\perp}, \mathbf{B}^{\perp}, \mathbf{X} \otimes \mathbf{Y} \quad \dfrac{\dfrac{\vdash \mathbf{X}^{\perp}, \mathbf{Y}^{\perp}, \mathbf{Z}}{\vdash X^{\perp} \parr Y^{\perp}, Z}\ \parr}{\vdash (X \otimes Y)^{\perp}, Z}\ neg\_eq}{\vdash \mathbf{A}^{\perp}, \mathbf{B}^{\perp}, \mathbf{Z}}\ Cut$$

(c) Proof tree

**Fig. 5.** The JOIN action between service $P$ with a composite output and service $Q$ with two matching inputs

of the JOIN action, as well as the corresponding proof tree[2] for the various cases are presented by means of a few examples next.

- If the output of $P$ is atomic e.g. $X$ and corresponds to the only input of $Q$, then the JOIN action corresponds to a trivial application of the *Cut* rule in CLL (Fig. 4).
- If the output of $P$ is composite, for example $X \otimes Y$ and:
  - $Q$ has composite inputs that accept both $X$ and $Y$ as input, the action will connect $P$ and $Q$ on both $X$ and $Y$ (Fig. 5).
  - $Q$ only has $X$ as input then a connection happens with $P$ on $X$, while $Y$ is simply buffered through $Q$ (Fig. 6).
- If the output of $P$ is optional, for example $X \oplus Y$, and $Q$ has $X$ as input then $Q$ is modified so that it can accept and properly handle the optional output. If $Y$ is received by this modified version of $Q$, it is buffered through together with any unused resources (from other inputs) of $Q$ (Fig. 7).

[2] We note that the `neg_eq` step in some of these proofs is an abbreviation of the *Cut* rule used with a lemma involving linear negation.

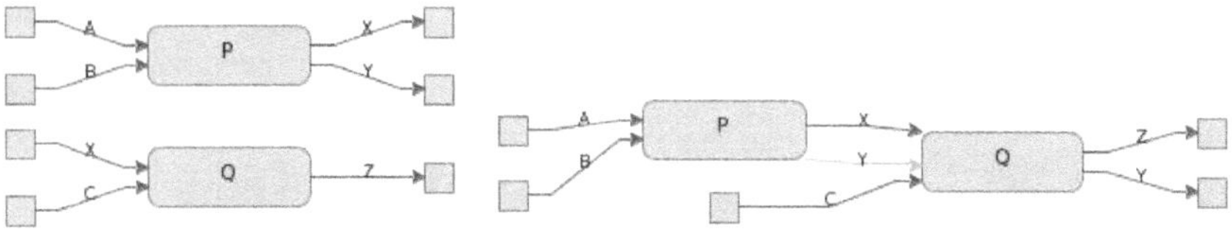


(a) Before (b) After

$$
\dfrac{\vdash \mathbf{A}^{\perp}, \mathbf{B}^{\perp}, \mathbf{X} \otimes \mathbf{Y} \quad \dfrac{\dfrac{\dfrac{\vdash \mathbf{X}^{\perp}, \mathbf{C}^{\perp}, \mathbf{Z} \quad \dfrac{}{\vdash Y^{\perp}, Y}\ Id}{\vdash X^{\perp}, C^{\perp}, Y^{\perp}, Z \otimes Y}\ \otimes}{\vdash X^{\perp} \parr Y^{\perp}, C^{\perp}, Z \otimes Y}\ \parr}{\vdash (X \otimes Y)^{\perp}, C^{\perp}, Z \otimes Y}\ neg\_eq}{\vdash \mathbf{A}^{\perp}, \mathbf{B}^{\perp}, \mathbf{C}^{\perp}, \mathbf{Z} \otimes \mathbf{Y}}\ Cut
$$

(c) Proof tree

**Fig. 6.** The JOIN action between service $P$ with a composite output and service $Q$ with only one matching input

In the special case where $Q$ has a single input $X$ and its output is $Y$, then the composition of $P$ and $Q$ will always generate $Y$ (if $P$ generates $X$ then $Q$ converts $X$ to $Y$, else $Y$ is generated directly from $P$) (Fig. 8).

– Finally, if the output of $P$ is a more complex combination of multiple composite and/or optional outputs, the above strategies are applied in a recursive, bottom-up way.

### 3.2 The TENSOR Action

The TENSOR action corresponds to the parallel composition of two services. This is particularly useful in cases where each of the components of a composite output of a service needs to be handled by a different service. Composing these handlers in parallel allows all the involved outputs to be handled simultaneously.

At the diagrammatic level, the user can accomplish the TENSOR action with the simple gesture of clicking on a service and then right-clicking on another. In the interface this creates a new outer box, representing the parallel composition of the original services into a new, composite service, whose inputs and outputs correspond to those of the original services. Note that this new box can be collapsed, thus hiding its component services, in order to make the diagram more concise.

In our CLL representation the TENSOR action can be easily verified based on the appropriate application of the tensor ($\otimes$) inference rule. The diagrammatic representation and proof tree for a simple example involving the TENSOR action on two services $P$ and $Q$ are given Fig. 9.

### 3.3 The WITH Action

The concept of the WITH action is similar to that of the TENSOR action. It corresponds to the optional composition of two services. This type of composition

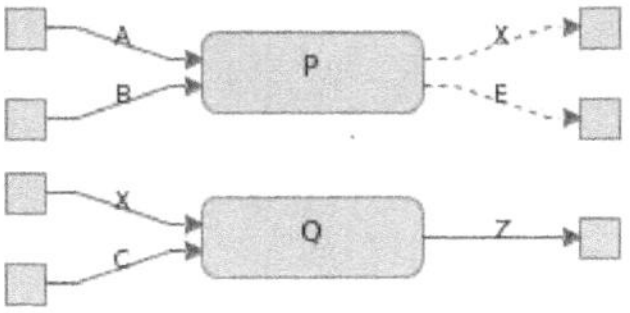


(a) Before

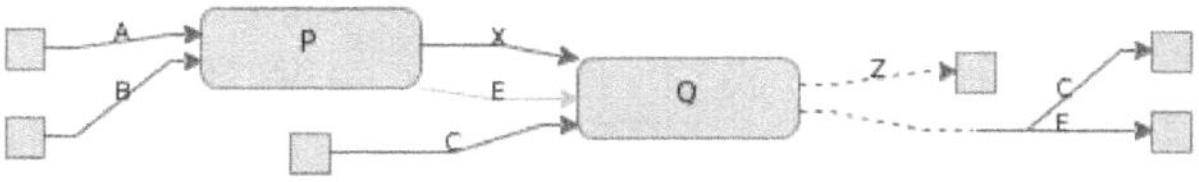


(b) After

$$\dfrac{\vdash \mathbf{A}^{\perp}, \mathbf{B}^{\perp}, \mathbf{X} \oplus \mathbf{E} \qquad \dfrac{\dfrac{\dfrac{\vdash \mathbf{X}^{\perp}, \mathbf{C}^{\perp}, \mathbf{Z}}{\vdash X^{\perp}, C^{\perp}, Z \oplus (C \otimes E)}\oplus \qquad \dfrac{\dfrac{\dfrac{}{\vdash C^{\perp}, C}Id \quad \dfrac{}{\vdash E^{\perp}, E}Id}{\vdash E^{\perp}, C^{\perp}, C \otimes E}\otimes}{\vdash E^{\perp}, C^{\perp}, Z \oplus (C \otimes E)}\oplus}{\vdash X^{\perp} \& E^{\perp}, C^{\perp}, Z \oplus (C \otimes E)}\&}{\vdash (X \oplus E)^{\perp}, C^{\perp}, Z \oplus (C \otimes E)}neg\_eq}{\vdash \mathbf{A}^{\perp}, \mathbf{B}^{\perp}, \mathbf{C}^{\perp}, \mathbf{Z} \oplus (\mathbf{C} \otimes \mathbf{E})}Cut$$

(c) Proof tree

**Fig. 7.** The JOIN action between service $P$ with an optional output and service $Q$ with a matching input

is useful in cases where each of the components of an optional output of a service needs to be handled by a different service.

For example, assume a service $S$ has an optional output $X \oplus E$ where $E$ is an exception. We want $X$ to be handled by some other service $P$ while another service $Q$ plays the role of the exception handler for exception $E$. For this to happen, we need to compose $P$ and $Q$ together using the WITH action so that $X \oplus E$ from $S$ can be dealt with in one go.

At the diagrammatic level, the user accomplishes the WITH action that will compose $P$ and $Q$ by clicking on input $X$ of $P$ and then right-clicking on input $E$ of $Q$. In the interface, similarly to the TENSOR action, a new collapsible, dashed, outer box is then created, representing a new composed service where either $P$ *or* $Q$ will be executed (contrast this to the solid outer box of the TENSOR action which, as explained before, consists of services that run in parallel).

Verifying this action in CLL is more challenging than for the TENSOR action. The main reason is that in this case we need to account for (unused resources from) other inputs. For example, if $P$ has another input $B$, then the service formed by the conditional composition of $P$ and $Q$ will also have this input. Therefore, if exception $E$ occurs and $B$ is provided by another source, $B$ will not be consumed (since $P$ will not be invoked). Thus, it needs to be buffered through together with the output of $Q$ that will handle $E$.

A more complicated example (where both $P$ and $Q$ have extra inputs $B$ and $D$ respectively) and the corresponding proof tree, as well as the diagrammatic representation are shown in Fig. 10.

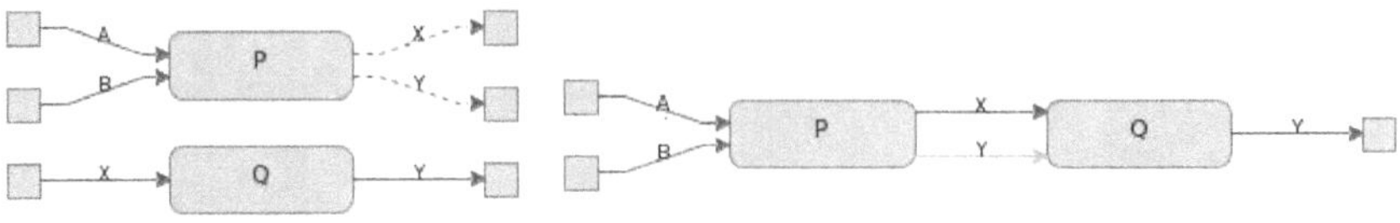


(a) Before (b) After

$$\dfrac{\vdash \mathbf{A}^{\perp},\ \mathbf{B}^{\perp},\ \mathbf{X} \oplus \mathbf{Y} \qquad \dfrac{\dfrac{\vdash \mathbf{X}^{\perp},\ \mathbf{Y} \quad \dfrac{}{\vdash Y^{\perp},\ Y}\ Id}{\vdash X^{\perp} \,\&\, Y^{\perp},\ Y}\ \&}{\vdash (X \oplus Y)^{\perp},\ Y}\ neg\_eq}{\vdash \mathbf{A}^{\perp},\ \mathbf{B}^{\perp},\ \mathbf{Y}}\ Cut$$

(c) Proof tree

**Fig. 8.** The JOIN action between service $P$ with an optional output $X \oplus Y$ and service $Q$ that only has $X$ as input and $Y$ as output

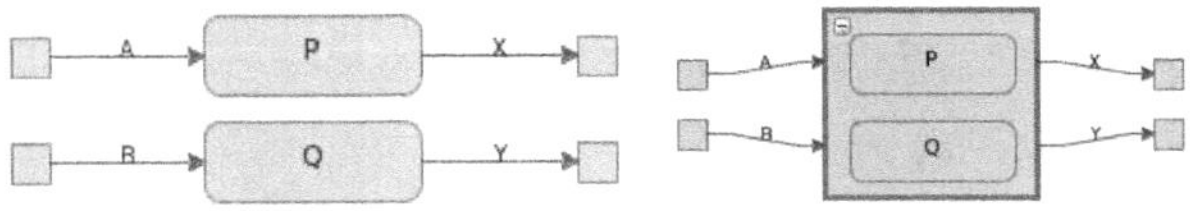


(a) Before (b) After

$$\dfrac{\vdash \mathbf{A}^{\perp},\ \mathbf{X} \quad \vdash \mathbf{B}^{\perp},\ \mathbf{Y}}{\vdash \mathbf{A}^{\perp},\ \mathbf{B}^{\perp},\ \mathbf{X} \otimes \mathbf{Y}}\ \otimes$$

(c) Proof tree

**Fig. 9.** The TENSOR action on two services $P$ and $Q$

# 4 Use Case: Home Purchasing

We consider the case of home purchasing as presented by Papapanagiotou and Fleuriot [7]. In this, a set of ten web services representing both automated and human agents involved in the various subtasks of purchasing a home is presented. These can be summarized as follows:

- The Buyer service represents the client making an offer for a home he wishes to purchase. For simplicity, in this example, we assume the point of view of the buyer, in which case an offer is always made (otherwise the execution is interrupted at that point).
- The HomeDir (home directory) service provides a list of available homes on the market based on a set of user-defined criteria.
- The CriminalService provides the level of average criminal activity in a region.
- The HouseAlert service generates alerts to notify about homes that match the given criteria.
- The EstateAgentSeller service represents the decision of an estate agent on a purchase offer.
- The MortgageService may generate a mortgage preapproval for a client or an *exception* if he is not eligible.
- The ContractService represents the notary that generates the contract between the estate agent and the buyer.
- The TitleSearch directory is used to retrieve the title and insurance information of a home.

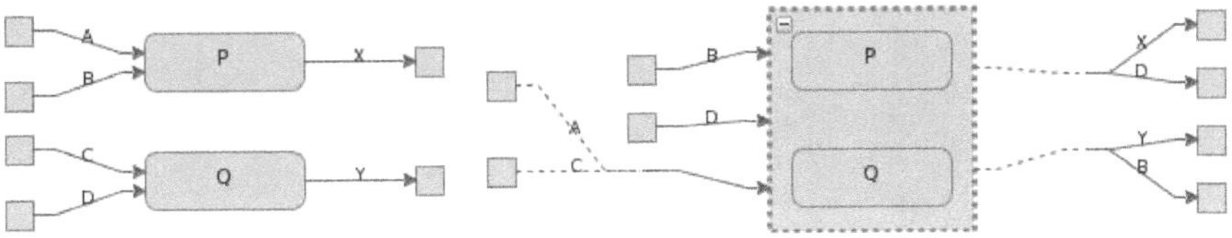


(a) Before (b) After

$$\dfrac{\dfrac{\dfrac{\vdash \mathbf{A}^{\perp},\ \mathbf{B}^{\perp},\ \mathbf{X} \qquad \dfrac{}{\vdash D^{\perp},\ D}\,Id}{\vdash A^{\perp},\ B^{\perp},\ D^{\perp},\ X \otimes D}\,\otimes}{\vdash A^{\perp},\ B^{\perp},\ D^{\perp},\ (X \otimes D) \oplus (Y \otimes B)}\,\oplus \qquad \dfrac{\dfrac{\vdash \mathbf{C}^{\perp},\ \mathbf{D}^{\perp},\ \mathbf{Y} \qquad \dfrac{}{\vdash B^{\perp},\ B}\,Id}{\vdash C^{\perp},\ B^{\perp},\ D^{\perp},\ Y \otimes B}\,\otimes}{\vdash C^{\perp},\ B^{\perp},\ D^{\perp},\ (X \otimes D) \oplus (Y \otimes B)}\,\oplus}{\dfrac{\vdash A^{\perp} \,\&\, C^{\perp},\ B^{\perp},\ D^{\perp},\ (X \otimes D) \oplus (Y \otimes B)}{\vdash (\mathbf{A} \oplus \mathbf{C})^{\perp},\ \mathbf{B}^{\perp},\ \mathbf{D}^{\perp},\ (\mathbf{X} \otimes \mathbf{D}) \oplus (\mathbf{Y} \otimes \mathbf{B})}\,neg\_eq}\,\&$$

(c) Proof tree

**Fig. 10.** The proof tree and diagrammatic representation of the WITH action on input $A$ of service $P$ and input $C$ of service $Q$. A new outer box is created with optional input $(A \oplus C)^{\perp}$ allowing either $P$ or $Q$ to be executed.

– The HomeInsurance service can provide home insurance contracts.
– Finally, the Settlement service represents the agreement between the estate agent and the buyer and the signing of the contracts.

Our goal, in this example, is to combine (graphically) this set of services in a single, fully verified, composite service that acts as a single point with which the end-user can interact and exchange information. In the composed service, the information flow between the composed components is handled automatically, thereby alleviating the need for the user to interact with each of the services in the set individually. Our underlying proof-based methodology, which only allows correct composition steps that eventually yield a verified result, can thus provide trust to all involved parties that the finished composition is correct and any exception or issue that arises will be handled or forwarded appropriately. Furthermore, it is worth noting the possibility of verifying a number of properties of the composed service by running model-checking tasks on the constructed process calculus term that is extracted from the finished proof.

In this example, the requested service can be represented by the following CLL statement:

$$\vdash HC^{\perp},\ RE^{\perp},\ DCL^{\perp},\ CI^{\perp},\ S \oplus ?E$$

This corresponds to a composite service that given the user's home criteria ($HC$), selected region ($RE$), desired maximum criminal activity level ($DCL$), and client information $CI$, generates a settlement ($S$) or an exception ($?E$). Note that $?E$ is a metavariable that becomes instantiated during the proof and is not expected to be known in advance. The binding for $E$ will be automatically propagated to the diagram and can simply be read-off the appropriate output edges by the user.

The entire proof of this statement, given that it involves ten services, is fairly complicated and the resulting proof tree (not given here due to space limitations) is large and convoluted to follow. The diagrammatic result, however, gives a much

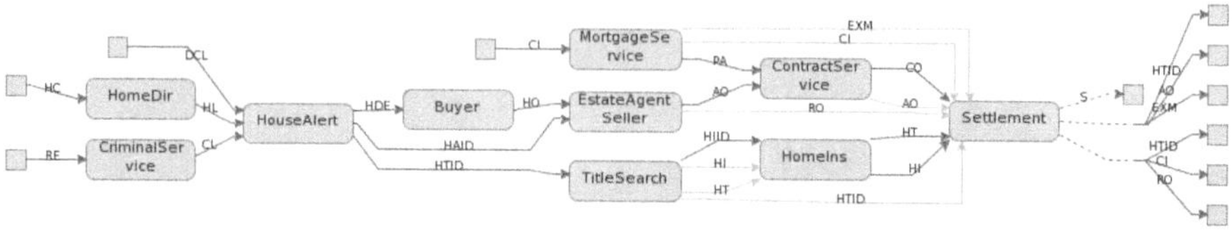


**Fig. 11.** The diagrammatic representation of the resulting composition in our Real Estate use case

clearer picture of how the services are composed together (see Fig. 11). The user obtains a clear view of the information flow in the composite service, an aspect that is completely obscured by the proof script.

In order to demonstrate the complexity of the proof and how it is abstracted from and made easy for the user in our diagrammatic interface, we focus on two particular examples of subproofs. Note that we only present enough details of these subproofs to demonstrate the benefits gained from the diagrammatic approach. For a more detailed discussion of the actual theorem proving, we refer the reader to the relevant paper [7].

### 4.1 Example Sub-proof: Composite I/O

In our first example, we attempt to join the HouseAlert service to the Buyer service. Given a home listing $HL$, the criminal activity level $CL$ in its region, and the desired criminal activity level $DCL$, the HouseAlert service produces a triple output $(HTID \otimes HAID \otimes HDE)$ involving the home's title database ID ($HTID$), its estate agent database ID ($HAID$), and its description ($HDE$). This can be described in CLL as follows:

$$\vdash HL^{\perp},\ CL^{\perp},\ DCL^{\perp},\ (HTID \otimes HAID \otimes HDE) \tag{1}$$

The Buyer expects the home description in order to decide if they want to make an offer $HO$. Therefore, we would like to connect output $HDE$ from HouseAlert to Buyer.

Accomplishing this in CLL is not as simple as it may sound. When connecting the two services, all of the involved resources need to be accounted for. In short, this means that the unused resources $HTID$ and $HAID$ need to be forwarded through the Buyer using buffers. The proof tree that constructs this composition is the following:

$$\dfrac{\dfrac{\dfrac{\dfrac{\dfrac{\dfrac{\dfrac{}{\vdash \mathbf{HDE}^{\perp}, \mathbf{HO}}\,Buyer \qquad \dfrac{}{\vdash HAID^{\perp}, HAID}\,Id}{\vdash HAID^{\perp}, HDE^{\perp}, (HAID \otimes HO)}\,\otimes \qquad \dfrac{}{\vdash HTID^{\perp}, HTID}\,Id}{\vdash HTID^{\perp}, HAID^{\perp}, HDE^{\perp}, (HTID \otimes HAID \otimes HO)}\,\otimes}{\vdash HTID^{\perp}, (HAID^{\perp} \parr HDE^{\perp}), (HTID \otimes HAID \otimes HO)}\,\parr}{\vdash (HTID^{\perp} \parr HAID^{\perp} \parr HDE^{\perp}), (HTID \otimes HAID \otimes HO)}\,\parr}{\vdash (HTID \otimes HAID \otimes HDE)^{\perp}, (HTID \otimes HAID \otimes HO)}\,neg\_eq}{\vdash \mathbf{HL}^{\perp}, \mathbf{CL}^{\perp}, \mathbf{DCL}^{\perp}, (\mathbf{HTID} \otimes \mathbf{HAID} \otimes \mathbf{HO})}\,Cut\ w/\ 1 \tag{2}$$

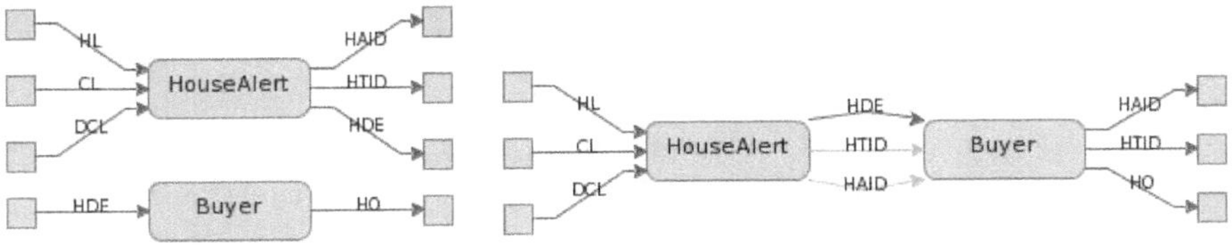


(a) Before (b) After

**Fig. 12.** The JOIN action between the HouseAlert and the Buyer services for the Home Purchase use case

Using our interface, this composition can be accomplished with a single JOIN action, by clicking on the dangling output edge labelled $HDE$ of HouseAlert, then right clicking on the corresponding input edge of Buyer. This will give us the result shown in Fig. 12. The complex task of performing the proof (2) is taken care of by the HOL Light tactic that corresponds to the JOIN action, thus verifying the correctness of the result in the background.

### 4.2 Example Sub-proof: Optional I/O

A similar situation is encountered when joining the MortgageService to the ContractService. The former generates a mortgage preapproval $PA$ for the client given the appropriate client information $CI$, but may throw an exception $EXM$ if the client is not eligible for mortgage. This service is expressed in CLL as follows:

$$\vdash CI^{\perp}, (PA \oplus EXM) \tag{3}$$

The ContractService expects a mortgage preapproval $PA$ and an accepted offer $AO$ from an estate agent in order to produce the contract $CO$.

In order to compose these two services, one needs to account for unused resources in the case of the exception. If $EXM$ is thrown, the ContractService will never be invoked because there will be no preapproval $PA$. Consequently, the resulting composite process needs to forward the unused resource of the the accepted offer $AO$ which is also part of its input. The CLL proof tree that performs this composition is shown below:

$$\cfrac{\cfrac{\cfrac{}{\vdash AO^{\perp}, AO}\,Id \qquad \cfrac{}{\vdash EXM^{\perp}, EXM}\,Id}{\vdash EXM^{\perp}, AO^{\perp}, AO \otimes EXM}\,\otimes}{\vdash EXM^{\perp}, AO^{\perp}, CO \oplus (AO \otimes EXM)}\,\oplus \tag{4}$$

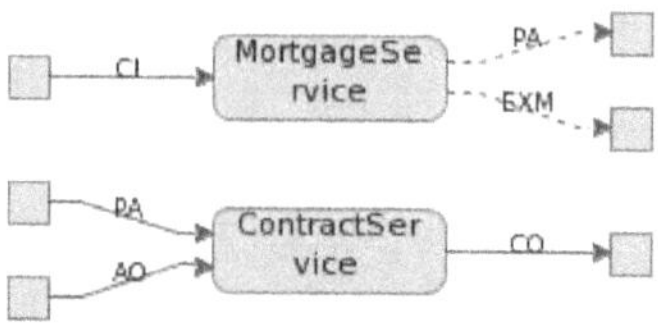


(a) Before

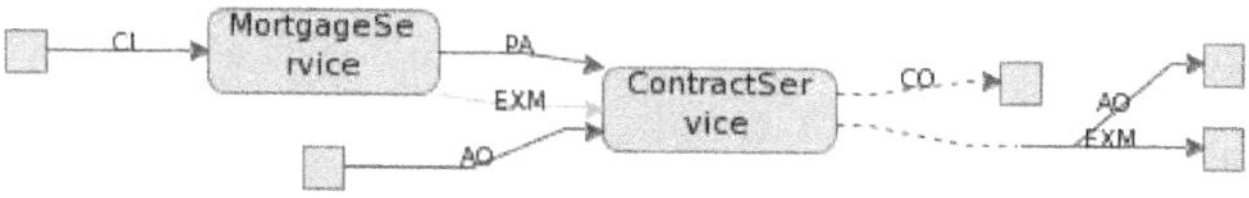


(b) After

**Fig. 13.** The JOIN action between the MortgageService and the ContractService for the Home Purchase use case

$$\dfrac{\dfrac{\dfrac{\dfrac{\dfrac{}{\vdash \mathbf{PA}^{\perp},\ \mathbf{AO}^{\perp},\ \mathbf{CO}}\ ContractService}{\vdash PA^{\perp},\ AO^{\perp},\ CO \oplus (AO \otimes EXM)}\ \oplus}{\vdash PA^{\perp} \,\&\, EXM^{\perp},\ AO^{\perp},\ CO \oplus (AO \otimes EXM)}\ \&\ w/\ 4}{\vdash (PA \oplus EXM)^{\perp},\ AO^{\perp},\ CO \oplus (AO \otimes EXM)}\ neg\_eq}{\vdash \mathbf{CI}^{\perp},\ \mathbf{AO}^{\perp},\ \mathbf{CO} \oplus (\mathbf{AO} \otimes \mathbf{EXM})}\ Cut\ w/\ 3 \qquad (5)$$

Similarly to the previous example, this proof is constructed in the background when the users joins the two services using a simple mouse gesture. The result of this action in our interface is shown in Fig. 13.

## 5 The Diagrammatic Interface

A screenshot of our diagrammatic tool is shown in Fig. 14. The GUI is implemented in Java where JGraph [6] is used for graph visualisation and layout. Behind the scenes, interactions with the GUI involves sending commands to an instance of HOL Light to construct proofs. Executed HOL Light commands communicate their results back to the GUI via JSON [3] so that the state of the current proof can be visualised. As all of the reasoning takes place within HOL Light, this guarantees that any proof we complete is correct.

Before beginning a proof, the user first defines the atomic services they want to use. A toolbar button is used to bring up a dialog for adding new services. The user then specifies the inputs and outputs, and name for the service by filling in text fields. The $\otimes$ and $\oplus$ operators are used to specify composite and optional outputs respectively. The HOL Light command for creating a new service is then executed in the background and an appropriate representation of the new service is added to the graph displayed by the GUI.

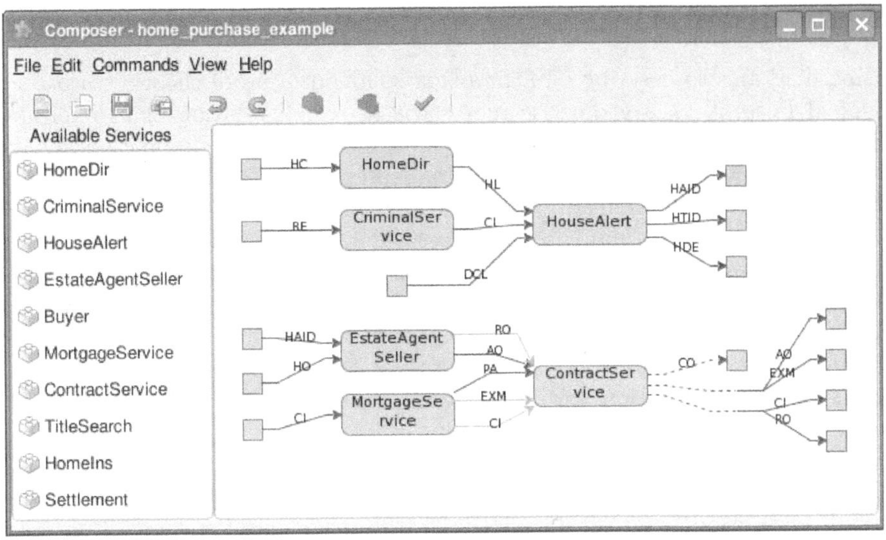


**Fig. 14.** A screenshot of our diagrammatic interface. The graph shows the current state of a proof in progress. The list on the left contains a list of all the atomic and composite services defined so far.

To start a proof, the same interface is used to specify a new composite service with the addition that metavariables are allowed for as yet unknown outputs (e.g. exceptions). The user must then manipulate the graph by applying actions (see Section 3) to create a graph that represents this composite service.

Each action is triggered using a different GUI gesture which executes a HOL Light tactic. When a tactic succeeds, the proof state of HOL Light is modified and the result of the tactic is communicated back to the GUI. The corresponding graph transformation to visualise this result is then applied.

For example, the tactic for the WITH action (see Section 3.3) is invoked when the user left-clicks one service then right-clicks a second service. If the tactic succeeds then 1) the graph nodes that represent these services are grouped together within a new outer node and 2) appropriate edges are added to this new node that represent the inputs and outputs specified by the tactic result. Note that should a service be required more than once in a proof, extra copies can be added to the graph by clicking its entry in a list that details the services defined so far (see Fig. 14).

Should the user wish to save their progress, a HOL Light proof script representing the proof state is created. This can be then be loaded to restore the GUI to the same state later. Using the same loading mechanism, the tool is thus also capable of visualising existing HOL Light proof scripts concerning web service composition that were written by hand directly in the theorem prover. For advanced users, this provides the flexibility for editing proof scripts using a combination of our interface and direct script manipulations.

## 6 Related Work

Using diagrams to describe CLL proofs is common in proof theory. Girard developed Proof-Nets, a geometric representation of CLL proofs that facilitates proof theoretic analysis, such as proof equivalence, and cut elimination [4]. We consider Proof-Nets to be a more general, complicated, and deeply theoretical construction than the simpler visualisation of composite web services that we are trying to achieve. Our aim is to construct a diagram that clearly demonstrates the intuitive interpretation of CLL proofs as web services composition. As mentioned earlier, the web services context restricts these proofs to specific patterns that our diagrams need to capture, as opposed to Proof-Nets that are designed to represent any possible CLL proof. Moreover, the diagrams constructed by our tool show the information flow in the resulting composition in a clear, straightforward way.

Web services are also often described using diagrams, and more specifically workflows. The Business Process Execution Language (BPEL) [1] and Business Process Modelling Notation (BPMN) [9] are among the most often used, workflow-based languages to describe web services. The main aims of such workflow-based specifications are to not only describe the web service information and control flow, but also provide an executable model. In our case, we focus mostly on the information flow, whereas the executable model is provided in the extracted process calculus term (upon which we have not elaborated here, but in the related papers [7,8]). We consider workflows as a more complete approach for the description of web services and consider an attempt to map our diagrams to such workflows as future work.

## 7 Future Work

Having implemented the core features for formally verified, interactive web services diagrammatic composition, there are several interesting topics and ideas to explore next. Primarily, we aim to introduce more actions so as to cover other specific cases of web services composition, such as the connection between a service with an atomic output and a service with an optional input (where one of the options matches the atomic output). Integrating the PiVizTool [2] for the visualisation of the executable process calculus term that is extracted from our proof is also among our immediate aims. As mentioned in the previous section, introducing links between our diagrammatic visualisation and existing workflow languages would help towards exporting a more concrete and reusable result. Moreover, there are various ways of improving our tool, such as guiding the user during the composition process, for example by highlighting matching inputs when the mouse hovers over a specific output. We also note that we are currently in the process of taking the framework beyond web services composition through some new work on the formal verification of collaborative healthcare workflows. This task is greatly facilitated by our abstract diagrammatic tool, enabling healthcare experts who have no prior knowledge of logic or theorem proving to use it and understand the resulting compositions.

## 8 Conclusion

We have described our efforts towards a diagrammatically-driven approach for the formal verification of web services composition. We use an intuitive graphical notation in which nodes (boxes) represent web services and edges correspond to inputs and outputs. Using simple mouse gestures, the user can apply predefined graphical rules to compose the available services and implicitly guide HOL Light into applying automatic procedures that verify the result of the composition steps in CLL. The interface provides a visualisation of the composition by creating a workflow-like graph representing the information flow between the component services. We evaluated our approach by developing a non-trivial example involving web services for home purchasing fully diagrammatically. Based on our previous experience on verifying this case-study directly in HOL Light, it is clear that the diagrammatic approach greatly simplifies the verification process by eliding the need for any direct proof derivation in CLL using (a wide variety of) HOL Light commands. We believe our tool demonstrates that an intuitive, diagrammatic approach to producing and visualising fully-verified composition of web services is not only feasible but also effective.

## References

1. Andrews, T., Curbera, F., Dholakia, H., Goland, Y., Klein, J., Leymann, F., Liu, K., Roller, D., Smith, D., Thatte, S., et al.: Business process execution language for web services, version 1.1. Standards proposal by BEA Systems, International Business Machines Corporation, and Microsoft Corporation (2003)
2. Bog, A., Puhlmann, F.: A Tool for the Simulation of $\pi$-Calculus Systems. Open. BPM (2006)
3. Crockford, D.: The application/json media type for JavaScript Object Notation (JSON). Internet RFC 4627 (July 2006)
4. Girard, J.Y.: Proof-nets: the parallel syntax for proof-theory. Logic and Algebra, 97–124 (1995)
5. Harrison, J.: HOL Light: A Tutorial Introduction. In: Srivas, M., Camilleri, A. (eds.) FMCAD 1996. LNCS, vol. 1166, pp. 265–269. Springer, Heidelberg (1996)
6. JGraph Ltd.: The JGraph homepage, `http://www.jgraph.com/`
7. Papapanagiotou, P., Fleuriot, J.: Formal verification of web services composition using linear logic and the pi-calculus. In: 2011 Ninth IEEE European Conference on Web Services (ECOWS), pp. 31–38. IEEE (September 2011)
8. Papapanagiotou, P., Fleuriot, J.: A theorem proving framework for the formal verification of web services composition. In: WWV 2011, vol. 61, pp. 1–16. EPTCS (2011)
9. White, S., Miers, D.: BPMN modeling and reference guide. Future Strategies Inc. (2008)

# The Diagram of Flow: Its Departure from Software Engineering and Its Return

S.J. Morris[1] and O.C.Z. Gotel[2]

[1] Department of Computing, City University London, UK
sjm@soi.city.ac.uk
[2] Independent Researcher, New York, NY, USA
olly@gotel.net

**Abstract.** The first diagrammatic notation used in software engineering represented the concept of flow. This paper considers the factors that affected the apparent departure of the flowchart from software engineering practice during the 1970s and 1980s and its subsequent return in the 1990s. A new emphasis on hierarchy (as level of abstraction) and on data structure meant that the general concept of flow was completely superseded, only to re-emerge later as a new duality of control flow and data flow. This reappearance took a variety of forms with varying semantics until its stabilisation in the latest version of the Unified Modeling Language. Flow is there re-instated as a fundamental concept in software engineering although its importance, and that of the activity diagram used to represent it, diminished as a consequence of its becoming just one among a wider set of paradigms for software systems development, each associated with its own diagrams.

**Keywords:** Activity Diagram, Diagrammatic Notation, Flowchart, History, Representation, Software Engineering.

## 1 Introduction

Earlier papers [1], [2] traced the history of flow diagrams from their early beginnings as representations of flows in nature and industry, through the introduction by Goldstine and von Neumann of a symbolic representation of sequence as a means of machine control, and through the use of their new version of the flowchart as an essential tool for programming and the automation of applied mathematics during the 1940s and 1950s, to the falling of the flowchart into disrepute as an unwieldy and counterproductive assistant for programming. This paper continues this history by considering the demise of the flowchart and its resurrection in the latest software engineering modelling language in an altered form and less crucial role.

Section 2 contrasts examples of programming practice before and after the initial demise of the flowchart and outlines the many factors leading to attempts to structure what were becoming increasingly large and complex programs and systems design projects. Section 3 considers the restructuring of representations of both program and system design, at first by variants of the flowchart concepts, then by more radical

P. Cox, B. Plimmer, and P. Rodgers (Eds.): Diagrams 2012, LNAI 7352, pp. 256–269, 2012.

hierarchical and data-centric approaches to the entire process. Section 4 deals first with the emergence of the object-oriented paradigm for programming and systems design and then considers two versions of one diagram in the Unified Modeling Language (UML), now the prevailing standard for development support in an object-oriented environment. Examination of this activity diagram shows how the concept of flow remained, although concealed, only to re-emerge quite unmistakably following significant revision of its original version. Section 5 presents general historical and practical conclusions.

## 2 Replacement of Flow as Program Paradigm

### 2.1 Contrasting Practices

In 1958 Mike Woodger defined the essential stages of program development in manuscripts that survive in the archive of the National Physical Laboratory (NPL) where he worked from 1945 to 1983. These manuscripts include fully detailed flowcharts of complex algorithms, one of which is reproduced in part in [1] and in full in [2]. Woodger defined programming practice in an era when there was only the particular code of a specific machine available to control it and when the programmer, almost always a mathematician or scientist, was also solely responsible for memory allocation and its manipulation. The flowchart then had a crucial role in the representation of algorithms and hence of programs. The production of a flowchart was an essential step following the analysis of the mathematical problem to be solved and the choice of a suitable procedure. This approach had developed in part because the most frequent tasks then involved complex mathematical tasks only made possible by automated computation at high speed [2].

Another program written by Woodger, not later than 1976, "to allow a user with VDU and keyboard to interrogate an existing stored body of information (about fungi)" survives in its entirety with a full program listing and documentation [3]. He uses it as an example in an article in which he again defines the steps of programming, but now without any mention of flowcharts. He sets out as programming principles the "separation of concerns", the separation of what is to be achieved from how it is to be achieved, and an "hierarchy of virtual machines" containing objects at an appropriate level of abstraction with their associated operations.

The process that Woodger defines and illustrates makes no use of diagrams. It begins with a "rough planning stage" whose purpose is to define "what is to be done". The second stage consists of "Further detail: how is this purpose to be achieved", again principally defined as text. The process then switches from what had become known as a top-down approach to the opposite bottom-up process. The final implementation stage involves direct use of the available "programming language and its implementation ... within the constraints of the OS".

An eyewitness account of the practices in a major UK company, ICI (Imperial Chemical Industries), during the intervening period indicates that it was a severe shock for many that excellent programs could be written using a completely new method. This writer, Ken Ratcliff, criticises such practices as leading to "indulgence into flights of bit fiddling, or unnecessary pirouetting on a recently learned technique of pointer arithmetic, cross-sectional definition of arrays, programmer controlled allocation of storage independent of program block structure and other excesses of the do-it-yourself type" [4]. The solution for ICI at the time was the rapid and comprehensive introduction of the data-centric program design principles of Michael Jackson with their emphasis on structure (See Section 3.4 below). This account also indicates how important project and staff management issues had already become.

## 2.2 Structuring the Unstructured

By the early 1970s, the unstructured complexity of increasingly large programs gave rise to reactions, at first little related to each other, under the titles of structured programming and structured systems development. To place developments in programming techniques and diagrammatic representations in context, the conspicuous causes included a wide variety of factors having different effects:

- The ability to write increasingly complex programs using higher level languages which used the original level of machine code as their basis and translators or compilers to produce code executable on a particular machine;
- The introduction of operating system programs to remove much of the effort from memory allocation, file management and the mechanics of input and output;
- The availability of mass storage devices (e.g., magnetic tape, drum or disk);
- The shift in commercial domains (and in other domains already exploiting rapid mathematical computation) towards mass data transformations and manipulations;
- The general lack of experience in dealing with the problems of scaling;
- The craft nature of programming;
- The indications of a continuing and rapidly expanding demand for machines and programming.

As discussions continued, the characteristics of structured programming coalesced around a number of issues, exemplified by a list published in 1978 by Infotech International [5]: structured analysis and design; structured coding; top-down implementation and testing; Hierarchy, plus Input, Process, Output technique (HIPO); team operations; project support libraries; structured walkthroughs; and project management systems. The latter four issues concern improvements to the organisation of programming personnel and the management of the programming process, and are not associated with any particular diagrammatic forms. It was expected that the solution of the former issues would draw on the increasing amount of material on formal, procedural methods for the design and construction of programs.

There is a clearly perceived switch from sub-division of the total problem into manageable parts to the use of formal criteria and techniques for decomposition and design. In the code itself, elimination of the uncontrolled use of GOTO statements was the benefit gained from the use of only three basic constructs (sequence, iteration and selection), credited to Böhm and Jacopini [6]. Top-down implementation and testing accepted the principle that software development should proceed from the highest control level modules downwards.

# 3 Restructuring the Representation of Program and System

## 3.1 Initial Responses

As they manifest themselves initially in program and systems design techniques, the responses to these issues focused on two issues both in procedures and in graphical representations: hierarchy (as level of abstraction) and data structure. The consequence was the complete supersession of flowcharts as the principal basic program design device (as used by Woodger in the 1950s) and its demotion to, at best, a small low-level supporting role.

These responses appeared in a number ways: alterations to the basic flowchart to accommodate some indication of hierarchy (e.g., Dill, Hopson and Dixon); more radically different representations of nested structures (e.g., Chapin); completely data-centric views of process (e.g., Jackson); and combinations of both data and hierarchical approaches in complete system descriptions (e.g., HIPO).

## 3.2 Tree Chart

As a preamble to examining these new diagrams it is necessary to first review another, the tree diagram or tree chart (Figure 1) to show the paradigmatic representation of hierarchy and how it came to be interpreted in the context of structured programming. The tree chart represents the results of the parsing of functions, or functional decomposition, as usually credited to Knuth [7].

The execution or control sequence represented by the diagram of Figure 1 cannot be, for example, A, C, J, P, A. The executing machine may or may not invoke the execution of the connected functions on the next lower subordinate level (B, C, D or E) in any sequence and any number of times. The essential proviso is that control always returns to the invoking function or module. Likewise with B and its invocation of F, G and H, and so on. This notion of nesting fundamentally altered program structure and introduced hierarchical levels not present in earlier generation programs, even if they made extensive use of routines and sub-routines. The value of decomposition of this type to overcome problems of size and complexity was a fundamental influence on the development of other diagrammatic representations.

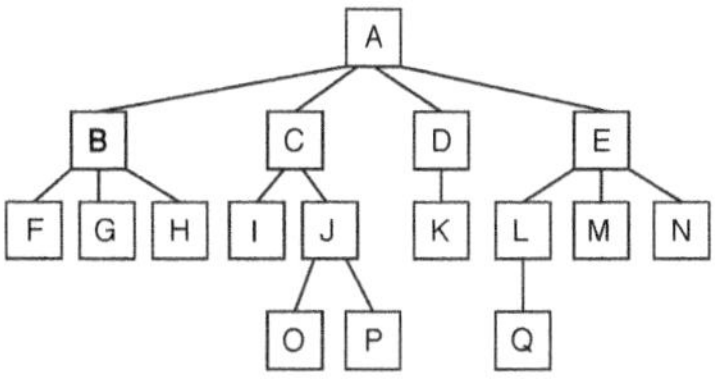


**Fig. 1.** Hierarchical tree structure for program structure and control

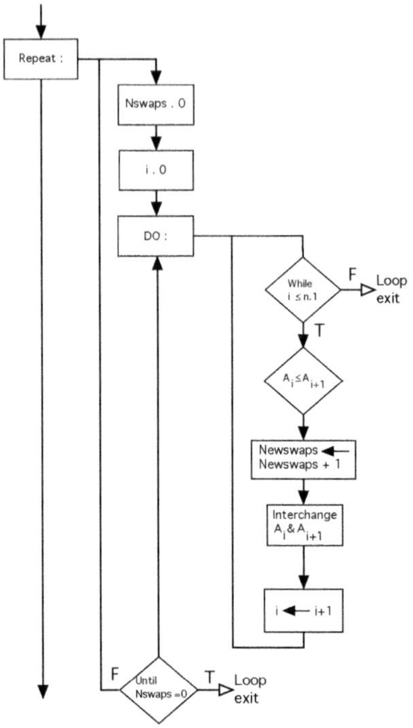


**Fig. 2.** Flowchart with left-to-right levels of nesting

### 3.3 Flowcharts Adjusted and Block-Like Variants

The standard ANSI and ISO flowchart [8], [9] imposed no configuration constraints other than a basic vertical sequence from top to bottom plus a reverse parallel path to represent returns to earlier positions in the sequence. Subsidiary horizontal paths served only to connect the single main parallel to its reverse companion(s). Such a structure served well until decomposition demanded some representation of hierarchy. Early attempts to do this, typified by the approach of Dill et al. [10], involved creating a horizontal left-to-right hierarchy, an early version of the graphical trope now called swim lanes. Loop exits now lead not directly back to a point above but to one to the left in a level higher in a hierarchical rather than topographical sense. The diagram shown in Figure 2 represents an exchange sort algorithm that repositions the numbers in an array in ascending order.

An alternative view of nesting, which came to be known as the Chapin chart, sacrificed all direct representation of flow and sequence in order to show nesting and selection. The diagram of Figure 3 [11] represents the same exchange sort algorithm as that described above. This innovative use of a two-dimensional space did not however solve any of the problems of complexity and size in a conclusive manner.

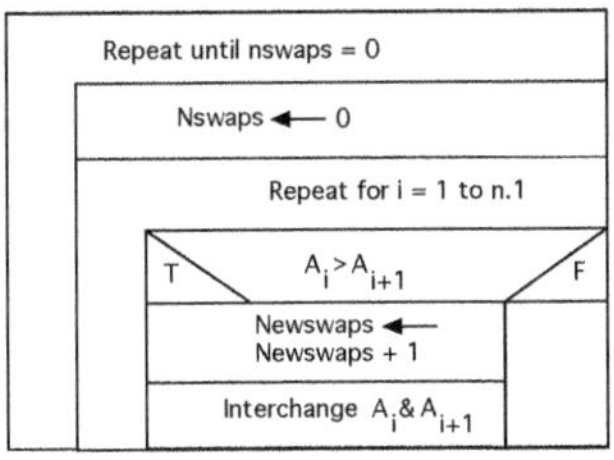


**Fig. 3.** Internally nested representation of loops in a Chapin chart

### 3.4 Data-Centric Views

The most radical approach was to switch completely to a program structure not only based on data but also independent of programming language, as exemplified by the initial work of Jackson [12]. The force of its impact comes over clearly in the same account from ICI UK [4]: "(He) casts many existing highly developed techniques to the winds, making a clean start with the explicit principle that program structure should be based on the structure of the data on which the program operates. To facilitate this development he defines a simple but clear notation for the basic components of structure, corresponding to the four constructs of structural coding and called sequence, iteration, selection and the elementary components. Because of the clear structure of stages, and the structure outlined, the design proceeds without the help (or rather hinderance) of flow charts."

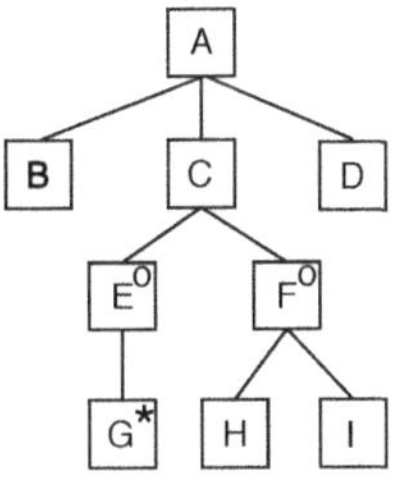


**Fig. 4.** Data structure in JSD notation

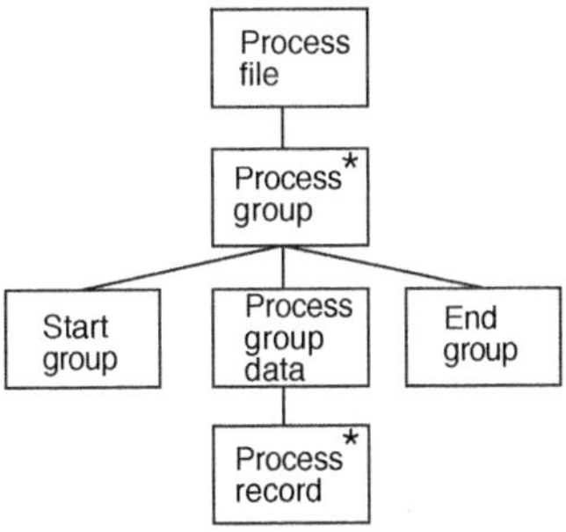


**Fig. 5.** Data-centric program structure

Figures 4 and 5 show the data and program structures as represented in a contemporary paper of Jackson [13]. In both cases, these represent tree structures in which elements annotated with a '0' are alternatives and those annotated with a '*' may be repeated 1 to n times.

### 3.5 Whole System Representation - HIPO

The Hierarchy, plus Input, Process, Output (HIPO) technique was intended to describe a whole system in terms of inputs, outputs and constituent, intervening processes. Originally developed at IBM for its own use [14], this takes the separation of concerns and top-down development as fundamental principles. It was not intended as a replacement for older techniques, rather as a means of system description at a level that was not possible to achieve previously when using them.

Figure 6 shows the basic notation used to represent a hierarchy of process components belonging to a main process P (in this case read at the subordinate level from left to right with $P_2$ following $P_1$) and the data inputs A and B to process P and its outputs C, D and E.

With the advent of techniques such as HIPO, and the many others that appeared during the 1970s (see Davis [15] for a review), the displacement of the flowchart as an essential programming tool was complete. Development of alternative programming paradigms demanding completely different representations meant a lapse into obscurity for the flowchart *per se* as a programming tool.

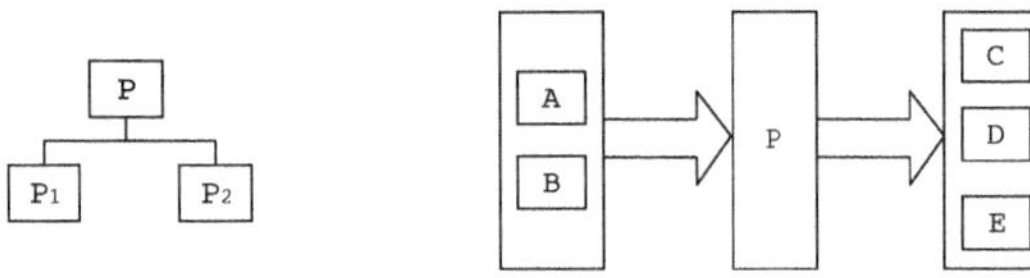


**Fig. 6.** Hierarchy of components in HIPO (left) Inputs to and Outputs from process P (right)

## 4 Flow Survives Object-Orientation

### 4.1 New Object-Oriented Approaches

From the earliest discussions of structured programming [16], basic concepts of object-oriented programming were emerging in the form of discussion of data types and their associated operations. The development of a completely different object-oriented paradigm for programming, derived in part from such notions of abstract data types, meant that a wholly new set of diagrams was created. In addition, a whole new class of diagrams for data and database structuring were emerging, which are outside the scope of this paper.

In a repeat of what had occurred in the 1960s and 1970s, there was again an explosion of diagrammatic forms required to assist the development process and an attempt to clarify complexity, leading to an effort to create initially a Unified Method

[17] and then a Unified Modeling Language (UML) [18] for object-oriented development. The UML has subsequently gone through many revisions, the latest major revision being Version 2.4 [19].

Flow, in the sense of a sequence of activities needed to achieve a particular goal, had not disappeared, nor had the need to represent notions of the selection of alternative routes and the possible repetition of segments in such sequences. Hence the many various uses of flowcharts continued outside the restricted domain of algorithmic expression and computation.

Within the evolution of the UML the role of flow has gone through three phases. In the first phase, flow played no recognised role at all in the semantics of the closest relative of the flowchart, known from the outset as the activity diagram. In the second phase, object flow became significant. In the third phase, object flow and the original concept of control flow form a new duality and mark the re-emergence of a full role for flow in the UML.

During the first and second phases, the principal underlying paradigm for behaviour was the finite state machine (FSM): "a hypothetical machine that can be in only one of a given number of states at any specific time. In response to an input, the machine generates an output and changes state. Both the output and the next state are purely functions of the current state and the input". [15]. Its particular development in the form of the state machines of Harel emphasises the importance of "the hierarchical decomposition of finite state machines and a mechanism for communication between concurrent finite state machines" [15]. This variant of the FSM provided the initial formal semantic basis for the initial activity diagram while another variant, the Petri net [20], contributed to its visual syntax. In the third phase, this essentially transitional view of behaviour was displaced in the activity diagram by the sequential view always implicit in the visual syntax. In the following sub-sections we examine the details of UML development and show how this occurred.

### 4.2 Initial Version of the UML Activity Diagram

When what was to became known as the Unified Modeling Language (UML) was proposed in 1995 [17], the software engineering community was presented with seven different types of diagram for specifying and designing a software-intensive system that would be implemented in an object-oriented manner (i.e., class, use case, message trace, object message, state, module and platform diagrams). Notably absent was any diagram specifically claiming to model the concept of flow.

The addendum to the initial version of the UML, known as Version 0.91 [18], remarked that: "Sometimes it is useful to show the work involved in performing an operation by an object." Activity diagrams were introduced to fill this gap in the emerging UML and therefore to provide a way to show the method for implementing an operation visually (i.e., the steps that occur in the procedural implementation of an operation). The activity diagram was not labelled as a flow diagram or flowchart; rather, it was described as: "a special kind of state machine that describes the implementation of an operation in terms of its sub-operations." These sub-operations were essentially the procedural activities internal to the object owning the activity diagram and referred to as activity states. The semantics of the initial activity diagram were explicitly grounded in those of state machines.

Examining the exemplar activity diagram in this addendum document, shown in Figure 7, reveals a visual notation with distinct graphical icons for the following:

- ***Activity states*** - Rounded rectangles that contain the name of a single activity inside. These model the individual steps (i.e., activities) in the implementation of an operation.
- ***Transitions*** - Solid lines, with a single directional arrow on one end. These model the sequencing between the activity states (i.e., the flow of control). The transitions are implicitly triggered by the completion of the preceding activity state.
- ***Synchronization bars*** - Solid thick horizontal lines. These model either the initiation or merging of concurrent control, the former when there are multiple arrows leaving a synchronization bar and the latter when there are multiple arrows entering it.
- ***Dummy nodes*** - Small hollow circles that chain guard conditions to model complex conditions on a transition.

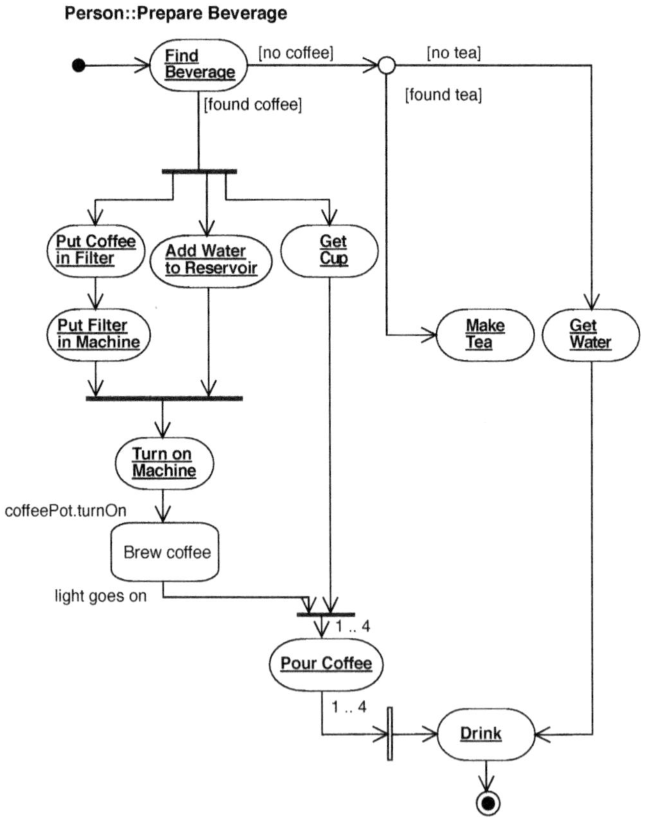


**Fig. 7.** Earliest version of the UML activity diagram [18]

- ***Start state*** - Small solid circle modelling the starting activity state in the implementation of an operation.
- ***End state*** - Small bulls-eye circle modelling the ending activity state in the implementation of an operation.

The activity diagram is read from the start state to the end state, and the exemplar diagram in the addendum document shows the steps in the implementation of the operation flowing down the page. The accompanying text also refers to the ability to model interrupts to the normal procedural flow of control and to model wait states in order to show operations external to the object but essential to the completion of the modelled operation, though no additional graphical icons are provided.

## 4.3 A Flow Diagram by Any other Name?

While the first appearance of the UML activity diagram clearly does depict the flow of control, it does not claim to use any of the standard flowcharting icons. Instead, the activity diagram claims to borrow from the notation used by the early UML state machine diagrams for representing its activity states, transitions, start state and end state. It also adopts the bars of the Petri net notation for modelling synchronisation [20], although Petri nets were not part of the syntax or semantics of the early UML.

However, in comparing the earliest incarnation of the activity diagram and the flowchart standards it is clear that there are considerable similarities between the concepts they model. Both activity diagrams and flowcharts model:

- Where the flow of control starts from and where the flow of control ends, though using different icons;
- Discrete processing steps, the processing being decomposed in the activity diagram into activities and represented in a homogeneous way using rounded rectangle icons, while in the flowchart processing types are differentiated with generic processing steps represented using rectangles, input/output represented using parallelograms, and preparatory work represented using hexagons;
- The flow of control, both using solid lines with arrow heads, control passing to the icon to which the arrow points;
- Parallel processing and concurrent control synchronisation, with activity diagrams using a thick single bar icon and flowcharts using two parallel horizontal lines;
- Interrupts in the flow of control;
- Processing steps running down the page, with the ability to express a sequence of operations, concurrent operations and branches in the flow of control, thus both having the potential to express the algorithms of program design and workflow;
- Nested diagrams.

While the activity diagram was described as a specialised form of state machine diagram in Version 0.9, it was evidently to be used to model a progression (or flow), something that was more the remit of flowcharts than state machine diagrams. Moreover, there was no explicit notion of progression in state machine semantics. Given the similarity in the underlying concepts which the icons of both the early activity diagrams and standard flowcharts were able to model, one could argue that it was the denigrated status of the flowchart, along with the paradigm shift from procedural to object-oriented development, that led to the redesign and renaming of a diagram intended to model familiar concepts.

### 4.4 Incorporating More Features From Flowcharts

With the release of Version 1.0 of the UML in 1997 [21], the activity diagram was firmly positioned as one of the core diagram types for modelling behaviour. The activity diagram was expanded in scope and intended to be attached not only to the implementation of operations, but also to classes and use cases, focusing on the flows driven by internal processing (i.e., the procedural flow of control), and not on external events. With a statement suggesting the "use (of) ordinary state diagrams in situations where asynchronous events occur" the activity diagram became the *de facto* choice for modelling all forms of synchronous behaviour.

In Version 1.0, the activity state was now relabelled as action state and the transitions were elaborated to add optional actions in addition to guards. The original icon for the dummy node, the small hollow circle, was replaced with a diamond shaped icon to represent a decision point, the exact same symbol that had long been used in traditional standard flowcharts. The action states could further be organised into swim lanes, depicted by solid vertical lines running down the length of a page, to allocate the actions to the objects responsible for their undertaking, reflecting the growing popularity of the use of activity diagrams for business modelling. Such swim lanes also reflect the structure of the much older flowchart variant shown in Figure 2.

In addition to the control flow indicated by the solid line transitions, object flow also began to be modelled using dashed lines to show those objects responsible for performing an action and those objects whose values are determined by an action. This was effectively the modelling of object message passing. The state of an object at any one time could further be described within a rectangular box. Additional icons to model the sending and receipt of signals were also added to the visual notation as its catalogue of icons began to grow. These were referred to as control icons from Version 1.1 onwards and were intended to elaborate the information that could be specified on a transition.

Thus the activity diagram steadily became more elaborate with each successive revision of the UML. By Version 1.4.2 in 2004 (and an ISO standard in 2005 [22]) the main distinction had become a focus on the modelling of nested structures, with both action states and sub-activity states being specified, along with new icons to show this nesting. The diamond decision icon was also now being used to merge decision branches, in addition to simply initiating them, introducing junction pseudo states. The control icons also continued to multiply, in particular increasing support for modelling concurrency. Throughout all these specification revisions, activity diagrams remained defined as a specialised form of state machine diagram while, according to anecdotal evidence, practitioners and tool vendors were coming to refer to them as the flowchart of the UML.

## 4.5 Acceptance of Flow in the UML

When UML Version 2.0 was released in 2005, activity diagrams were completely reformulated based on the Petri net semantics of token flow. "By flow, we mean that the execution of one node affects, and is affected by, the execution of other nodes, and such dependencies are represented by edges in the activity diagram." [23]. The change in the base semantics was made to increase the number of flows that could be modelled by the UML and perhaps to reflect increased interest in business process and workflow modelling for systems development.

Version 2.0 specifically delineated the concepts of action ("the fundamental unit of behaviour specification") and activity, which provides the conditions and sequencing information for coordinating the lower-level behaviours. The primary modelling artefacts of the earlier activity diagram were renamed as activity nodes and activity edges. Distinct icons were provided to indicate the various types of node (e.g., action nodes with a rounded rectangle icon, control nodes with a regular rectangle icon and five types of control node each with their individual associated icons). Concepts were introduced to depict iteration and data storage, both common to flowcharting, and a whole series of icons were introduced to model the concept of containment in activity diagrams. Moreover, the traditional flowchart icons for collation and summing junction were now being used to model accept event actions and final flow nodes.

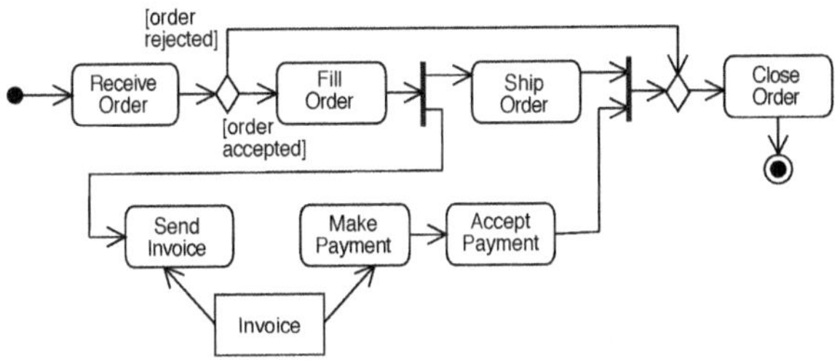


**Fig. 8.** Example of a UML activity diagram from UML Version 2.4 [19]

With UML Version 2.0, not only was the modelling of control flow and object flow made quite explicit for the first time, but an unequivocal way to model data and information flow was also provided. The latest version [19] defines an object flow as "an activity edge that can have objects or data passing along it … which models the flow of values to and from object nodes" and defines a control flow as "an edge that starts an activity node after the previous one is finished … objects and data cannot pass a control flow edge". The consequence of this change is a diagram that represents, in both a semantic sense and in visual appearance, much of the essence of the original flowchart. Figure 8, extracted from Figure 12.35 of the UML Superstructure Specification Version 2.4 [19], provides an illustration.

As the UML continues to evolve the popularity of activity diagrams in industrial practice is in no doubt, evidenced by its prominence in leading commercial tools. These tools also show, however, that the status of flow has altered drastically. The taxonomy of diagrams provided in UML Version 2.4 includes seven structure diagrams (i.e., class, component, object, composite structure, deployment, package and profile diagrams) and four behaviour diagrams (i.e., activity, use case, state machine and interaction diagrams, for which there are four variants). While it is quite clear that the concept of flow has re-emerged and that its value in the specification, design and development of software systems has unequalled longevity, it has completely lost the supremacy that it initially acquired and maintained for so long.

## 5 Conclusions

Understanding the concept of flow remains basic to software engineering. Data flow has joined control flow to form a pair of concepts fundamental to the understanding of programs and the design of software systems. The history of the representation of these concepts shows how diagrams can reveal shifts of technology and changes in the manner in which it is exploited. The succession of flow diagrams created and used prior to the reinvention of Goldstine and von Neumann shows their importance to industrial processes. The new form of flowchart invented to represent the flow of control in the earliest automated computational machines then come to dominate programming practice for more than two decades. Its use also spread to every field where the flow of a sequence of decisions would benefit from being shown in a simple diagram.

While this vernacular use continued, to the extent of misuse and caricature, the original diagram acquired the status of an international standard but fell into disrepute as a programming tool. The complexities of programs and the necessity to represent other concepts, in particular data structure and other views of machine behaviour, required the introduction of alternative concepts and diagrams and their promotion. As a consequence, any assessment of the flowchart made now (if it is given at all), emphasises its importance only outside the specialist field of software engineering and programming practice.

Examination of contemporary practice and its conceptual support shows, however, that flow and flow representations remain firmly entrenched albeit in a new role. The history of diagrammatic notations has in this case revealed the continuing significance of an important concept with the possible consequence of improving understanding and practice. Further work will examine the introduction, development and use of other concepts and diagrams that have been central to the history and progress of software engineering.

## References

1. Morris, S.J., Gotel, O.C.Z.: Flow Diagrams: Rise and Fall of the First Software Engineering Notation. In: Barker-Plummer, D., Cox, R., Swoboda, N. (eds.) Diagrams 2006. LNCS (LNAI), vol. 4045, pp. 130–144. Springer, Heidelberg (2006)
2. Morris, S., Gotel, O.: The role of flow charts in the early automation of applied mathematics. BSHM Bulletin. Journal of the British Society for the History of Mathematics 26(1) (2011)

3. Woodger, M.: The aims of structured programming. In: Structured Programming. Infotech State of the Art Report. Infotech International Limited, Maidenhead (1976)
4. Infotech Management Report Structured Programming Practice and Experience. Vol. 1: Management Guide to Techniques and Implementation Vol. 2: Management Report on Implementation Practice. Infotech International Limited, Maidenhead (1978)
5. Infotech International Survey Structured Programming Practice and Experience. Vol. 1: Overview of Structured Programming Vol. 2: Structured Programming Methodologies and Techniques Vol. 3: International Survey and Analysis of User Experience. Infotech International Limited, Maidenhead (1978)
6. Böhm, C., Jacopini, G.: Flow diagrams, Turing machines and languages with only two formation rules. Communications of the ACM 9(5) (May 1966)
7. Knuth, D.E.: The art of computer programming - fundamental algorithms. Addison Wesley, Reading (1968)
8. Chapin, N.: Flowcharting with the ANSI standard. A tutorial. Computing Surveys 2(2), 119–146 (1970)
9. International Organization for Standardization. Information Processing - Flowchart symbols. ISO 1028 (1973)
10. Dill, J.M., Hopson, R.W., Dixon, D.F.: Design and documentation standards. Brown University, Providence (1975)
11. Nassi, I., Shneiderman, B.: Flowchart techniques for structured programming. SIGPLAN Notices 8(8) (August 1973)
12. Jackson, M.A.: Principles of Program Design. Academic Press, Orlando (1975)
13. Jackson, M.A.: Data structures as a basis for program design. In: Structured Programming. Infotech State of the Art Report. Infotech International Limited, Maidenhead (1976)
14. HIPO - A Design Aid and Documentation Tool. Poughkeepsie, NY, IBM Corporation, Form SR20 - 9413 (1973)
15. Davis, A.M.: Software Requirements Analysis and Specification. Prentice-Hall International, Englewood Cliffs (1990)
16. Dahl, O.-J., Dikstra, E.W., Hoare, C.A.R.: Structured Programming. Academic Press, London (1972)
17. Unified Method V0.8. Rational Software Corporation, Santa Clara (October 1995)
18. Booch, G., Jacobson, I., Rumbaugh, J.: The Unified Modeling Language for Object-Oriented Development. Documentation Set Version 0.91 Addendum. Rational Software Corporation, Santa Clara (1996)
19. Object Management Group. OMG Unified Modeling Language™ (OMG UML), Version 2.4 (January 2011), `http://www.omg.org/spec/UML/2.4/` (accessed July 26, 2011)
20. Peterson, J.: Petri Nets. ACM Computing Surveys 9(3) (September 1977)
21. Unified Modeling Language V1.0. Rational Software Corporation, Santa Clara (January 13, 1997)
22. International Organization for Standardization. Open Distributed Processing - Unified Modeling Language (UML) Version 1.4.2. ISO/IEC 19501 (2005)
23. Unified Modeling Language Version 2.0. Object Management Group (August 2005)

# DDA\Repository: An Associative, Dynamic and Incremental Repository of Design Diagrams

Bharat Dave and Gwyllim Jahn

University of Melbourne, Parkville VIC 3010 Australia
b.dave@unimelb.edu.au, gwyllo@gmail.com

**Abstract.** This paper describes implementation of an online prototype that, on the one hand, offers interactive diagramming support to externalize thinking about design compositions and, on the other hand, acts also as an incremental repository of diagrams that can be dynamically interrogated to find other proximate compositional thinking and ideas related to a particular position. The prototype helps both notate design thinking and draw out associations between separately notated design compositions.

**Keywords:** design configurations; constructional logic; compositional diagrams; knowledge accretion; associative thinking.

## 1 Introduction

Geometry serves a pivotal role in architectural design however the specific nature of relationship between the two continues to be redefined over time. March and Steadman (1971) highlighted one such shift from the use of geometry "... to measure [metric] properties of space such as area, volume, angle ..." to descriptions of "structural relationships" of space such as 'adjacent to', 'contained by'. As an example, March and Steadman analyse three very dissimilar architectural house plans by the American architect Frank Lloyd Wright to reveal the presence of exactly the same spatial organization in all three houses, i.e. they are topologically equivalent.

The diffusion of computing in architectural design has fostered renewed exploration of structural relationships as generators of spatial configurations. To facilitate such higher level thinking among architecture students while designing by scripting, we describe here DDA\Repository, an online repository of diagrams that is associative, incremental and dynamic.

## 2 Prototype Overview

The prototype DDA\Repository supports different functionalities and user profiles. The stored diagrams and associated data are browsed via a dynamically generated interface with image tiles of stored design projects (Fig. 1, left). Each tile is clickable and leads to a second level display comprising two interface panes (Fig. 1, right). The

P. Cox, B. Plimmer, and P. Rodgers (Eds.): Diagrams 2012, LNAI 7352, pp. 270–276, 2012.

left-hand pane displays image(s) of the currently selected design project, textual annotations and diagram(s) associated with a specific design developmental history. The right-hand pane displays image tiles of the all other projects in the repository.

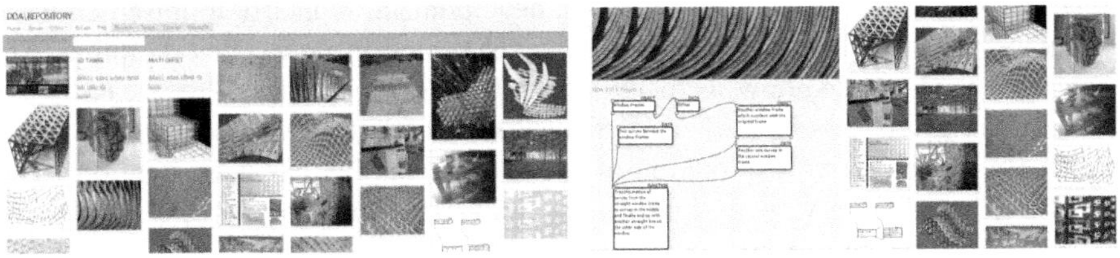

**Fig. 1.** Browsing in DDA\Repository

# 3 Motivating Context

The diffusion of computing in architectural design fosters cultural practices in which representational tools demand even more attention than before. The greater expressive power of contemporary modelers with multiple interaction modes including direct manipulation, scripting, visual programming, and external plug-in architectures comes at a cost. Users now need to cultivate mental models of more than one interaction mode, each of which may operate in one or more representational spaces including graphical, numeric, or symbolic.

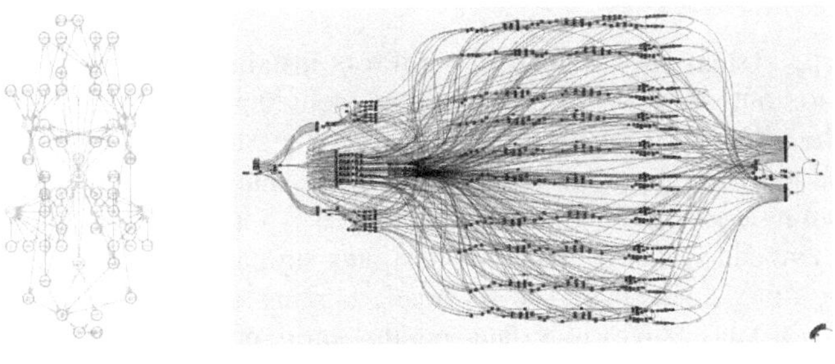

**Fig. 2.** Spaghetti of symbolic representations

There is resurgent interest in visual programming to support interactive three dimensional modelling environments. Although these developments help flatten the initial learning curves, anything but simple spatial compositions rapidly turn into dense spaghetti of symbolic graphs (Fig. 2). Interdependencies become hard to follow and trace, conflicts harder to resolve, and generalisable knowledge harder to distil.

These issues become more problematic when visual scripting environments become embedded in existing design cultures. One central tenet of design is to imagine that which does not yet exist. The process rarely follows a linear, uni-directional unfolding; sometimes it revisits and retrieves discarded design fragments and paths. The possibility to record sequential state histories in contemporary interactive systems might encourage divergent design exploration. However, complexities and effort required to manage visual representations of design scripts very often thrust projects along a uni-directional developmental trajectory.

Similar problems arise when architecture students embrace scripting and computational thinking. Whereas traditional design development implicitly followed an intuitive blackbox model, now we require students to design by explicit scripting. The challenges then become how to foster higher level compositional design knowledge without being too closely tied to particular software, one that supports divergent thinking and associative exploration of ideas, and one that can incrementally grow over time. Further, we view design compositions as a *process of making* and not just as a *final state description* as in traditional design drawings.

### 3.1 Related Ideas

Our work draws upon recurrent themes in development of computational approaches suited to the designerly modes of thinking. The notion of design patterns appears in vernacular architecture (Alexander, 1977) and formalized rule books (Durand, 2000). The notion of levels of abstractions in design appears in cognitive studies of designers (Gero, 1998) to pedagogical traditions in design (Fischer et. al., 2000). The notion of internalizing and using design moves of various grain sizes appears elsewhere in how designers think and develop models of software systems to effectively and efficiently use those tools (Coyne el. al., 1993; Pantazi, 2008; 2010; Woodbury 2010).

## 4 Prototype Architecture

The prototype system uses standard WordPress installation as the primary backend component to compose posts of different data including text, images, video, etc. Diagrams can be added to this collection of media using an interactive diagram builder implemented in Flash. Diagrams can be grouped (displayed as tabbed pages) based on a simple naming convention. The diagram builder is integrated with PhP which turns text strings embedded within diagrams into tags similar to how they are used in the standard WordPress metadata. The automatic tagging of text chunks in diagrams allows specific instances of various data and also entire posts to be associated with each other. The front end uses a combination of PhP scripts and JQuery functions to generate dynamic CSS/HTML code to format and render user-level information display.

## 5 Prototype Functionalities

The prototype system revolves around three kinds of users. The 'reader' role allows anyone, anywhere on the net to browse and query different information chunks including diagrams. The 'creator' role enhances 'reader' privileges with creation and modification of diagrams and posts generated by registered users. The 'admin' role has all the privileges including moderation and publishing of diagrams and posts.

## 5.1 Diagram Creation

The interactive diagram builder supports a small number of container nodes (represented as rounded rectangles): *data, function, object, image* and *fabricate*. Data nodes contain parameter(s) or variable(s) used by other nodes. Function nodes represent a series of operations, possibly using data received from and passed to other nodes. Object nodes represent repeatable elements, e.g. a panel or a rib, with variable properties. Image nodes point to network-accessible image/video data to support visual references within diagrams. Fabricate nodes represent specialised chunks of operations to accommodate translation of generated geometry to the needs of specific fabrication technologies, e.g. laser cutter, 3D printing, etc. Each node has two connectors at top left and bottom left corners to represent input and output to/from nodes respectively.

To create a new diagram, nodes are dragged out from a toolbar onto the blank canvas. Flow of information between nodes is represented by joining output connector of one node to input connector of another node. Free form text annotations can be added to a node and displayed inside its bounding rectangle. Nodes can be repositioned anywhere on canvas; the connecting edges between nodes are maintained by the system.

## 5.2 Uses and Benefits

The prototype implementation supports declaration of compositional intents prior to, during and retrospective to a design project. It is facilitated by interactive composition of a series of diagrams which are associated with other design diagrams via 'tags'.

- *Modular thinking*. The use of diagrams fosters modular design moves in terms of required design elements and operations on them (Fig. 3, left). It encourages a degree of higher level meta-designing instead of being caught up with all the details simultaneously at a small grain size. The explicit recognition of input and output data reinforces the notion that design compositions are materialized outcomes of a temporal assembly process and that correct sequence matters!
- *Idea histories*. A series of diagrams enables occasional look back and (re)tracing of paths followed and perhaps discarded. The use of multiple, alternate diagrams (Fig. 3, right) encourages what-if explorations and provides a record of design history that mitigates premature idea fixation and encourages reflective assessment.
- *Proximate ideas*. The tagged associations between diagrams bring to fore proximate design ideas. When used as a search term (Fig. 4), the system highlights all projects that are similarly tagged. When used from within diagrams by clicking on a string, all similarly tagged projects are highlighted.
- *Grain size of representations*. Diagrams enforce choices about grain size of representations and help generalize or disambiguate specific details. Since the diagram builder provides only a limited canvas, users become selective about what to describe in diagrams with brevity and clarity. It also reinforces encapsulation of ideas leading to clearer organization and definition of scripts.

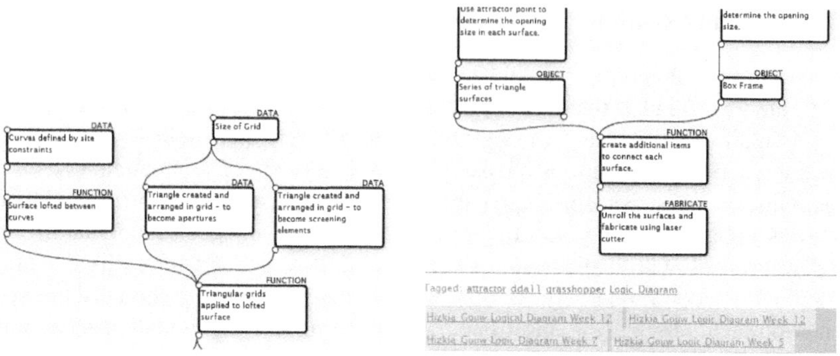


**Fig. 3.** Left: Sequenced and modular design moves. Right: Tabbed idea histories

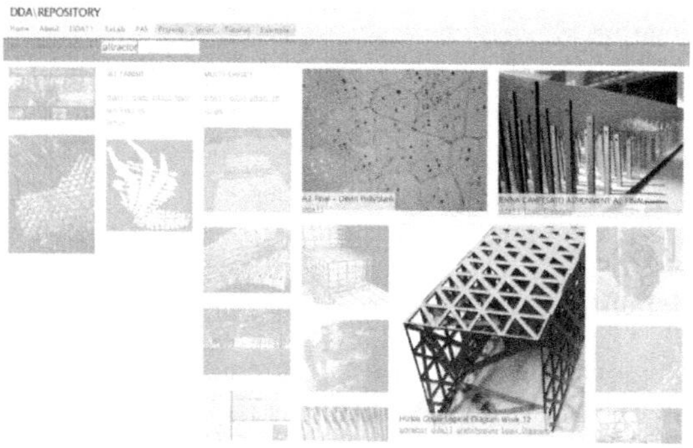


**Fig. 4.** Query and display of proximate design ideas

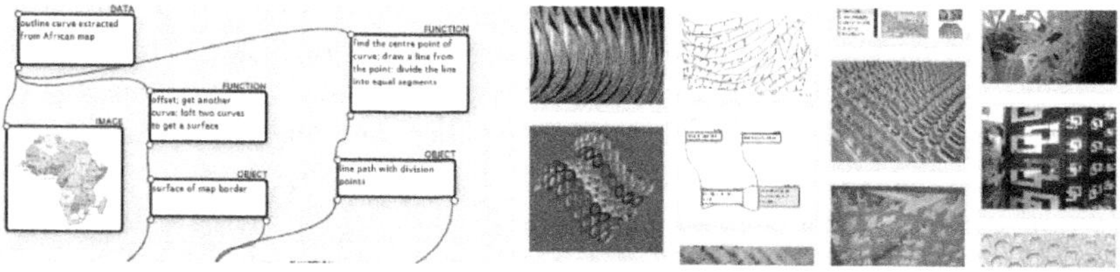


**Fig. 5.** Pointers to local and remote data

- *Recombinant fragments.* Using tagged searches (Fig. 5), diagrams of design intents and operations make possible creative exploration in which retrieved fragments may suggest new compositions and novel design outcomes on the fly.

### 5.3 Evaluation

DDA\Repository has been used in a graduate subject that introduces scripting and fabrication technologies to design purposely small scale projects, e.g. design of a secondary skin over a surface. The students typically face two simultaneous demands: learning to script while also developing design ideas that are to be generated using scripts, and then fabricated and assembled. The students use DDA\Repository as an ongoing part of project development. The following discussion draws upon feedback provided by students using semi-structured interviews in the use of DDA\Repository.

There was a variable pattern to the use of diagrams by students. In the beginning, most students employed diagrams retrospectively, closer to the end of their projects rather than in the beginning. Other students found it useful to begin with graphical sketches which were then transcribed into diagrammatic representations.

Some students tentatively explored the available range of operations in the scripting environment first, scoping out possible design configurations. Here the scripting operations appear to act as symbolic 'sketches' though keeping a persistent record in some form was not a priority for most students. As one student put it: "*... I was new to [*scripting environment*] ... I didn't know and needed to work out what I didn't know*". As a result, she used diagrams as "a description of process" rather than as a "planning tool". However, with experience, she also noted that "*... thought was given as to how to better plan the scripting process by using the logic diagram tools.*"

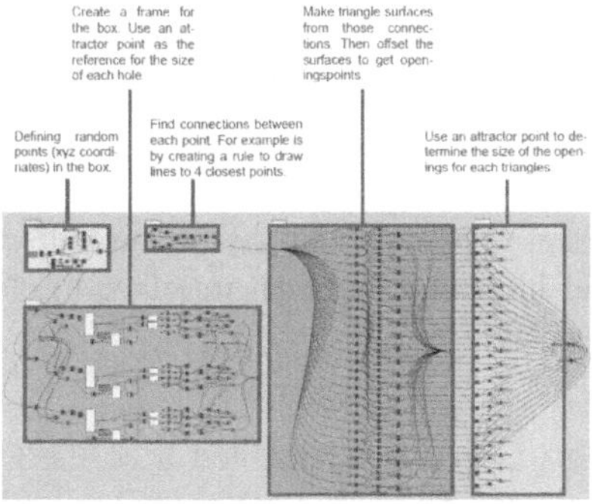


**Fig. 6.** Abstracting and navigating different representational spaces

Having generated a series of diagrams, she "... understood a lot more" about both the range of design possibilities she could explore and also alternative computational procedures through which to realize them. This retrospective abstraction highlighted that her initial efforts were limited by the handful of scripting functions she knew which, in turn, limited her design ambitions. The use of diagrams allowed her to step back from these limitations and reframe design intentions in a way that was independent of what she knew or the limits imposed by a particular scripting environment.

Another student employed groups of nodes distinguished by different colors to translate back and forth between diagram and executable scripts (Fig. 6). The visual correspondence between the two representations allowed him to develop a better documented diagram and executable code with generalised, reusable design operations.

The use of "free-form" text in diagrams and its translation in executable scripts made another student reflect on different ways to achieve similar geometric outcomes by following different pathways. For example, operations of lofting and extrusion of geometric primitives may result in visually similar shapes but each operation involves very different parameters and intermediate processing steps.

The use of diagrams offered others with scaffolding for design search. A student with limited design and scripting experience noted: "... *in the beginning, I have no idea what I am doing or why*". The use of diagrams and free form text gave him "*goals ... a direction*", helped by looking at others' pseudo-code and diagrams.

## 6 Future Extensions

DDA\Repository is accretive, exploratory, and independent of particular software environments. It is accretive and associative, i.e. accommodates work of multiple authors with tagged connections between them. It is exploratory in that it fosters divergent and lateral thinking rather than rapidly converging on to a solution. It fosters higher-level compositional skills instead of being constrained by software.

We are aware of the potential trade-offs between formal syntactic specificity of diagrams and designerly creativity as we consider the following future extensions.

- Enable levels of details for nodes to support embedding of diagrams within diagrams as a way to encourage problem subdivision.
- Allow diagram nodes to refer to structured pseudo-code. It will help explore alternative computational sequence of operations in a given design context.
- Add contextual pop-ups for interactive input of freeform text strings in diagram nodes to suggest alternative visual examples and pseudo-code fragments.
- Develop a multi-level hierarchy in visual interface to cluster and organise design examples and their display.

**Acknowledgements.** This project has benefited from the work and feedback from a number of students enrolled in the subject *Digital Design Applications* during 2008-2011. The current version of DDA\Repository is accessible online at http://scripts.crida.net/gh/dda11.

## References

1. Alexander, C.: A Pattern Language. Oxford University Press, New York (1977)
2. Coyne, R.F., Flemming, U., Piela, P., Woodbury, R.: Behavior Modeling in Design System Development. In: CAAD Futures 1993, Pittsburgh, pp. 335–354 (1993)
3. Durand, J.-N.-L.: Précis of the lectures on architecture; with, Graphic portion of the lectures on architecture. Getty Research Institute, Los Angeles (2000)
4. Fischer, T., Burry, M., Woodbury, R.: Object-Oriented Modelling Using XML in Computer-Aided Architectural and Educational CAD. In: The Problem of Interoperability Exemplified in Two Case Studies, CAADRIA 2000, Singapore, pp. 145–155 (2000)
5. Gero, J.: Concept formation in design. Knowledge-Based Systems 10(7-8), 429–435 (1998)
6. March, L., Steadman, P.: The geometry of environment. RIBA, London (1971)
7. Pantazi, M.: Using Patterns of Rules in the Design Process. In: Critical Digital: What Matters(s)?, pp. 349–356. Harvard University, Cambridge (2008)
8. Woodbury, R.: Elements of Parametric Design. Routledge, London (2010)

# Structure, Space and Time: Some Ways That Diagrams Affect Inferences in a Planning Task

David L. Mason[1], James E. Corter[1], Barbara Tversky[1], and Jeffrey V. Nickerson[2]

[1] Teachers College, Columbia University, New York, USA
{dlm2153,jec34}@columbia.edu,
btversky@stanford.edu
[2] Stevens Institute of Technology, Hoboken, USA
jnickerson@stevens.edu

**Abstract.** An efficient way to notify a set of people is to use a calling tree, where one person calls a few people who call others until everyone has been notified. Calling trees are typical of a large class of planning tasks that entail considering both the structure of agents and tasks in time. Participants were asked to choose the optimal diagram for a calling tree problem, and to compute the time needed to call everyone. Participants computed more accurately when the tree diagrams were scaled to represent elapsed time as well as the connection structure of the callers. In addition to efficiency, both gestalt factors and social equity considerations biased selection of the best diagram.

**Keywords:** diagram understanding, planning, inference, comprehension, representing time.

## 1 Introduction

Planning requires representing actors, tasks, and their relationships in both space and time. For complex planning tasks like business decisions, people often rely on external representations to aid them in making, modifying, and using plans. These external representations can be as simple as calendars, or as complex as charts of multi-year plans for implementation of corporate mergers.

One class of planning problems is a distribution plan. A familiar example of this class of problems is a calling tree. In a typical calling tree problem, information must be distributed quickly and reliably to a predetermined group of people. The message must be delivered in one-on-one conversations, to insure that the message has been delivered, and perhaps because of technological limitations or concerns of security or privacy. When more than a few people are involved, designing a calling tree is a non-trivial task, in part because there are often multiple objectives that may be in opposition. Some goals make a tree with many callers more desirable, for example when all people should be notified as quickly as possible or when sharing the burden of calling

P. Cox, B. Plimmer, and P. Rodgers (Eds.): Diagrams 2012, LNAI 7352, pp. 277–290, 2012.

seems desirable. By contrast, other goals, such as concerns about security or possible degradation of the message, favor a tree with few callers.

Many knowledge structures can be represented as network diagrams. Networks consist of nodes and links, where the nodes represent concepts like agents, objects, places, or ideas, and the links represent the relationships among them. A calling tree can be represented as a specific kind of network diagram, a directed tree, where a single message is to be propagated outwards from a single source. The nodes represent the callers and/or callees and the directed links represent the direction of calls. As such, the tree represents the structure of connections among the callers and callees, explicitly showing the set of people called by each successive caller.

Two types of trees, seen in Fig. 1, can be used to represent a calling tree plan. Calling Tree A, termed a *uniform* tree, emphasizes the structure of the calling plan, but also shows the sequence of calls made. In the uniform tree the height of nodes on the page is proportional to depth in the tree, or "generation" of each caller/callee. However, because the time to complete all the calls can be important, a network diagram that represents elapsed time directly would be useful. Calling Tree B, termed a *time* tree, uses node height to explicitly represent the time elapsed before each callee is notified (assuming that each call takes a constant amount of time). This tree gives more emphasis to the temporal structure of the calls, and demonstrates that calls by a single caller happen sequentially, while calls by different callers can happen in parallel.

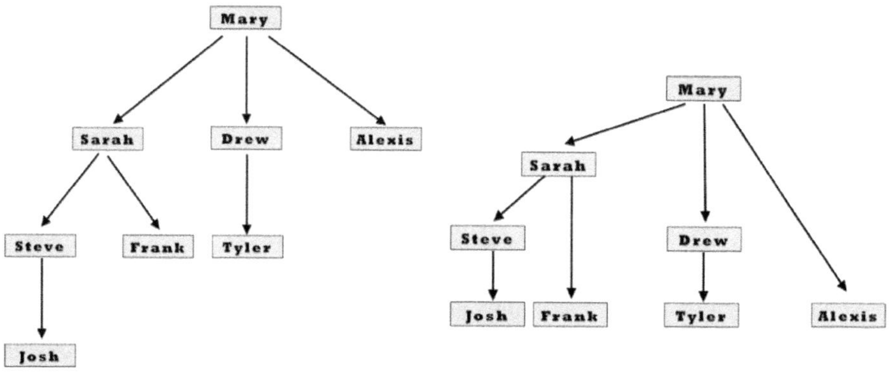


*A* *B*

**Fig. 1.** Two ways to represent the same eight-node calling plan as a tree: the "uniform tree" in panel *a* lines up nodes on the vertical dimension according to generation or depth in the tree; the "time tree" in panel *b* lines up nodes on the vertical dimension according to elapsed time under the assumption that calls have equal durations

Representing elapsed time directly, as with the time tree in Figure 1b, should facilitate reasoning about the time to complete a task – in the case of calling trees, informing everyone. In a time tree, the position of nodes along the vertical dimension represents elapsed time. If all calls are assumed to take a constant amount of time, then the set of people called by the same elapsed time are lined up horizontally, and the sum of the number of units of time elapsed is directly indexed by the sum of the levels of the tree. Metric time can be computed directly from a time tree by multiplying the number of levels by the time required for a call. Calculating metric time from a uniform tree is indirect and onerous. However, time trees are more complex than uniform trees because the nodes and lines not only code spatial structure, the links from caller to callee, but also code metric time.

Perhaps because of the extra complexity of time trees, when participants are asked to produce representations of calling trees and use them to compute the time needed to complete the calls, many of them draw networks that represent the structure and order of calls, but do not represent time directly [1]. Participants for that previous study, as for this one, were novices, with little if any experience constructing representations for and solving planning problems. However, even novices may recognize and understand the use of vertical position on the page to represent elapsed time. If so, when time is directly represented, they should be better able to select optimal designs among alternative calling tree designs and they should be better able to compute the total time to complete the calling tree. Thus, in the present study we turn from production to preference and performance. Participants will be asked to select alternative calling tree designs either for a set of time trees or a set of uniform trees, and then to compute total time. Although time trees allow more efficient calculation of total time, they are more complex, hence harder to comprehend, so they might not be preferred, and they might not facilitate computation.

The second goal of the present study is more exploratory. Diagrams can be a boon to reasoning because they extract the essential information and present it directly, using place in space and spatial relations to represent crucial aspects of problems [2]. However, diagrams are visual objects in themselves, and extraneous characteristics of the spatial array may mislead participants [2-4]. A secondary goal of the present research is to investigate whether a visually salient, rapidly recognized but irrelevant feature of diagrams affects preference, namely, symmetry [5-8]. The significance of symmetry in the human visual system is undoubtedly related to its importance as a cue to recognizing biological entities, notably people, as well as artifacts designed to serve people. More abstractly, symmetry connotes *balance*, the same features, the same weight, on each side of the axis of symmetry. In the case of network diagrams, a symmetric tree might suggest a balance of burden or responsibility. Another visual aspect of a calling tree diagram that may affect preferences is the depth of the tree (in the sense of the number of levels of the hierarchy). People may believe that the depth of a tree can be used as a reliable cue to find the minimal-time plan, although because of parallelism this is not always true. Finally, extraneous social-pragmatic concerns might play a role in people's judgment of an optimal calling plan [12]. Will symmetry and other visual features of diagrams affect choice, even when irrelevant?

### 1.1 Selecting and Using Planning Trees

The experiment was designed to investigate how specific aspects of the diagram representing the calling tree plan would affect preferences and performance. Would participants provided with diagrams representing time directly ("time trees") more often choose the diagram representing the best plan and calculate the time to affect the plan more accurately, compared to participants provided with diagrams that only represented the structure of the calling plan ("uniform trees")? Next, would visual aspects of the diagram irrelevant to the task, or related to measure of equity, affect diagram selection? One group of participants selected from among five *uniform* trees representing calling plans, then calculated the time required to complete the selected plan; another group of participants selected among plans represented by *time* trees, then calculated required time. We used three different sets of uniform trees and three different sets of time trees (between participants) to ensure generalizability of our results. Also, we were able to vary aspects of the trees such as symmetry, depth, and equity across the distractor diagrams of each problem.

## 2 Method

### 2.1 Participants

Participants were recruited from a crowdsourcing website, Amazon's Mechanical Turk (AMT). This website is designed to allow "requesters" to pay volunteer workers to complete tasks for which computers are ill equipped, so-called "human intelligence tasks" or HITs. These tasks range from tagging images and websites for Internet search optimization to simple psychological experiments and may take as little as 30 seconds to complete. There were N=139 participants. Based on participant answers to several demographic questions, the sample was 54% female, and the mean age was 34 (with a range from 20 to 84). 75% of them reported English as their first spoken language, and 91% of respondents had attended college. Some programming experience was reported by 29%; only 9% had more than four years of such experience. Each worker in AMT has a unique ID number; these were used to ensure that no worker did the study more than once.

### 2.2 Stimuli

Each participant was presented with five diagrams that depicted different plans for calling friends. These diagrams were presented either as "Uniform" trees or "Time" trees. In the Uniform tree condition, the five calling-plan trees were presented so the nodes in the tree were lined up vertically based on "generational" relationships. That is, all the people called by a particular caller are represented as aligned on a horizontal line. And all the callees called by callers in that row are represented on the next level, so the relationships between callers and callees resemble a genealogical tree. In the Time tree condition, the same calling plans were presented but the nodes were aligned vertically according to the time they were actually called (participants were asked to assume that all calls took 1 minute). That is, calls that occurred simultaneously are represented by aligning them on the vertical axis so calls during the first minute are aligned on the first horizontal line, those in the second minute on the second line, and so on (please see Fig. 2 for examples of both conditions). The connectivity between nodes and position on the horizontal axis remained identical between these conditions – only the position on the vertical axis differed.

We designed these calling plan problems with seven, eight, or nine actors (factor Nodes). Thus there were six conditions for the experiment: 7-Node Uniform, 7-Node Time, 8-Node Uniform, 8-Node Time, 9-Node Uniform, 9-Node Time. Each of the corresponding six problems presented five alternative diagrams to participants, making thirty diagrams in total.

### 2.3 Procedure

Participants were randomly assigned to one of the six conditions. Each calling plan problem was presented in the form of text that described the need to distribute information and the task objectives. The problem text for the 8-node problem was:

> *Mary has invited 7 friends to a party tonight. But she just came down with the flu and needs to contact them all to cancel. She (and her friends) have only cell phones to communicate with. Assume that her friends are willing to help her make the calls, and each call will take approximately 1 minute. Design a plan to notify everyone as quickly as possible by selecting which illustrates the fastest plan from the 5 diagrams below. (You may mark more than one)*

This problem text was identical across conditions except for the number of friends Mary had to call. Additionally, to assure that participants interpreted both types of tree as representing ordinal time relationships by the horizontal ordering of links, the participants were informed that:

> *In the diagrams below, assume that the first call each caller makes is represented by the left-most arrow in their branch and the last call a caller makes is represented by the right-most arrow in their branch.*

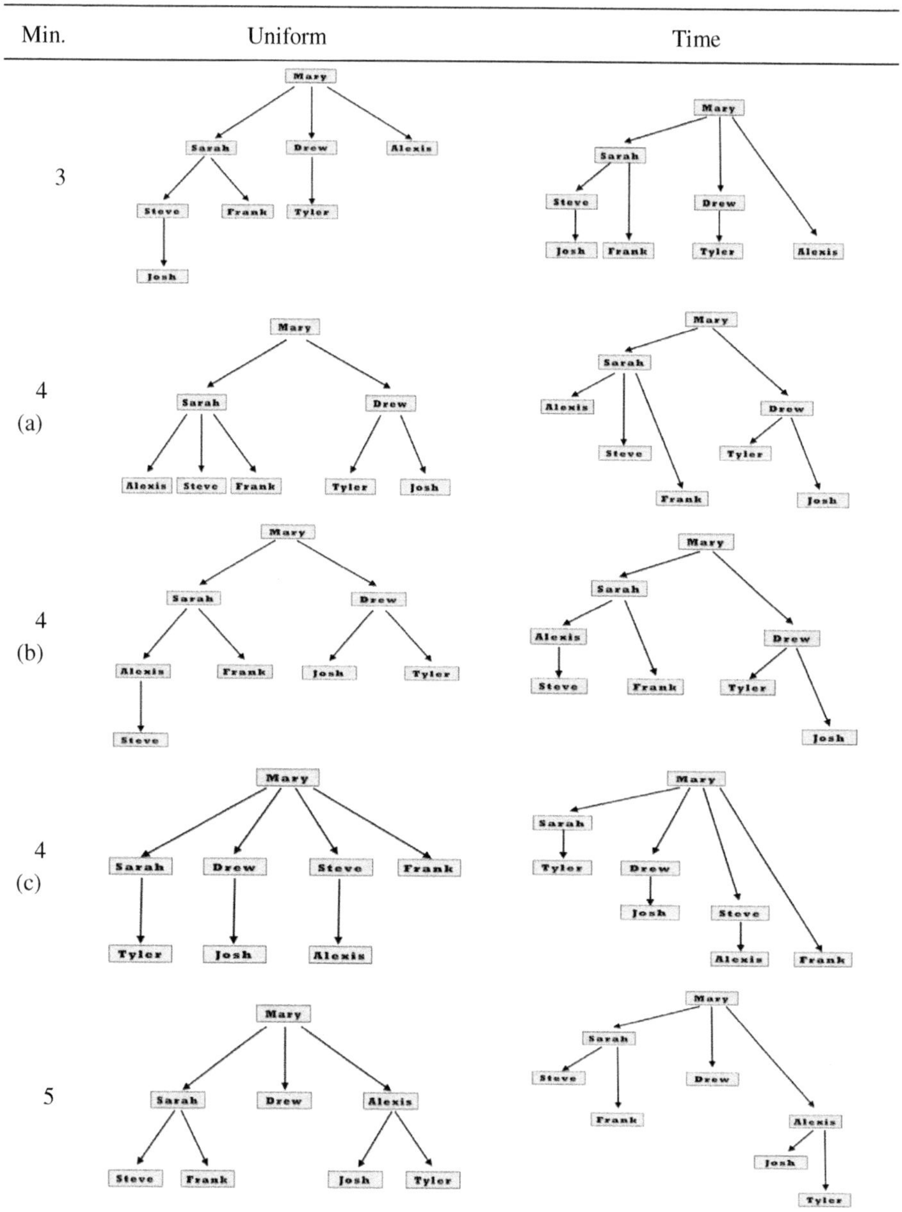


**Fig. 2.** Alterative diagrams used for the 8-node problem. Left column: diagrams used in the Uniform trees condition; Right column: diagrams used in the Time trees condition. The time for all calls to be completed is in the Minutes column.

This description was followed by presentation of five randomly ordered tree diagrams that depicted the different plans for calling friends. Participants were not given explicit instruction in the meaning of alignment along the vertical axis; we were interested in whether they would spontaneously adopt the appropriate interpretation (as evidenced by correct task performance).

After participants selected which of the five diagrams represented a calling plan that would notify all the friends the fastest, they were asked to specify how long in minutes the plan would take to complete (assuming that each call lasted approximately one minute). Lastly we asked participants the demographic questions. The entire task took an average of three and a half minutes. Participants were paid $.25 for their participation. Feedback from the participants, when offered, was entirely positive, to the effect that the task was engaging and enjoyable.

## 3 Results

Two dependent variables are of interest: choice of the correct tree diagram (i.e., the one that represents the plan with shortest completion time) and correctness of the answer given to the time computation question (the actual time required for completion of the calling plan). Because we allowed participants to choose more than one of the diagrams, a correct answer to the tree choice entails choosing only the correct diagram. Results for the six conditions are shown in Table 1.

**Table 1.** Performance by condition: percentage correct for choice of best tree ("choice") and time computation ("time"), with percentage of respondents getting both questions correct ("both"). Total N = 139

| | 7-node | 8-node | 9-node | Overall |
|---|---|---|---|---|
| Uniform | (N=36) | (N=22) | (N=20) | (N=78) |
| choice | 36 | 27 | 40 | 35 |
| time | 39 | 36 | 45 | 40 |
| both | 28 | 23 | 25 | 26 |
| Time | (N=22) | (N=16) | (N=23) | (N=61) |
| choice | 64 | 38 | 57 | 54 |
| timc | 55 | 44 | 61 | 54 |
| both | 55 | 38 | 39 | 44 |

It can be seen from Table 1 that the time tree diagrams facilitated correct computation of the time needed for completion of all calls, greatly increasing the rate of choosing the optimal tree ("choice"), of correctly answering the time inference question ("inference"), and of correctly answering both questions ("both"). The advantage of the time tree diagram was confirmed using a generalized linear model with a logit link function, with Type of Tree (= uniform, time) and Nodes (= 7, 8, 9) as factors. In the analysis predicting whether participants chose the correct diagram and correctly answered the time question, the effect of Type of Tree was significant, Wald $X^2$(d.f.=1) = 4.850, p=.028, while the effect of Nodes and the Type*Nodes interaction were not significant.

Figure 3 shows that for all three calling tree problems (those with 7, 8, and 9 nodes) performance was improved when participants selected among time tree diagrams, compared to when they selected among uniform tree diagrams. Specifically, in the Time Trees conditions more people chose the diagram illustrating the optimal plan: the 7-node 3-minute plan (diagram "7_3), the 8-node 3-minute plan ("8_3"), and the 9-node 4-minute plan ("9_4"). Furthermore, the Time Trees conditions decreased the probability of selecting the incorrect diagrams (including the incorrect response of selecting two or more diagrams as optimal, denoted "2+" in Fig. 3). However, there were several exceptions, where incorrect alternatives were more often chosen with time trees. These exceptions tend to be cases where the uniform tree is symmetric, and the time tree is symmetric (but on a bias). One example is the 8-node 5-minute plan (see Fig. 2), which was selected more often in the Time Trees condition. Another example is the 9-node 5-minute plan shown in Fig. 4.

These results suggest that people are choosing diagrams without formally (or at least correctly) optimizing the calling plan completion times: only 54% of respondents select the optimal tree even in the Time Tree condition (and only 35% in the Uniform Tree condition). If people are not correctly optimizing, on what basis do they make their diagram choices? The finding mentioned in the previous paragraph, that people can be misled by symmetric trees, suggests that people may sometimes use heuristics to select a tree. In other words, people may be biased to select symmetric trees, simply because a symmetric tree seems a good design that might be presumed to be relatively efficient. Other aspects of the tree diagrams might also play a role in choice heuristics: the overall depth of the tree (i.e., the length of the longest chain from the root to any leaf) might influence participants, in that they might be disposed to choose the shallowest trees (even though a shallow tree may correspond to a very slow calling plan, as when the "root" caller calls all the friends one by one). Finally, the rooting of this problem in a situated pragmatic context (of a calling tree to be created among friends), may bias people towards certain types of solutions. As an example, in this domain people might see an additional optimization criterion to be taken into consideration, namely social equity. That is, it might be that people think, consciously or unconsciously, that the burden of making the calls should be shared across the members of the group, distributed as evenly as possible.

We investigated the idea that participants might be influenced in their choices by these visual aspects of the diagrams and social-pragmatic aspects of the underlying problem. To do so, we created a database of all 30 diagrams used in the present study, plus another 25 diagrams used in our associated pilot studies (the pilot studies had only trivial differences in stimuli, procedures, or instructions). First, we recorded the optimality criterion that can be used to order the trees objectively (i.e., total completion time for the corresponding calling plan). Then we coded each diagram on certain key visual properties, including the degree of symmetry of the graph, total depth of the tree, and two alternative measures intended to capture the notion of social equity (i.e., how evenly or unevenly the burden of making calls was spread among the friends).

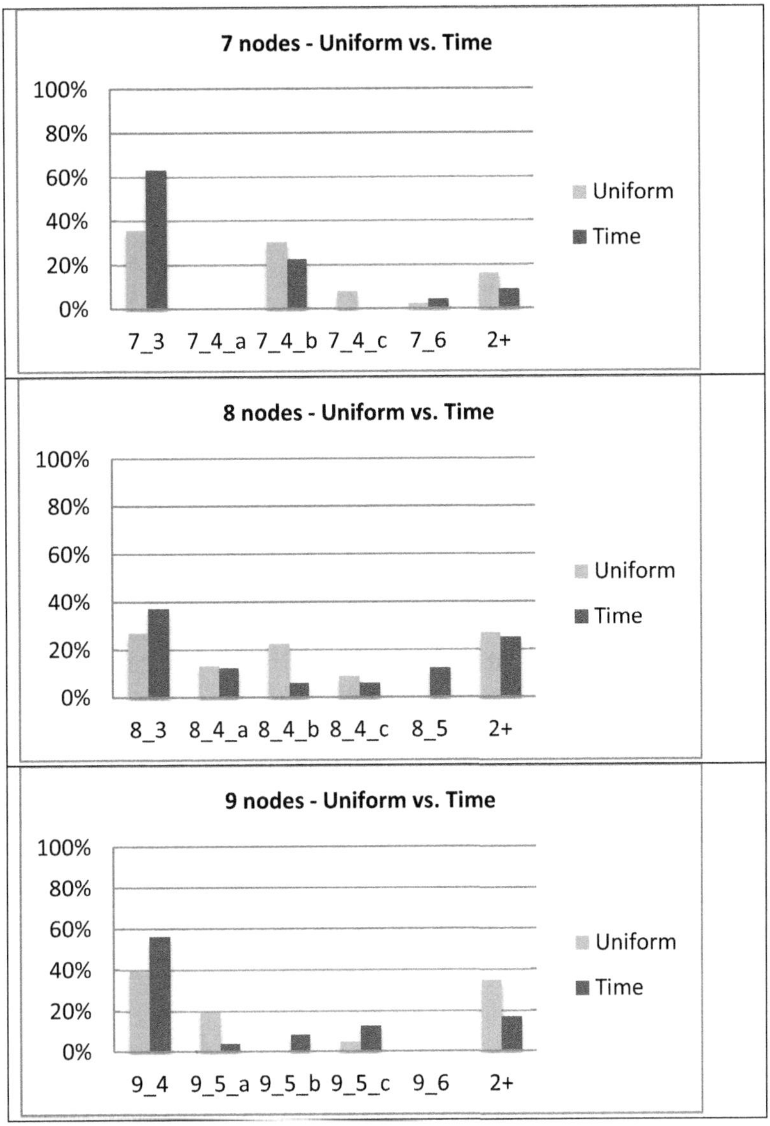


**Fig. 3.** Proportion of participants in each condition choosing each specific diagram (correct diagram shown on left; participants choosing more than one diagram as "optimal" shown at right, labeled "2+"). Diagrams are labeled as <Nodes_Time>, so Diagram "7_4_a" means the 7_node 4 minute tree (version a).

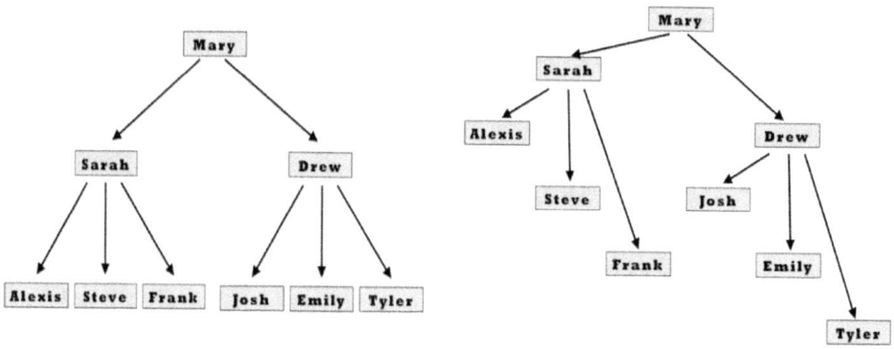


9-node 5-minute (version c) Uniform tree 9-node 5-minute (version c) Time tree

**Fig. 4.** An incorrect diagram for the 9-node problem, chosen as optimal more often in the Time Trees condition than in the Uniform condition

To explore reasons why participants err in selecting the optimal tree/plan, we deleted the optimal tree (i.e., the correct answer) for each problem, leaving N=44 incorrect diagrams in the database, then correlated the proportion of participants choosing each incorrect diagram with certain visual characteristics of the tree diagram. The results are shown in Table 2. Because the choice probabilities are not independent across the 44 coded diagrams, the p-values in Table 2 cannot be interpreted as inferential tests – they are shown for descriptive purposes only.

**Table 2.** Simple and partial correlations (N=44) between the proportion of participants choosing an incorrect diagram and certain coded visual aspects of diagrams (see text). The partial correlations were obtained in multiple regression analyses predicting the choice proportion for each (incorrect) diagram from the named visual aspect and actual time required for completion of the plan.

| Aspect: | Simple corr. | | Partial corr. | |
|---|---|---|---|---|
| | R | p | R | p |
| Time | -.266 | .081 | - | - |
| Symmetry | .390 | .009 | .564 | <.001 |
| Depth | .408 | .006 | .320 | .036 |
| Equity | .398 | .007 | .347 | .022 |
| N_callers | .343 | .022 | .263 | .088 |

The correlation of -.266 between Time and choice proportion shows that a shorter completion time for a plan is only marginally correlated (p=.081) with the probability that the corresponding (incorrect) diagram is chosen. Thus the objective criterion of completion time does not seem to capture all, or even most, of the variance in the choice proportions.

Symmetry was defined as a topological property of the tree, and was not defined to be all-or-none, rather it was assessed separately at each level of the tree (below the root), then the overall proportion of levels that showed symmetry was computed. For example, in Fig. 4 both trees are balanced or symmetric at the two bottom levels, thus they are given symmetry score 2/2 = 1.0. In Fig. 2, the 3-minute diagram has three levels below the root. At the first level, the pattern of nodes is symmetric or balanced around the vertical axis, while at the bottom two levels it is not, thus the symmetry score for this diagram is 1/3 = 0.33. In Table 2, the positive correlation (r=.390) between choice probability and degree of symmetry of the diagram means that more symmetric "distractor" diagrams are more often chosen; people seem to be attracted to symmetric diagrams in choosing what they believe to be efficient distributions plans. This positive relationship is not due to a correlation of symmetry with the objective criterion of required time – in fact, the relationship becomes even stronger (r=.564, p<.001) when completion time is statistically controlled for in a multiple linear regression predicting the choice proportion for each diagram (Table 2).

The depth of the tree, defined by the number of (vertical) levels of the uniform tree, is also positively correlated with choice probability. The direction of the relationship means that "deeper" trees are more often chosen, or alternatively that participants avoid choosing broad shallow trees. This may happen because participants are aware that parallelism is good: i.e., when a caller calls several friends (say, A then B then C), callee A has time to make further calls while B and C are being contacted, thus deeper trees (at least in these context sets) may tend to be the most efficient from the standpoint of total elapsed time.

Two potential measures of social equity of the calling plan were devised. First, note that the maximum number of callees of any caller (denote this quantity C) can be taken as a rough measure of the social inequity of a calling plan. For example, diagram 4(c) in Fig. 2 (C=4) seems more inequitable than diagram 4(b) (C=2), while the 7-node 6-minute tree in Fig. 5 (C=7) seems more inequitable than either. Thus, one possible measure of social equity is simply: Equity = 1 - C. This measure is positively correlated with a diagram's probability of being chosen. Another rough measure of equity, based on how widely distributed is the burden of making calls, can be calculated as simply the number of friends who take on the role of making calls. This measure, denoted "N_callers") is also positively correlated with the probability that a particular incorrect diagram is chosen.

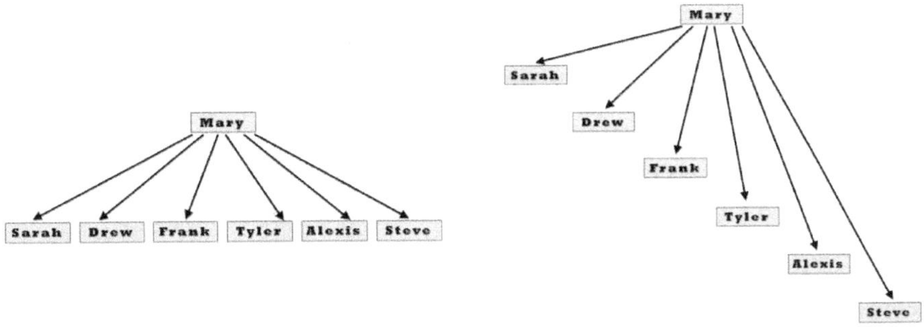


7-node 6-minute Uniform tree 7-node 6-minute Time tree

**Fig. 5.** An incorrect diagram for the 7-node problem, representing a highly inequitable plan

## 4 Discussion

The problem of disseminating information quickly and accurately to a set of people, a calling tree, is representative of many planning tasks. Designing an effective distribution plan involves reasoning about many components that may not be readily apparent. In the case of calling trees, designers need to take account of the structure of callers as well as the elapsed time. Designing effective distribution trees can be facilitated by expressing the problem in diagrams that use elements and their spatial relations to represent key aspects of the problem. When asked to produce diagrams of specific calling trees, novice but educated designers typically create diagrams that represent the structure of callers, but only rarely represent elapsed time explicitly [12]. Diagramming spatial structures seems to take precedence over diagramming temporal ones. Even young children readily diagram spatial structures earlier than temporal ones [13]. Moreover, adjusting line lengths to represent both spatial and temporal structure is complicated, and, as noted, few novices represented time.

Here we asked whether novices would recognize and use explicit representation of time in order to select optimal designs and to compute the time to complete the calling tree. The answers to both questions were positive. Participants more frequently selected optimal designs and computed total time accurately when they chose among diagrams that represented time directly than when they chose among diagrams that did not represent time directly. This held despite the added complexity of the diagrams that represented time directly.

Performance in both tasks was far from perfect, even for diagrams that represented time directly. The errors were revealing. Many appeared to stem from irrelevant but salient visual features of the diagrams, notably symmetry. The bias toward symmetric calling trees could be a purely visual preference driven by perceptual biases or esthetic leanings, or it might be a more abstract inference that symmetry is a consequence of desirable properties of a calling tree, for example, that effort is more evenly distributed among the callers/callees. Support for the idea that participants were using this abstract property of symmetry is provide by the finding that incorrect diagrams were selected more often when they had a more equal, thus more equitable, distribution of effort across actors. Thus, equity considerations seem to bias people towards plans that distribute work responsibilities more broadly and more fairly, even though the present task asked participants only to find the fastest plan. People were also biased toward "deep" trees over broader shallower trees, even when the actual time required by plans was statistically controlled in a multiple regression. This bias might again reflect equity considerations, since very broad shallow trees are typically those in which many calls are placed by the original source, or by the first few people contacted. Shallow plans may be more efficient in terms of time, but may seem inequitable in some social/pragmatic contexts.

Our previous studies showed that novice designers fail to produce good designs for representing and solving problems because of failures of imagination; they represent the spatial structure of participants in a calling tree but rarely represent elapsed time. The present findings show that people do not always evaluate or use designs correctly, a failure of comprehension or judgment. The failures of judgment may be due to the sheer complexity of the diagrams, using lines for spatial structure and node position on the page to represent elapsed time, but it is also seems to be due to invoking irrelevant criteria, namely, equity considerations. However, the findings also show that selecting the optimal diagram improves accuracy of problem solving. The implication is that ways to induce designers to select optimal designs must be sought, that is, ways to insure comprehension of the relevant features of the diagrams and ignoring of irrelevant features.

As such, these findings appear to have implications for other classes of problems and for even other classes of diagrams. Networks and trees are used to represent countless situations, traffic patterns on roadways and the internet, spread of disease and rumors, distributing supplies and information. Salient but irrelevant visual factors like symmetry and depth of tree are likely to affect reasoning in those situations as well. More generally, many other types of planning and monitoring problems involve multiple agents who must coordinate in both space and time. Sometimes the space is physical space, sometimes virtual; the relevant structure may involve only spatial relationships or connectivity as well. In these tasks one challenge is to find effective diagrams that can represent structure and time simultaneously. Choosing an effective representation is a crucial step in design, one that offers opportunities to improve problem understanding and performance.

**Acknowledgments.** The authors are grateful to grants National Science Foundation HHC 0905417, IIS-0725223, IIS-0855995, and REC 0440103, the Stanford Regional Visualization and Analysis Center, and Office of Naval Research NOOO14-PP-1-O649, N000140110717, and N000140210534 for partial support of the research reported.

## References

1. Nickerson, J.V., Tversky, B., Corter, J.E., Yu, L., Mason, D.: Thinking with Networks. In: Carlson, L., Hoelscher, C., Shipley, T.F. (eds.) Proceedings of the 33rd Annual Conference of the Cognitive Science Society, Austin, TX, pp. 2662–2667 (2011)
2. Tversky, B.: Visualizations of Thought. Topics in Cognitive Science 3, 499–535 (2011)
3. Corter, J.E., Mason, D.L., Tversky, B., Nickerson, J.V.: Identifying Causal Pathways With and Without Diagrams. In: Carlson, L., Hoelscher, C., Shipley, T.F. (eds.) Proceedings of the 33rd Annual Conference of the Cognitive Science Society, Austin, TX, pp. 2662–2667 (2011)
4. Landy, D., Goldstone, R.L.: How Abstract is Symbolic Thought? Journal of Experimental Psychology: Learning, Memory, and Cognition 33(4), 720–733 (2007)
5. Freyd, J., Tversky, B.: The Force of Symmetry in Form Perception. American Journal of Psychology 97, 109–126 (1984)
6. McBeath, M.K., Schiano, D.J., Tversky, B.: Three-dimensional Bilateral Symmetry Bias in Judgments of Figural Identity and Orientation. Psychological Science 8, 217–223 (1997)
7. Wagemans, J.: Detection of Visual Symmetries. Spatial Vision 9(1), 9–32 (1995)
8. Wagemans, J.: Characteristics and Models of Human Symmetry Detection. Trends in Cognitive Sciences 1(9), 346–352 (1997)

9. Locher, P.J., Nodine, C.F.: Influence of Stimulus Symmetry on Visual Scanning Patterns. Attention, Perception, & Psychophysics 13(3), 408–412 (1973)
10. Carmody, D.P., Nodine, C.F., Locher, P.J.: Global Detection of Symmetry. Perceptual Motor Skills 45(3 part 2), 1267–1273 (1977)
11. Roddy, G., Gurnsey, R.: Mirror Symmetry is Subject to Crowding. Symmetry 3, 457–471 (2011)
12. Yu, L., Nickerson, J.V., Corter, J.E., Tversky, B.: The Shifting Shape of Collaboration: The Effect of Hierarchy on the Topology of Communication Plans. In: The Ninth Annual SIG IS Cognitive Research Exchange Workshop (2010)
13. Tversky, B., Kugelmass, S., Winter, A.: Cross-cultural and Developmental Trends in Graphic Productions. Cognitive Psychology 23, 515–557 (1991)

# What Can Concept Diagrams Say?

Gem Stapleton[1], John Howse[1], Peter Chapman[1], Ian Oliver[2], and Aidan Delaney[1]

[1] Visual Modelling Group, University of Brighton, UK
{g.e.stapleton,john.howse,p.b.chapman,a.j.delaney}@brighton.ac.uk
[2] Nokia Services, Finland
ian.oliver@nokia.com

**Abstract.** Logics that extend the syntax of Euler diagrams include Venn-II, Euler/Venn, spider diagrams and constraint diagrams, which are first-order. We show that concept diagrams can quantify over sets and binary relations, so they are second-order. Thus, concept diagrams are highly expressive compared with other diagrammatic logics.

Many diagrammatic logics that have been developed to date are restricted to a fragment of monadic first-order logic [1,3,5,6], but some include two-place predicates symbols, such as constraint diagrams [2] which are equivalent to some unknown fragment of dyadic first-order logic. Here, we focus on the expressiveness of concept diagrams [4]. We demonstrate that concept diagrams can quantify over sets and binary relations, as well as elements, thus significantly advancing the state-of-the-art in terms of what can be expressed diagrammatically.

As an example, suppose that the individual *Erica* is a *Person* who is *marriedTo* only the *Person Peter* and that *Erica* owns exactly two pets (identified by the role *ownsPet*), both of which are *Dogs*, including a *Beagle* called *Minnie*. A concept diagram asserting this information uses three closed curves to represent the concepts *Person*, *Dog*, and *Beagle*; see figure 1. The concepts *Person* and *Dog* are disjoint and *Beagle* is a subsumed by *Dog*. The named individuals are represented by labelled dots, with their location telling us of which concepts they are instances; for example, *Minnie* is located inside the curve labelled *Beagle*. The fact that *Erica* owns a set of *Pets* is visualized by the use of the arrow, labelled *ownsPet*, hitting an unlabelled curve. This unlabelled curve is drawn inside *Dog*, to assert that the image of the relation *ownsPet*, when its domain is restricted to *Erica*, is subsumed by *Dog*. The fact that this unlabelled curve contains two trees, called spiders, tells us that *Erica* owns two *Dogs*. *Erica*'s dog

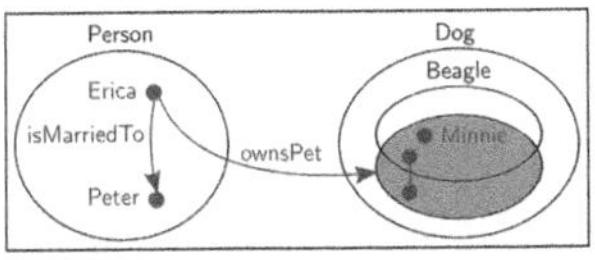


**Fig. 1.** Visualizing roles

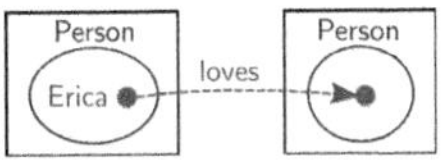


**Fig. 2.** Arrows between boxes

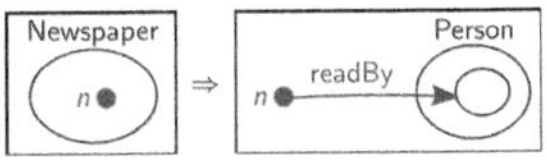


**Fig. 3.** Free variables

P. Cox, B. Plimmer, and P. Rodgers (Eds.): Diagrams 2012, LNAI 7352, pp. 291–293, 2012.

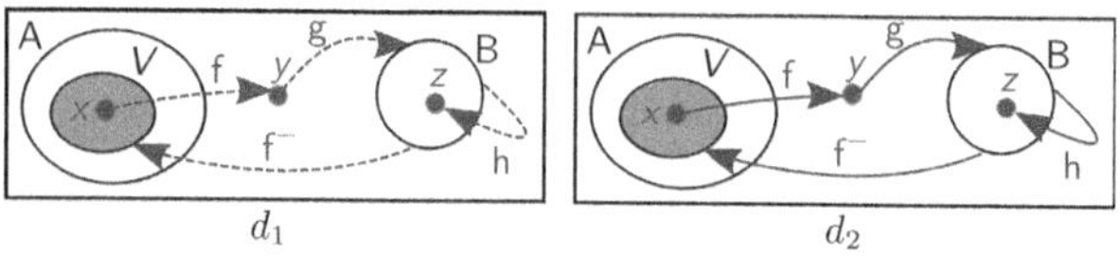


**Fig. 4.** Translating concept diagrams to second-order predicate logic

that is not called *Minnie* could be either a *Beagle* or not a *Beagle*. Shading is used to assert that the only dogs owned by *Erica* are represented by the spiders: in a shaded region, all elements must be represented by spiders.

The arrows convey information about the image of a role under a domain restriction, such as *Erica isMarriedTo Peter* and only *Peter*. We use dashed arrows to represent partial information, such as *Erica loves* some *Person*; *Erica* may love many things, but all we know that she loves at least one *Person*. Further, we do not know whether the *Person Erica loves* is distinct from herself. A concept diagram expressing this is in figure 2. Note that the arrow connects diagrammatic syntax placed in different boxes. This ensures that we have not made any assertion about whether the *Person Erica loves* is distinct from *Erica*.

A concept diagram asserting that every *Newspaper* is *readBy* only a subset of *Person* can be seen in figure 3. Here, the labelled dot inside the curve labelled *Newspaper* is a free variable, and so is equivalent to a universally quantified variable. The diagram expresses that if $n$ is a *Newspaper* then the set of things $n$ is read by is subsumed by *Person*.

We now compare the expressiveness of concept diagrams with predicate logic. The specific logic that we consider has as its terms, $x = y$ where $x$ and $y$ are either constants (individuals) or variables that represent elements. Formulae are of three forms, where $x$ and $y$ are as previously stated: (a) $c(x)$ where $c$ is either a monadic predicate symbol or a variable acting as a monadic predicate, (b) $r(x, y)$ where $r$ is either a dyadic predicate symbol or a variable acting as a dyadic predicate, and (c) formulae joined by the usual logical logical connectives or quantified over by $\exists$ or $\forall$ using any one of the three types of variable.

We demonstrate how to translate concept diagrams into this second-order predicate logic using figure 4. The lefthand diagram translates to the conjunction of the following formulae:

1. All elements must lie in the sets represented by zones, so all elements must 'lie in' one of the disjuncts of the following:

$$\forall i\Big((\neg A(i)\wedge\neg B(i)\wedge\neg V(i))\vee(A(i)\wedge\neg B(i)\wedge\neg V(i))\vee(A(i)\wedge\neg B(i)\wedge V(i))\vee(\neg A(i)\wedge B(i)\wedge\neg V(i))\Big).$$

2. The shading tells us that all elements in shaded zones must be represented by spiders: $\forall i\big((A(i) \wedge V(i)) \Rightarrow (i = x \vee i = y \vee i = z)\big)$.
3. Each spider represents an element in the set represented by the region in which it is placed: $A(x) \wedge \neg B(x) \wedge V(x) \wedge \neg A(y) \wedge \neg B(y) \wedge \neg V(y) \wedge \neg A(z) \wedge B(z) \wedge \neg V(z)$.
4. No two spiders represent the same element: $x \neq y \wedge x \neq z \wedge y \neq z$.
5. The four arrows translate as follows:

(a) the dashed arrow from $x$ to $y$ only tells us that $x$ is related to (at least) $y$ under $f$: $f(x,y)$
(b) the dashed arrow from $y$ to $B$ tells us that $y$ is related to at least the elements in $B$ (equivalently, everything in $B$ is related to by $y$) under $g$: $\forall i\big(B(i) \Rightarrow g(y,i)\big)$
(c) the dashed arrow from $B$ to $z$ tells us that at least one element in $B$ is related to $z$ $\exists i\big(B(i) \wedge h(i,z)\big)$
(d) the dashed arrow from $B$ to $V$ tells us that every element in $V$ is related to by at least one element in $B$ under $f^{-}$: $\forall i\big(V(i) \Rightarrow \exists j(B(j) \wedge f(i,j))\big)$

Diagram $d_2$ has a translation similar to that of $d_1$, differing only when we translate the arrows because they are dashed rather than solid:

(a) $f(x,y) \wedge \forall i\big(f(x,i) \Rightarrow i = y\big)$
(b) $\forall i\big(B(i) \Rightarrow g(y,i)\big) \wedge \forall i\big(g(y,i) \Rightarrow B(i)\big)$
(c) $\exists i\big(B(i) \wedge h(i,z)\big) \wedge \forall i \forall j\big((B(i) \wedge h(i,j)) \Rightarrow j = z\big)$
(d) $\forall i\big(V(i) \Rightarrow \exists j\big(B(j) \wedge f(i,j)\big)\big) \wedge \forall i \forall j\big((B(i) \wedge f(j,i)) \Rightarrow V(j)\big)$.

Second-order predicate logic formulae can also be translated to concept diagrams. The formula $\exists r \forall x \exists y\Big(\neg(x = y) \Rightarrow \big(A(x) \wedge X(x) \wedge r(x,y)\big)\Big)$ translates to:

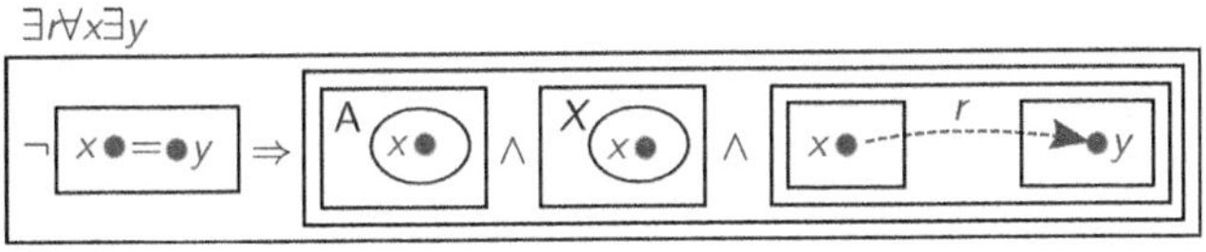


**Theorem.** Concept diagrams and second-order predicate logic with quantification over elements, sets, and binary relations are equivalent in expressiveness.

This result demonstrates that concept diagrams are highly expressive. They are more expressive than any other Euler diagram based logic.

## References


1. Hammer, E.: Logic and Visual Information. CSLI Publications (1995)
2. Kent, S.: Constraint diagrams: Visualizing invariants in object oriented modelling. In: Proceedings of OOPSLA 1997, pp. 327–341. ACM Press (1997)
3. Mineshima, K., Okada, M., Sato, Y., Takemura, R.: Diagrammatic Reasoning System with Euler Circles: Theory and Experiment Design. In: Stapleton, G., Howse, J., Lee, J. (eds.) Diagrams 2008. LNCS (LNAI), vol. 5223, pp. 188–205. Springer, Heidelberg (2008)
4. Oliver, I., Howse, J., Stapleton, G., Nuutila, E., Törma, S.: A proposed diagrammatic logic for ontology specification and visualization. In: International Semantic Web Conference (2009)
5. Shin, S.-J.: The Logical Status of Diagrams. Cambridge University Press (1994)
6. Swoboda, N., Allwein, G.: Using DAG transformations to verify Euler/Venn homogeneous and Euler/Venn FOL heterogeneous rules of inference. Journal on Software and System Modeling 3(2), 136–149 (2004)

# CDEG: Computerized Diagrammatic Euclidean Geometry 2.0

Nathaniel Miller

University of Northern Colorado
nat@alumni.princeton.edu

**Abstract.** This paper briefly describes **CDEG** 2.0, a computerized formal system for giving diagrammatic proofs in Euclidean geometry.

**Keywords:** diagrams, computer formal systems, Euclidean geometry.

This paper briefly describes the computer proof system **CDEG**, version 2.0. **CDEG** stands for "Computerized Diagrammatic Euclidean Geometry." This computer proof system implements a diagrammatic formal system for giving diagram-based proofs of theorems of Euclidean geometry that are similar to the informal proofs found in Euclid's *Elements* [1]. It is based on the diagrammatic formal system **FG**, which is described in detail in my book, *Euclid and his Twentieth Century Rivals: Diagrams in the Logic of Euclidean Geometry* [10]. That book also describes an earlier version of **CDEG**; however, **CDEG** has evolved significantly since the publication of the book. In particular, a beta version of **CDEG** is now publicly available, and can be downloaded from `http://www.unco.edu/NHS/mathsci/facstaff/Miller/personal/CDEG/`. I encourage interested readers of this paper to download **CDEG** and to try it out for themselves.

When we say that **CDEG** is a diagrammatic computer proof system, this means that it allows its user to give geometric proofs using diagrams. It is based on a precisely defined syntax and semantics of Euclidean diagrams. To say that it has a precisely defined syntax means that all the rules of what constitutes a diagram and how we can move from one diagram to another have been completely specified. The fact that these rules are completely specified is perhaps obvious if you are using the formal system on a computer, since computers can only operate with such precisely defined rules. However, it was commonly thought for many years that it was not possible to give Euclidean diagrams a precise syntax, and that the rules governing the use of such diagrams were inherently informal.

To say that the system has a precisely defined semantics means that the meaning of each diagram has also been precisely specified. In general, one diagram drawn by **CDEG** can actually represent many different possible collections of lines and circles in the plane. What these collections all share, and share with the diagram that represents them, is that they all have the same topology. This means that any one can be stretched into any other, staying in the plane. So, for example, a diagram containing a single line segment represents all possible single line segments in the plane, since any such line segment can be stretched

P. Cox, B. Plimmer, and P. Rodgers (Eds.): Diagrams 2012, LNAI 7352, pp. 294–296, 2012.

into any other. See [10] for more details concerning the syntax and semantics of Euclidean diagrams, and see [9] for more details of how to **CDEG** can be used, including a tutorial.

**CDEG** is essentially a computer implementation of the formal system **FG** described in [10]. Actually implementing the formal system on a computer was a highly non-trivial matter that took several years worth of work. Why would we want to implement an existing formal system on a computer?

The first reason that we might want a computer implementation is to demonstrate that this system really is completely formal: that the diagrams that are being manipulated are, indeed, completely specified as formal objects, and that the rules of the system are completely specified on these objects. With traditional, sentential formal systems, we do this by writing our axioms in a formal language, and then carefully writing rules of inference as typographical manipulations of sentences in this formal language. However, when our formal objects are diagrams, it is difficult to achieve this level of specificity without a computer implementation. Diagrams are complicated formal objects, and we have very strong informal intuitions about how they should work that may cloud our ability to judge if our rules have been completely formally specified.

Furthermore, even if our rules are completely formally specified, without a computer implementation, it will be quite difficult to play with the formal system to see what derivations are like, and to make sure that they really work the way that we think they will. This is particularly true in geometry, where constructions can lead to case branching, with a large number of cases that are virtually impossible to keep track of without using a computer. Thus, we may not be able to prove everything we think we can.

This worry is not just academic. Several other diagrammatic formal systems have been proposed by other researchers and have appeared in print but have later turned out to have ill-defined and/or unsound rules. For example, Isabel Luengo's formal system **DS1**, described in [4] and [3], turned out to be unsound, as explained in [10, Appendix C]. Likewise, John Mumma's **Eu**, described in [6], [7], [8], and, at a previous conference in the Diagrams series, in [5], is also unsound, as described in [11]. Neither of these proposed formal systems for geometry was implemented as a computer system, and neither worked quite in the way that their designers intended. Furthermore, both systems were examined by quite a number of article referees and dissertation committee members who failed to notice their significant problems. Thus, we should approach any proposed diagrammatic formal system with a certain amount of healthy skepticism. A working computer system is one way to allay some of this skepticism.

Secondly, a computer system is the only way to make a formal system widely available. Many potential users will not be able to make sense out of a formal system that is just specified mathematically, but will be able to try out a computer implementation.

The third reason for a computer implementation is to be able explore exactly what the formal system is able to prove. **CDEG**'s rules of inference are closely modeled on those found in Euclid's *Elements*[1], and I therefore claim that **CDEG** should be able to duplicate the first four books of Euclid's *Elements*. The only way to verify this claim is to systematically go through each of Euclid's proofs, and to see how to duplicate it within **CDEG**. To date, I have done this with many different proofs from Euclid's Book I, but have not yet gone systematically through all of Euclid's proofs. This is a future project of mine, and one that would be essentially impossible without the computer implementation.

As mentioned above, a previous version of **CDEG**, version 1.0, was discussed in [10], but was never made publicly available, because it did not include a stand-alone means of drawing its diagrams. The new version of **CDEG** relies on OGDF (the "Open Graph Drawing Framework") to lay out its diagrams, using the mixed model algorithm of Gutwenger and Mutzel [2].

Other significant changes to **CDEG** include the addition of the triangle congruence rules and rules for deleting pieces of diagrams, as well as numerous bug fixes. It also now has the ability to draw its own output diagrams rather than relying on an external program to do this. However, the version of **CDEG** that is now available is a beta version and most likely still contains bugs. If you try out **CDEG** and discover any bugs, please let me know by sending an email to `nat@alumni.princeton.edu`.

## References

1. Euclid: The Elements, 2nd edn. Dover, New York (1956); translated with introduction and commentary by Heath, T.L.
2. Gutwenger, C., Mutzel, P.: Planar Polyline Drawings with Good Angular Resolution. In: Whitesides, S.H. (ed.) GD 1998. LNCS, vol. 1547, pp. 167–182. Springer, Heidelberg (1999)
3. Luengo, I.: Diagram. In: Geometry. Ph.D. thesis, Indiana University (1995)
4. Luengo, I.: A Diagrammatic Subsystem of Hilbert's Geometry. In: Allwein, G., Barwise, J. (eds.) Logical Reasoning with Diagrams. Oxford University Press, New York (1996)
5. Mumma, J.: Ensuring Generality in Euclid's Diagrammatic Arguments. In: Stapleton, G., Howse, J., Lee, J. (eds.) Diagrams 2008. LNCS (LNAI), vol. 5223, pp. 222–235. Springer, Heidelberg (2008)
6. Mumma, J.: Intuition formalized: Ancient and modern methods of proof in elementary geometry. Ph.D. Dissertation, Carnegie Mellon University (2006), `http://www.contrib.andrew.cmu.edu/~jmumma/list.html` (retrieved May 20, 2010)
7. Mumma, J.: Proofs, pictures, and Euclid. Synthese 175(2), 255–287 (2010), doi:10.1007/s11229-009-9509-9
8. Mumma, J.: Review of Euclid and his twentieth century rivals: Diagrams in the logic of Euclidean geometry. Philosophia Mathematica 16(2), 256–264 (2008)
9. Miller, N.: **CDEG**User's Manual, `http://www.unco.edu/NHS/mathsci/facstaff/Miller/personal/CDEG/`
10. Miller, N.: Euclid and his twentieth century rivals: Diagrams in the logic of Euclidean geometry. CSLI Press, Stanford (2007)
11. Miller, N.: On the Inconsistency of Mumma's Eu. Notre Dame Journal of Formal Logic (in press, 2012), preprint available at `http://www.unco.edu/NHS/mathsci/facstaff/Miller/personal/diagrams`

# Design and Implementation of Multi-camera Systems Distributed over a Spherical Geometry

Hossein Afshari, Kerem Seyid, Alexandre Schmid, and Yusuf Leblebici

Ecole Polytechnique Fédérale de Lausanne (EPFL), Lausanne, Switzerland
first_name.last_name@epfl.ch

**Abstract.** The current trend in constructing high-end computing systems consists of parallelizing large numbers of processors. A similar trend is observed in digital imaging where multiple camera inputs are utilized to obtain multiple images of a scene and thus enhance the performance envelope of the image capture. A methodology based on Voronoi diagrams is presented for coverage analysis of multi-camera systems mounted on spherical geometry. Interconnected network of camera concept is introduced for the purpose of the application development of multi-camera systems.

## 1 Introduction

Inspired from insects' compound eyes a multi-camera system is devised by distributing camera modules over a spherical surface. This multi-camera system is referred to as the Panoptic camera [1].

An arrangement of camera modules over a spherical surface is desired which considers the mechanical spacing of the camera modules as major constraint. Each camera position is modeled as a circular face with a constant radius. The surface of a unit hemisphere is divided into latitude floors. Each floor is populated with identical camera positions to its maximum extent. A seven-floor built prototype is shown in Fig. 1(a).

## 2 Omnidirectional Vision

The Panoptic camera can be used to emulate the omnidirectional vision of a virtual observer located anywhere inside the hemisphere by combining the light information collected by each camera. The principle method used in this process is the interpolation of light information in the light-ray space domain (*i.e.*, light field [4]). For this purpose the omnidirectional view on a discretized sphere $\mathcal{S}_d$ is estimated. The surface of the sphere can be discretized into an equiangular grid with $N_\theta$ latitudes and $N_\phi$ longitudes pixels. An example of a discretized sphere surface with sixteen spherical pixels for $N_\theta$ and $N_\phi$ is shown in Fig. 1(b).

P. Cox, B. Plimmer, and P. Rodgers (Eds.): Diagrams 2012, LNAI 7352, pp. 297–299, 2012.

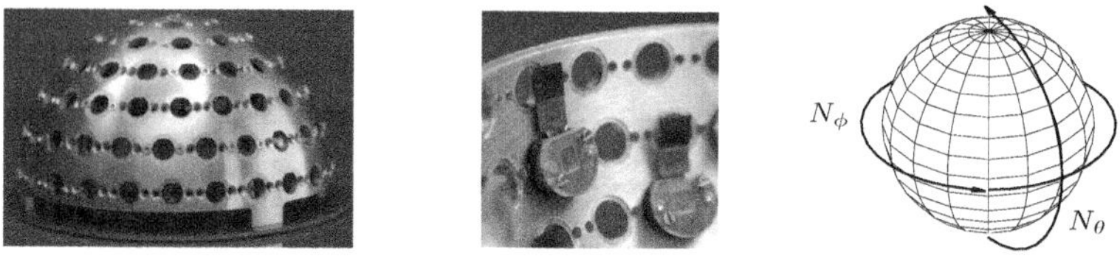


(a) (b) (c)

**Fig. 1.** Fabricated seven-floor Panoptic camera (a) side view (b) and internal view. (c) Discretized sphere surface $\mathcal{S}_d$ with $N_\theta = 16$ latitude and $N_\phi = 16$ longitudes pixels.

## 3 Coverage Analysis

The surface of the Panoptic device hemisphere can be partitioned into set of cells centered on the camera locations. Each cell defines the set of all points on the hemisphere surface which are closer to its camera location than to any other camera position. This definition falls into the category of Voronoi diagrams. The farthest point of the largest cell (in terms of radius) is the direction of the virtual view that is less covered by the Panoptic cameras. The top view of the Voronoi diagram of the hemisphere structure with five-floors is shown in Fig. 2(a). The utilization of individual imagers with limited angle-of-view implies that full-view coverage of the surrounding with the Panoptic device is achievable from a minimal distance and beyond. This distance is referred to as the Full-view coverage Distance (FCD) of the Panoptic device. Using spherical trigonometric identities and the position of the least covered direction on the Panoptic device, the FCD of the Panoptic device is calculable.

## 4 Interconnection Network

An interconnection network is a programmable system capable of transporting data between terminals [3]. A multi-camera system can be realized through an interconnected network of cameras. An interconnected network of cameras, where each of its cameras is provided processing and intelligence capability, is a platform intended for distributed and parallel (*i.e.*, running at the same time) implementation of multi-camera system's applications. Applications of multi-camera system demand exchange of (image) information among the cameras. Hence an interconnected network provides the means for this purpose.

The assignment of cameras to a target interconnected network nodes can be defined in the context of a facility allocation problem known as the Quadratic Assignment Problem [2]. Selection of cameras for direct access to a central unit which is in charge of commanding and controlling the interconnected network is mapped to another facility allocation problem known as the vertex p-center problem. The utilization of these techniques also improve access time and overall performance requirement of the interconnected network.

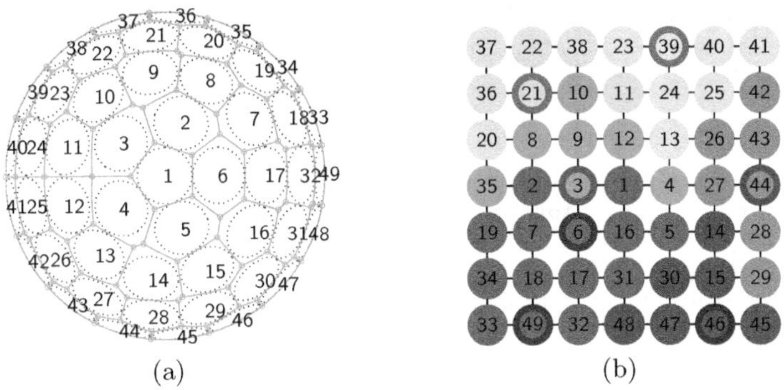


**Fig. 2.** (a) Top view of the Voronoi diagram of a five-floor Panoptic system containing 49 camera positions (b) The assigned $7 \times 7$ mesh topology interconnected network with 7 vertex p-centers

A QAP problem has been solved for assigning the cameras of the five-floor Panoptic system containing 49 cameras to a $7 \times 7$ mesh topology graph as the target interconnection network topology. The assigned camera numbers of Fig.2(a) is represented on the mesh graph shown in Fig. 2(b). A vertex-7 center problem has also been applied on the mesh topology. The vertex centers are indicated with bold edges in Fig. 2(b).

## 5 Conclusion

A method for estimating the minimum angle-of-view requirement and full coverage distance of the Panoptic cameras is presented. To this aim the Voronoi diagram and the are utilized as useful geometrical tools. Interconnection network of camera concept is introduced as a novel implementation of multi-camera systems. Facility allocation methods of Quadratic Assignment Problem and vertex $p$-center are shown as usefull optimization techniques for improving the access time performance of interconnection network of cameras.

## References

1. Afshari, H., Jacques, L., Bagnato, L., Schmid, A., Vandergheynst, P., Leblebici, Y.: Hardware implementation of an omnidirectional camera with real-time 3d imaging capability. In: 3DTV Conference: The True Vision - Capture, Transmission and Display of 3D Video (3DTV-CON), pp. 1–4 (May 2011)
2. Burkard, R.E., Karisch, S., Rendl, F.: Qaplib-a quadratic assignment problem library. European Journal of Operational Research 55(1), 115–119 (1991)
3. Dally, W., Towles, B.: Principles and Practices of Interconnection Networks. Morgan Kaufmann Publishers Inc., San Francisco (2003)
4. Levoy, M., Hanrahan, P.: Light Field Rendering. In: SIGGRAPH 1996, Proceedings of the 23rd Annual Conference on Computer Graphics and Interactive Techniques, pp. 31–42. ACM (1996)

# Algebraic Aspects of Duality Diagrams

Lorenz Demey*

University of Leuven, Belgium
lorenz.demey@hiw.kuleuven.be

**Abstract.** Duality phenomena are widespread in logic and language; their behavior is visualized using square diagrams. This paper shows how our recent algebraic account of duality can be fruitfully used to study these diagrams. A duality cube is constructed, and it is shown that 14 duality squares can be embedded into this cube (two of which were hitherto unknown). This number is also an upper bound.

**Keywords:** duality, negation, logic, linguistics, logical geometry.

Logicians and linguists use the term 'duality' to describe pairs of notions such as conjunction/disjunction ($\wedge/\vee$), universal/existential quantification ($\forall/\exists$), and necessity/possibility ($\Box/\Diamond$). Duality behavior is often visualized by means of *duality diagrams*;[1] Fig. 1 shows three examples.[2] Duality is also connected with a well-known object from group theory: the *Klein four-group* $\mathbb{V}_4$ [1,4].

We have recently shown that this connection with $\mathbb{V}_4$ can be developed into a full group-theoretical account of duality. Given expressions/operators $O_1$ and $O_2$, we use L to say that they are each other's external negation ($O_2 = \mathrm{L}(O_1) = \neg O_1$ and $O_1 = \mathrm{L}(O_2) = \neg O_2$), R to say that they are each other's internal negation ($O_2 = \mathrm{R}(O_1) = O_1\neg$ and $O_1 = \mathrm{R}(O_2) = O_2\neg$), and LR to say that they are each other's duals ($O_2 = \mathrm{LR}(O_1) = \neg O_1 \neg$ and $O_1 = \mathrm{LR}(O_2) = \neg O_2 \neg$). Adding the identity operation I, we obtain the Klein four-group $\mathbb{V}_4$, whose composition table is given in Fig. 2(a). It is well-known that $\mathbb{V}_4$ is isomorphic to the direct product of $\mathbb{Z}_2$ with itself ($\mathbb{V}_4 \cong \mathbb{Z}_2 \otimes \mathbb{Z}_2$). The group $\mathbb{Z}_2$ has domain $\{0,1\}$; its composition table is given in Fig. 2(b). The domain of $\mathbb{Z}_2 \otimes \mathbb{Z}_2$ is $\{0,1\} \times \{0,1\}$; its composition table is given in Fig. 2(c). A concrete isomorphism between $\mathbb{V}_4$ and $\mathbb{Z}_2 \otimes \mathbb{Z}_2$ is given by: $\mathrm{I} \leftrightarrow (0,0), \mathrm{L} \leftrightarrow (1,0), \mathrm{R} \leftrightarrow (0,1), \mathrm{LR} \leftrightarrow (1,1)$. This is syntactically meaningful: 0 and 1 represent the number of negations in a given 'position', and the left and right coordinates stand for the external and internal position, respectively. For example, LR corresponds to $(1,1)$, which represents 1 negation in the external position and 1 negation in the internal position, i.e. duality. Representing $\mathbb{V}_4$ as $\mathbb{Z}_2 \otimes \mathbb{Z}_2$ thus gives us a firm syntactic handle on duality behavior: it shows how this behavior arises out of the interplay of the independent behaviors of an external and an internal negation position.

* Thanks to Hans Smessaert for his extensive feedback. The author is financially supported by a PhD fellowship of the Research Foundation–Flanders (FWO).

[1] Despite some superficial similarities, these duality diagrams are very different from another type of diagrams in logic, viz. the *Aristotelian squares of oppositions* [3,6].

[2] L stands for *external negation*, R for *internal negation*, and LR for *duality*.

P. Cox, B. Plimmer, and P. Rodgers (Eds.): Diagrams 2012, LNAI 7352, pp. 300–302, 2012.

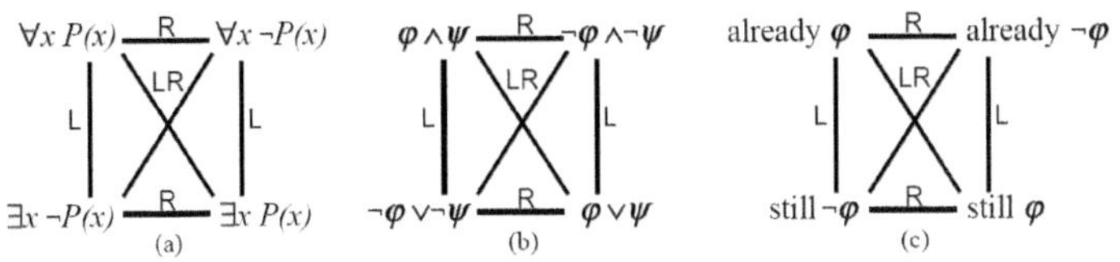


**Fig. 1.** Duality squares for (a) universal and existential quantification, (b) conjunction and disjunction, and (c) the natural language expressions 'already' and 'still'

We now turn to composed operators $O_2 \circ O_1$; e.g. $\Box \circ \wedge$: $(\varphi, \psi) \mapsto \Box(\varphi \wedge \psi)$. The duality behavior of these operators is not described by $\mathbb{V}_4$, but rather by an 8-element group $\mathbb{G}_8$, which is isomorphic to $\mathbb{Z}_2 \otimes \mathbb{Z}_2 \otimes \mathbb{Z}_2$. Compared to $\mathbb{V}_4$, one more copy of $\mathbb{Z}_2$ is thus added. This makes perfect syntactic sense: each copy of $\mathbb{Z}_2$ governs the behavior of one negation position, and by going from single operators to composed operators, we have added one more negation position: not only *external* $(\neg O_2 O_1)$ and *internal* $(O_2 O_1 \neg)$, but also *intermediate* $(O_2 \neg O_1)$. We will use the letter M to denote intermediate negation (so $\mathrm{M}(O_2, O_1) = O_2 \neg O_1$); the group $\mathbb{G}_8$ can then be presented as {I, L, R, M, LR, LM, RM, LRM}. This group is visualized by means of a *cube*; see Fig. 3(a) (diagonals are omitted for reasons of visual clarity). Duality squares *within* this cube correspond to 4-element *sub*groups of $\mathbb{G}_8$. There are exactly 7 such subgroups [2]:

1. {I, L, R, LR}, {I, R, M, RM} and {I, L, M, LM},
2. {I, L, RM, LRM}, {I, R, LM, LRM} and {I, M, LR, LRM},
3. {I, LR, RM, LM}.

Each subgroup corresponds to two squares; hence, there exist exactly 14 squares within the cube. The first three subgroups correspond to the cube's outer faces; for example, {I, L, M, LM} corresponds to the left and right face; see Fig. 3(b). The next three subgroups correspond to the cube's diagonal planes; e.g. {I, L, RM, LRM} corresponds to the two 'vertical' diagonal planes; see Fig. 3(c). Also the final group corresponds to two four-point 'clusters'. Consider one such cluster, e.g. $\{O_2 O_1 \neg, O_2 \neg O_1, \neg O_2 O_1, \neg O_2 \neg O_1 \neg\}$. Connecting all points with straight lines does not yield a square, but rather a *tetrahedron*; see Fig. 3(d). Such a tetrahedron can still be regarded as a square, albeit a 'twisted' one. To see this, start with one of the cube's three central symmetry planes; this plane intersects the cube along a square; see Fig. 4(a). Rotate one side of this intersection square 45° clockwise, and the opposite side 45° counterclockwise; the result

| (a) | I | L | R | LR |
|---|---|---|---|---|
| I | I | L | R | LR |
| L | L | I | LR | R |
| R | R | LR | I | L |
| LR | LR | R | L | I |

| (b) | 0 | 1 |
|---|---|---|
| 0 | 0 | 1 |
| 1 | 1 | 0 |

| (c) | (0, 0) | (1, 0) | (0, 1) | (1, 1) |
|---|---|---|---|---|
| (0, 0) | (0, 0) | (1, 0) | (0, 1) | (1, 1) |
| (1, 0) | (1, 0) | (0, 0) | (1, 1) | (0, 1) |
| (0, 1) | (0, 1) | (1, 1) | (0, 0) | (1, 0) |
| (1, 1) | (1, 1) | (0, 1) | (1, 0) | (0, 0) |

**Fig. 2.** Composition tables for (a) $\mathbb{V}_4$, (b) $\mathbb{Z}_2$, and (b) $\mathbb{Z}_2 \otimes \mathbb{Z}_2$

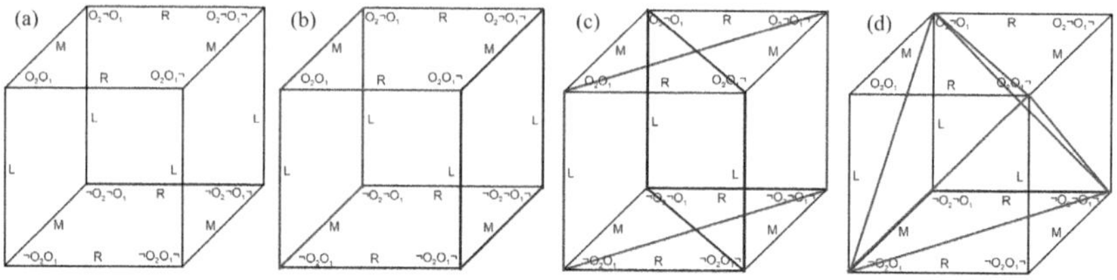


**Fig. 3.** (a) The duality cube for $\mathbb{G}_8$. (b) The two squares for {I, L, M, LM}, and (c) for {I, L, RM, LRM}. (d) One of the two tetrahedra for {I, LR, RM, LM}.

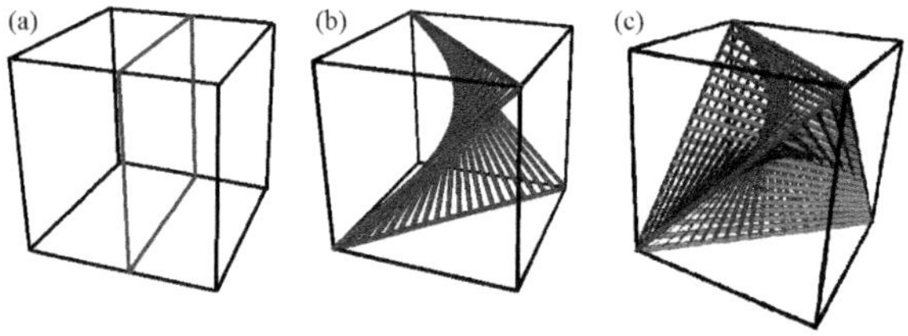


**Fig. 4.** (a) Intersection of the cube with one of its three central symmetry planes. (b) 'Twisting' the intersection square. (c) The three 'twisted' squares put together.

is a *twisted square* whose four corners are exactly our cluster; see Fig. 4(b). Doing this for each of the cube's three central symmetry planes yields three twisted squares, which together 'approximate' the tetrahedron; see Fig. 4(c).

Our algebraic account is thus a powerful tool to study duality diagrams. The first twelve squares were already known [5], but the final two are new discoveries (and turn out to be very interesting). The *limitative* result is useful as well: there cannot exist any squares within the cube beyond the 14 that we described.

## References

1. van Benthem, J.: Linguistic universals in logical semantics. In: Zaefferer, D. (ed.) Semantic Universals and Universal Semantics, pp. 17–36. Foris, Berlin (1991)
2. Călugăreanu, G.: The total number of subgroups of a finite Abelian group. Scientiae Mathematicae Japonicae 60, 157–168 (2004)
3. Demey, L.: Structures of oppositions in public announcement logic. In: Béziau, J.-Y., Jacquette, D. (eds.) Around and Beyond the Square of Opposition. Springer (2012)
4. Löbner, S.: Wahr neben Falsch. Duale Operatoren als die Quantoren natürlicher Sprache. Max Niemeyer Verlag, Tübingen (1990)
5. Moretti, A.: A cube extending Piaget-Gottschalk's formal square (ms.)
6. Smessaert, H.: The classical Aristotelian hexagon versus the modern duality hexagon. Logica Universalis (forthcoming), doi:101007/s11787-011-0031-8

# The Use of Diagrams in *Science*

## An Examination of Trends in Articles Published in Science between 1880 and 2010

Lillian P. Fanjoy, A. Luke MacNeill, and Lisa A. Best

University of New Brunswick, Psychology Department, Box 5050, Saint John, NB E2L 4L5
{d9za0,s8me9,lbest}@unb.ca

**Abstract.** Scientists use inscriptions, such as tables, graphs, and illustrations to provide readers with a visual representation of data. A sample of articles from the journal, *Science* was collected and a random selection of eight articles was drawn from each decade from inception to the present decade (2008 - 2010). Overall, we found different trends in the use of graphs, tables, and non-graph illustrations.

**Keywords:** Visual inscriptions, Trends over time, *Science*.

## 1 Scientific Inscriptions

A scientific inscription device is a specific type of visualisation aid that provides an illustrative display in a scientific text [1]. Collection of empirical data began in the middle ages [2], and, by the 17th century, scientists presented numerical information in data tables [3] and used early graphical methods, including anatomical drawings, geographical and astronomical maps, and geometric diagrams. Furthermore, mechanical recording devices were created that produced moving line graphs of natural events [2]. During this time, graphical methods were underutilized. Gross et al. [3] found that only 38% of articles published in the 17th century contained a visual representation.

Founded in 1880, *Science* is a scientific journal with widespread readership and appeal. *Science* reaches an estimated worldwide readership of more than 1,000,000 people, its articles consistently rank among the world's most cited research, and each year, less than 8% of submissions are published. Both the age and prestige of *Science* make it an ideal focus for the sampling of historical graph use [4]. The primary purpose of this study was to analyze the use of visual inscriptions in the *Science*, with a particular emphasis on how the use of these figures has changed over time.

## 2 Methods

A sample of eight articles was randomly selected from Science for 14 time periods (every ten years from 1880 to 2010). Book reviews, addendums, and errata were excluded from the sample, as the study focused solely on the use of visual inscriptions

P. Cox, B. Plimmer, and P. Rodgers (Eds.): Diagrams 2012, LNAI 7352, pp. 303–305, 2012.

in scientific writings (empirical articles, experiments, etc.). Inscription information was recorded, including the number and type of inscriptions, as well as the total area and fractional area for each inscription. Information was recorded about three types of inscriptions: visual inscriptions (included graphs and non-graph illustrations, NGI); non-visual inscriptions (included tables and equations); and, montages (included two or more types of inscriptions; see [5]).

## 3 Results

A total of 111 articles (only seven articles were selected from 1880, due to the dearth of empirical articles) were sampled. Averaged over all decades, 33.33% of articles contained at least one graph, 36.04% contained at least one table, and 35.14% contained at least one illustration. Figure 1 shows trends in graph use, NGI use, and table use.

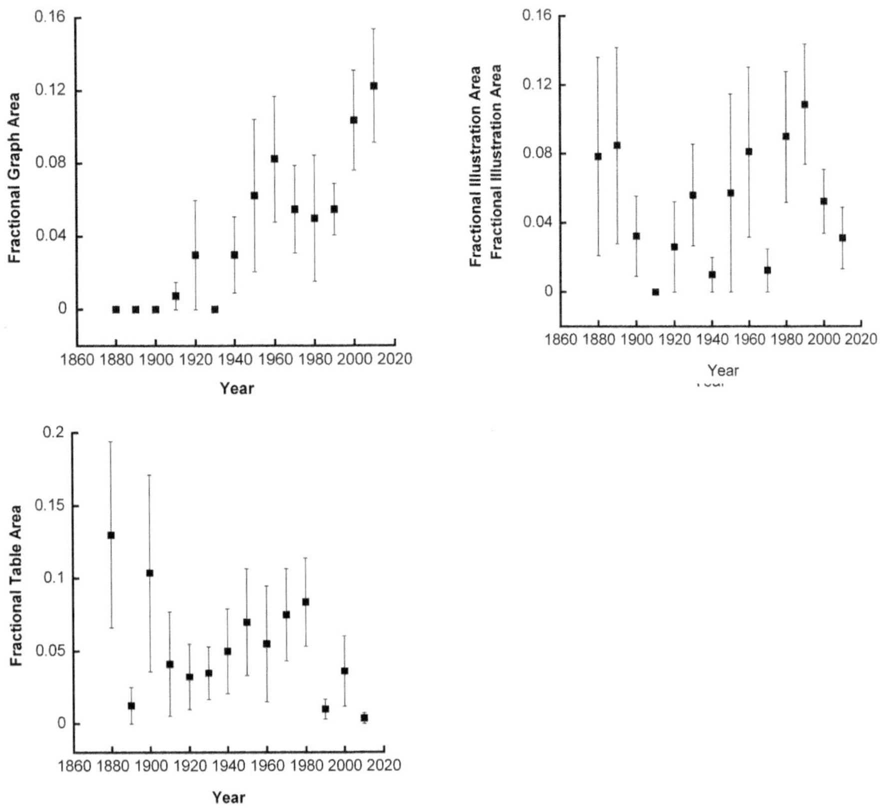


**Fig. 1.** Proportion of page space dedicated to tabular presentation (Panel A), equations used per journal page (Panel B), and overall non-visual inscription use per page (Panel C)

As can be seen in the figure, the percentage of article space dedicated to graph use increased significantly over time ($r(14)=.89$, $p<.0001$), the use of NGIs ($r(14)=.277$, $p=.39$) and tables ($r(14) = .06$, $p=.84$) has been relatively stable. The overall use of visual inscriptions has increased over time ($r(14)=.74$, $p=.03$), driven largely by the increase of graphical data displays. During the early years of publication, graphs were not used to present data but illustrations, tables, and equations were quite common. After 1940, graphs became more and more popular while the use of tables declined.

## 4 Discussion

Funkhouser [2] called the period from 1860 to 1890 the "golden age of graphs", but he may have been somewhat premature. If the articles sampled from *Science* are any indication, graph use did not see its sharpest increases until a full century later. In fact, although the overall use of scientific inscriptions has increased since 1880, this growth has been driven almost entirely by graph usage. The greatest increases in graph use have occurred in recent decades and, presumably, part of this growth can be attributed to the pervasiveness of computers and statistical software, which have made graphs more accessible to researchers.

Although graph use has increased over the decades, the same cannot be said for other scientific inscriptions. It appears that the use of illustrations has remained static over the lifespan of *Science*. We found that the use of non-graph illustrations has undergone very little fluctuation between 1880 and 2010. Table use, however, has decreased over time. These results support Bazerman [6] and Gross, Harmon, and Reidy [3] who found that, although graph use in certain natural sciences has increased over the past century, table use has remained stable or declined.

## References

1. Latour, B.: Drawing things together. In: Lynch, M., Woolgar, S. (eds.) Representation in Scientific Practice, pp. 19–68. MIT Press, Cambridge (1990)
2. Funkhouser, H.G.: Historical development of the graphical representation of statistical data. Isis 3, 269–404 (1937)
3. Gross, A.G., Harmon, J.E., Reidy, M.: Communicating science: The scientific article from the 17th century to the present. Oxford University Press, Oxford (2002)
4. Science. AAAS, `http://www.sciencemag.org/site/about/index.xhtml`
5. Arsenault, D.J., Smith, L.D., Beauchamp, E.A.: Visual Inscriptions in the Scientific Hierarchy: Mapping the "Treasures of Science". Sci. Comm. 27(3), 376–428
6. Bazerman, C.: Theoretical integration in experimental reports in twentieth-century physics: Spectroscopic articles in Physical Review, 1893-1980. In: Bazerman, C. (ed.) Shaping Written Knowledge, pp. 153–186. University of Wisconsin Press, Madison (1988)

# A User Study on Curved Edges in Graph Visualisation

Kai Xu, Chris Rooney, Peter Passmore, and Dong-Han Ham

Middlesex University, UK

## 1 Introduction

It seems that straight lines seldom occur in natural objects and that humans actually prefer curved lines [1]. Thus it may not seem surprising that in aesthetics, curved lines are often to be preferred over straight ones, as found for example in Hogarth's serpentine Line of Beauty [2]. More recently a number of "confluent drawings" [3] and "edge bundling" [4] methods have been proposed to reduce edge clutter by using curved edges. Inspired by the work of Mark Lombardi, there is also theoretical work [5] that uses curved edges to optimises *angular resolution*, i.e., keep the angles between adjacent edge uniform.

Many examples are available to demonstrate the results of graph visualization with curved edges. However, there has been little effort to empirically evaluate their effectiveness on common graph-related tasks. The only related experiment [6] we are aware of is a qualitative study comparing hierarchical edge bundling against node-link diagrams with five software developers. The data used were not general graphs (directed acyclic graphs) and the tasks were software engineering-specific. The results show that all participants strongly prefer hierarchical edge bundling to node-link diagrams.

In this paper we describe our experiment studying the impact of edge curvature on general graph readability. We wanted to avoid any confounding factors and limit the difference in visualization to edge curvature only. Therefore, we did not include edge bundling or confluent drawing methods, because they require special layout methods that can not be applied to straight-line graphs.

## 2 Experiment and Results

A total of twenty-eight subjects voluntarily participated in the study. They were from diverse social-economical background and included university students and staff (both academic and non-academic), and general public. The graphs used in the experiment were generated with the model proposed by Ware et al. [7]. The graphs have three sizes: 20, 50, and 100 nodes, and ten graphs were generated for each of the three sizes. The force-directed method [8] was then used to generate graph layout. After applying the layout algorithm, three visualizations were produced for each graph with straight, slightly curved, and heavily curved edges (Figure 1). A four-point Bezier curve was plotted for the slight curve, and three point Bezier curve was plotted for the heavy curve. The node positions

P. Cox, B. Plimmer, and P. Rodgers (Eds.): Diagrams 2012, LNAI 7352, pp. 306–308, 2012.

were kept unchanged. A within-subject design is used. During the experiment, participants were asked to identify whether a path of length two existed between two nodes (i.e., path-finding task).

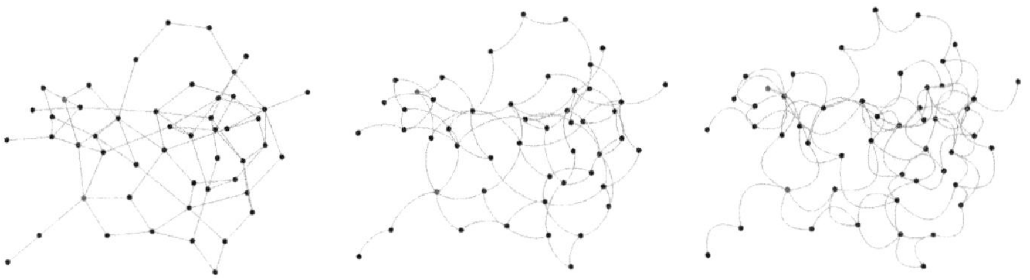

(a) Straight edge (b) Slightly curved edge (c) Heavily curved edge

**Fig. 1.** Examples of graph visualizations used in the experiment

**Hypothesis 1:** Steeper arcs will be detrimental to performance as participants will have to follow a longer path between nodes.
**Hypothesis 2:** Task time will increase with the graph size.
**Hypothesis 3:** Participants would prefer straight edge for effectiveness, but slightly curved edges for aesthetics.

This study used two objective measures: time to answer (TIME) and the number of correct answers (CORRECT)) and two subjective measures: user preference on effectiveness (PREF-EFFECTIVE) and look (PREF-LOOK) of line type. Table 1 summarizes the pairwise comparisons using Tukey test for TIME and CORRECT. TIME: The ANOVA results showed that all the main effects were

**Table 1.** Pairwise comparisons of two objective measures

| | Line Type | | | Number of Nodes | | |
|---|---|---|---|---|---|---|
| | STL vs. SCL | STL vs. HCL | SCL vs. HCL | 20 vs. 50 | 20 vs. 100 | 50 vs. 100 |
| $Log_{10}(TIME)$ | ** | ** | ** | ** | ** | ** |
| CORRECT | | ** | ** | | * | |

* Significant at the $\alpha = 0.05$ level; ** Significant at the $\alpha = 0.01$ level

statistically significant at the 0.01 significance level (line type (F(2, 54) = 16.31, $p < 0.01$) and number of nodes (F(2,54) = 26.64, $p < 0.01$). CORRECT: The main effect of line type was statistically significant at the 0.01 significance level (F(2,216) = 61.20, $p < 0.01$). The main effect of number of nodes was also significant at the 0.05 level (F(2,216) = 3.05, $p < 0.05$). PREF-EFFECTIVE and PREF-LOOK: For the two subjective measures, the Friedman test showed that there was a statistically significant difference among three line types (PREF-EFFECTIVE ($\chi_F^2 = 16.83$, $p < 0.01$; adjusted for ties) and PREF-LOOK ($\chi_F^2 = 16.83$, $p < 0.01$; adjusted for ties)). STL is the most preferable line type in both subjective measures.

## 3 Discussions

This study has found that edge curvature increase leads to longer task completion time, which is in agreement with our Hypothesis 1. It is also found that correct answer percentage decreases as curvature increases. These results indicate that using curved edges alone does not improve graph readability. The results also show that the time taken increases with the number of nodes for each curvature condition, which agrees with our Hypothesis 2. Participants preferred the aesthetics of straight lines to curved ones, and judged them to be more effective for the path finding task. This disagrees with Hypothesis 3 that curved lines would be preferred and the previously cited studies that show humans prefer curved contours. It is probable that any initial reaction is overridden by the requirements of our task. As a result, subjects prefer the look of straight lines when they find them easier to use for the task.

## References

1. Bar, M., Neta, M.: Humans prefer curved visual objects. Psychological Science 17(8), 645–648 (2006)
2. Hogarth, W.: The Analysis of Beauty. Yale University Press (1753)
3. Dickerson, M.T., Eppstein, D., Goodrich, M.T., Meng, J.Y.: Confluent Drawings: Visualizing Non-planar Diagrams in a Planar Way. In: Liotta, G. (ed.) GD 2003. LNCS, vol. 2912, pp. 1–12. Springer, Heidelberg (2004)
4. Holten, D.: Hierarchical edge bundles: Visualization of adjacency relations in hierarchical data. IEEE Transactions on Visualization and Computer Graphics 12, 741–748 (2006)
5. Duncan, C.A., Eppstein, D., Goodrich, M.T., Kobourov, S.G., Nöllenburg, M.: Lombardi Drawings of Graphs. In: Brandes, U., Cornelsen, S. (eds.) GD 2010. LNCS, vol. 6502, pp. 195–207. Springer, Heidelberg (2011)
6. Telea, A., Ersoy, O., Hoogendorp, H., Reniers, D.: Comparison of node-link and hierarchical edge bundling layouts: A user study. In: Keim, D.A., Pras, A., Schönwälder, J., Wong, P.C. (eds.) Visualization and Monitoring of Network Traffic. Schloss Dagstuhl - Leibniz-Zentrum fuer Informatik, Germany (2009)
7. Ware, C., Bobrow, R.: Supporting visual queries on medium-sized node-link diagrams. Information Visualization 4(1), 49–58 (2005)
8. Battista, G.D., Eades, P., Tamassia, R., Tollis, I.G.: Graph Drawing: Algorithms for the Visualization of Graphs. Prentice Hall (1999)

# Truth Diagrams: An Overview

Peter C.-H. Cheng

Department of Informatics, University of Sussex, Brighton, BN1 9QH, UK
p.c.h.cheng@sussex.ac.uk

**Abstract.** *Truth Diagrams*, TDs, are a new diagrammatic notation for propositional logic. TDs provide: (1) representations of logical states of affairs and relations; (2) operators on such relations; (3) a test of the validity of derivations. A proof of one of de Morgan's laws is given as an illustration of TDs.

Truth Diagrams (TDs) were invented as part of a programme of research on the *Representational Epistemic* approach to the study of how notational systems encode knowledge and the potential cognitive benefits that novel codifications of knowledge may confer [1-3]. The core principle of the approach claims that effective representational systems should directly encode the fundamental conceptual structure of their knowledge domains, using coherent notational schemes. TDs were designed as a further test of this idea. The purpose here is simply to give an informal overview of TDs by describing the derivation of one of de Morgan's laws: –(*P*v*Q*) ⊢ –*P*&–*Q*.

The derivation is shown in Fig. 1. The sequent to be derived is stated at the top; columns C to H in row 1, or {C-H,1}. The derivation has three main parts: (a) the construction of the TD for the assumption of the sequent, to the left of Fig. 1 {A-C,2-6}; (b) the construction of the conclusion TD, on the right {H-J,2-6}; (c) a test of that the assumption and the conclusion constitutes a tautology, as required for a valid derivation, at the bottom of the diagram {F,7-8}.

TDs are configurations of lines and symbols that may be interpreted in three different ways. First, TDs may represent **logical states of affairs**. There are ten such TDs in Fig. 1; {A,2}, {C,2}}, {H,2}, {J,2}, {B,4}, {H,4}, {J,4}, {B,6}, {I,6}, and {F,8}. The formula for the state of affairs represented by a TD is shown by the underlined expression at the top of each TD; e.g., in {B,4} the relation is <u>*P*v*Q*</u>. A TD possess one or more variables identified by the letter(s) in the middle of each diagram; in the unary TD {A,2} the variable is *P*, and in the binary TD {B,4} they are *P* and *Q*. The position of the *end* of a line next to a variable represents a truth-values assignment to that variable: the top position stands for True and the bottom position stands for False. For example, the top left of {B,4} is *P*=T and the bottom right is *Q*=F, and the top of {A,2} is *P*=T and the bottom is *P*=F.

Within a TD, each line stands for particular set of truth-value assignments to the variables, depending on the position of the ends of the line. For example, in {B,4} the top horizontal line stands for (*P*=T, *Q*=T). The bottom (dashed) horizontal line represents (*P*=F, *Q*=F). The descending and ascending diagonals represents (*P*=T, *Q*=F) and (*P*=F, *Q*=T), respectively. The style of the line indicates whether the set of

P. Cox, B. Plimmer, and P. Rodgers (Eds.): Diagrams 2012, LNAI 7352, pp. 309–311, 2012.

assignments is itself T or F. A solid line assigns True to the set and is called a *Tine*. A dashed line assigns False to the set and is called a *Faint*. For example, the descending diagonal Tine in {B,4} is (*P*=T, *Q*=F)=T and bottom Faint is (*P*=F, *Q*=F)=F. As all the lines of {F,8} are tines, all possible sets of assignments are true, so this TD stands for a tautology. The TDs in row {2} are simply unary variables, so by definition they have a Tine at the top and a Faint at the bottom.

The second interpretation is **TDs as operators**: {B,3}, {H,3}, {J,3}, {B,5} and {I,5} in Fig. 1, which are enclosed by dashed rectangles. The symbol above the TD identifies the operator. The bracketed formulas in the middle identify the argument TDs to which the operator is applied. In Fig. 1 the argument and result TDs are drawn just above and below the operator TD; e.g., {B,4} is the argument for operator {B,5} and the result is {B,6}. The configuration of the lines within an operator TD determines what the operator does. The position, either top or bottom, of the end of a line is an instruction to find either a Tine or a Faint, respectively, in the relevant argument TD; e.g., lines ending at the bottom left of {B,3} means locate a Faint in the TD for the expression [P]. In {B,5} the top position means find a Tine in the TD for [*P*v*Q*]. The two ends of a line in a binary operator constitute a pair of instructions to find the specified types of lines in each of the argument TDs; e.g., the ascending diagonal in {I,5} means find a Faint in the TD for [-*P*], {H,4}, and find a Tine in [-*Q*], {J,4}. The actual type of a line in an operator TD specifies the type of line to be drawn in the result TD; e.g., the negation operator {B,5} transforms all the Tines in {B,4} into Faints in {B,6}, as there is a top Faint in {B,5}; and vice versa for the Faint in {B,4} given the bottom Tine in {B,5}. The operator TD in {B,3} constructs {B,4} from all the four possible combinations of the Tines and Faints in {A,2} and {C,2}, with new Tines drawn whenever there is a Tine in either of the arguments.

On the left of the diagram, TDs for *P* and *Q* are disjunctively combined by the *or* operator {B,3} and the result is negated by {B,5}, to give a TD for –(*P*v*Q*), {B,6}. On the right, *P* and *Q* are individually negated before being conjunctively combined by the *and* operator {I,5}, to give –*P*&–*Q*, {I,6}.

The **tautology test** of the assumption and the conclusion of a derivation is the third interpretation of TDs. It applies the material implication TD in the solid rectangle {F,7} to {B,6} and {I,6}, in the manner of any operator. As there is a Tine in the same position in both {B,6} and {I,6}, a Tine is drawn in {F,8}, labeled 'TT'. Similarly, as Faints occur in all the same positions, Tines are also drawn for them in {F,8} 'FF', because the bottom line in {F,7} is a Tine. There are no combinations of a Tine in the assumption and a Faint in the conclusion, so no faints occur in {F,8}. Therefore, the test TD only has Tines, so it is a tautology and the derivation is thus valid.

The verbal descriptions of the interpretations of TDs given here do require some effort to follow, but graphical explanations in fuller treatments of TDs are easier to understand, because they can fully exploit the spatial structure of TDs.

Proofs that Truth Diagrams comprise a sound and complete system for propositional logic will be presented elsewhere, along with comparison of the relative benefits of TDs in comparison to the conventional formula notation and truth tables.

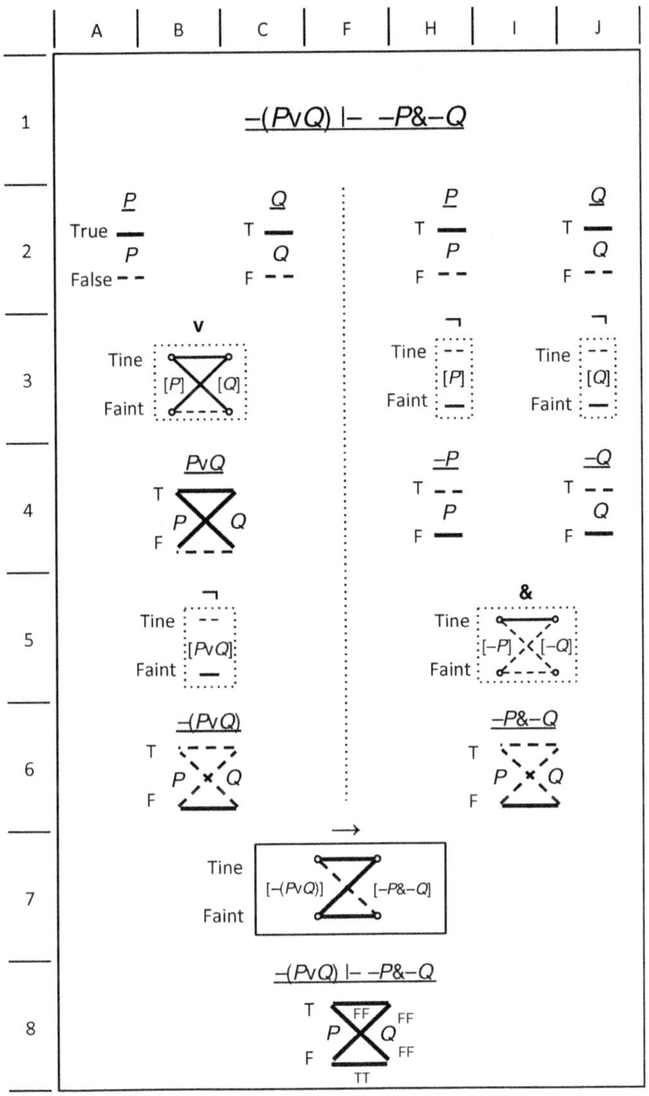


**Fig. 1.** TD derivation of one of de Morgan's Laws

# References


1. Cheng, P.C.-H.: Electrifying diagrams for learning: principles for effective representational systems. Cognitive Science 26(6), 685–736 (2002)
2. Cheng, P.C.-H.: Probably good diagrams for learning: Representational epistemic re-codification of probability theory. Topics in Cognitive Science 3(3), 475–498 (2011)
3. Cheng, P.C.-H., Barone, R.: Representing complex problems: A representational epistemic approach. In: Jonassen, D.H. (ed.) Learning to Solve Complex Scientific Problems, pp. 97–130. Lawrence Erlbaum Associates, Mahmah (2007)

# Are Teachers Aware of Students' Lack of Spontaneity in Diagram Use? Suggestions from a Mathematical Model-Based Analysis of Teachers' Predictions

Yuri Uesaka[1], Emmanuel Manalo[2], and Masanori Nakagawa[3]

[1] Graduate School of Education, The University of Tokyo, Japan
[2] Faculty of Science and Engineering, Waseda University, Tokyo, Japan
[3] Graduate School of Decision Science & Technology,
Tokyo Institute of Technology, Tokyo, Japan
y_uesaka@p.u-tokyo.ac.jp, emmanuel.manalo@gmail.com,
nakagawa@nm.hum.titech.ac.jp

**Abstract.** Although many studies have shown that diagrams are effective tools for problem solving, research evidence shows that students do not always use diagrams effectively. One of the most serious problems is their lack of spontaneity in diagram use. However, no previous studies have examined whether teachers are adequately aware of this problem. In this investigation, data were gathered on students' mathematics performance (including their spontaneous use of diagrams) and teachers' predictions of the students' performance. Using a mathematical model (Uesaka & Nakagawa, 2010) to analyze the data, it was found that the parameter representing the accuracy of teachers' prediction was lower for their assessment of spontaneous diagram use compared to other mathematical tasks. This suggests that spontaneity in diagram use is an overlooked aspect in teachers' view of student performance.

**Keywords:** spontaneous diagram use, math problem solving, teachers' awareness, mathematical model based analysis.

## 1 Introduction

Constructing diagrams in problem solving is considered by both educators and researchers to be an efficacious strategy [1]. Although teachers use a lot of diagrams during instruction, researchers have noted that in contrast students lack spontaneity in constructing and using diagrams [2]. Uesaka, Manalo & Ichikawa [2] suggested that this is one primary reasons for student failure in problem solving. The lack of spontaneity in diagram use basically means that students miss out on the benefits that diagram use brings to problem solving and other learning tasks in school. A previous study [3] using a newly developed assessment tool called COMPASS (a componential assessment of students' basic competence in mathematics) [4] has also revealed that students' performance was relatively poorer in tasks assessing the spontaneous construction and use of diagrams. A key finding in this study [3] was that students performed relatively better in a task that assessed their ability to interpret and construct diagrams *when they were given explicit instructions to do so*. Thus, the issue of spontaneity appears to be of particular concern. However, it is not clear whether teachers are aware of the prevalence of this problem in spontaneity.

P. Cox, B. Plimmer, and P. Rodgers (Eds.): Diagrams 2012, LNAI 7352, pp. 312–314, 2012.

Thus this study explored the question of whether teachers are adequately aware of diagram use problems that *their own* students might have. Examining this question is important because awareness is crucial to enabling teachers to take the necessary steps in formulating and using appropriate instructional strategies to address those problems. To achieve this goal, the accuracy of teachers' predictions about their own students' use of diagrams in designated tasks were examined and analyzed with the use of a mathematical model.

Four tasks in COMPASS were used to assess students' performance and examine the accuracy of teachers' predictions. If teachers' predictions in the four tasks distinguish between those that assess spontaneity in diagram use and those that do not, it would empirically show that there is a problem in teachers' awareness of the spontaneity problem. To analyze the accuracy of teachers' predictions, a model proposed by Uesaka and Nakagawa [5] was used. This model can be used with categorical data, and its advantage compared to other models is that it enables comparison of accuracy of predictions between different tasks.

## 2 Method

The participants were 18 junior high school mathematics teachers, and 682 8th-grade students (aged 13–14 years). The teacher and student participants came from four schools selected from different prefectures in Japan.

Four tasks in COMPASS were used. To assess diagram use spontaneity, a diagram self-construction task and a diagram utilization task were used. As comparison tasks (to assess diagram use *other than spontaneity*), a simple calculation task, and an interpretation and drawing task were used. The teachers were first asked to carry out predictions of their own students' performance in these tasks. After this, the students were administered the tasks.

## 3 Results and Discussion

The data were analyzed using the following mathematical model [5]. In this model, the teachers' prediction is considered as a function of their students' performance. When $Q_{ijk}$ is defined as the probability in category $k$ of task $i$ that teacher $j$ predicts as the percentage of students belonging to that category, and $P_{ik}$ is the real probability of students in category $k$ of task $i$, the mathematical model can be represented as follows (with $\alpha$ and $\beta$ as parameters to be estimated).

$$Q_{ijk} = \beta_{ij} * P_{ik}^{\alpha ij} \qquad [1\text{-}1]$$

Here, $\alpha$ represents the accuracy of the prediction: an $\alpha$ value closer to 1 suggests a greater level of prediction accuracy. A one-way ANOVA, in which $\alpha$ estimates were used as the dependent variable and task difference was the independent variable, was conducted. The effect of task difference was found to be significant ($F_{(3, 51)} = 3.05, p < .01$). The difference of $\alpha$ between tasks was contrasted by using multiple $t$-test. Firstly, $\alpha$ estimates for the two tasks assessing the spontaneous use of diagrams were found to be significantly lower than the other two ($t_{(51)} = 2.71, p < .01$). Secondly, no difference in $\alpha$ estimates between the two tasks assessing spontaneity in diagram use was found ($t_{(51)}$= .12, n.s.). Finally, no significant difference was found in comparing the two tasks assessing other competencies ($t_{(51)}$= .89, n.s.). In sum, the $\alpha$ estimates in the two tasks involving spontaneous diagram use were lower that the $\alpha$ estimates in the other two tasks. In addition, the $\alpha$ estimates in the two tasks relating to spontaneous diagram use deviated more from 1 (cf. when $\beta$ = 1, then $\alpha$ = 1 means perfect fit), so the teachers' predictions can be considered as significantly less accurate where predicting students' spontaneity in diagram use was concerned.

These results indicate that teachers were comparably poorer in predicting performance in tasks assessing spontaneity in diagram use than in other tasks assessing non-spontaneity aspects of diagram use. They suggest that the problem of students' lack of spontaneity in diagram use may be overlooked by teachers and thus may require particular attention in teacher professional development.

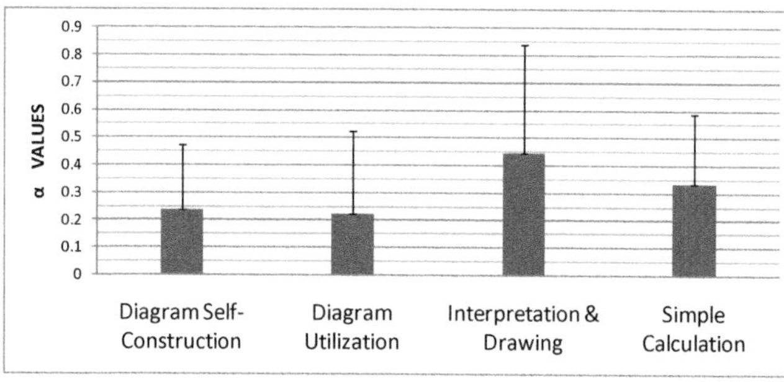


**Fig. 1.** Teacher Predictions of $\alpha$ Values for the COMPASS Tasks

## References

1. Cheng, P.C.H.: Electrifying Diagrams for Learning: Principles for Complex Representational Systems. Cognitive Science 26, 685–736 (2002)
2. Uesaka, Y., Manalo, E., Ichikawa, S.: What Kinds of Perceptions and Daily Learning Behaviors Promote Students' Use of Diagrams in Mathematics Problem Solving? Learning and Instruction 17, 322–335 (2007)
3. Uesaka, Y., Suzuki, M., Kiyokawa, S., Seo, M., Ichikawa, S.: Using COMPASS (Componential Assessment) to Reveal Japanese Students' Actual Competence in the Fundamentals of Mathematics: Is it True that "Students are Generally Fine with the Fundamentals, and that the Problems Exist Only in Applications"? (submitted)
4. Ichikawa, S., Haebara, T., Sugisawa, T., Seo, M., Kiyokawa, S., Inuzuka, M., Murayama, K., Uesaka, Y., Kobayashi, H., Shinogaya, K.: Development of COMPASS: Componential Assessment for Basic Competence and Study Skills in Mathematics. Cognitive Studies 16(3), 333–347 (2009)
5. Uesaka, Y., Nakagawa, M.: Development and Application of Mathematical Model Analyzing Teachers' Accuracy of Predication of Students' Performance: Proposing Empirical Methods Detecting Overlooked Competences and Learning Skills. In: Paper in Proceedings of the Annual Conference of the Japanese Cognitive Science Society, pp. 350–356 (2010)

# Modelling Delivery Information Flow: A Comparative Analysis of DSMs, DFDs and ICDs

Christopher Durugbo[1], Ashutosh Tiwari[2], and Jeffrey R. Alcock[2]

[1] Centre for Concurrent Enterprise, University of Nottingham, Nottingham, NG8 1BB, UK
christopher.durugbo@nottingham.ac.uk
[2] School of Applied Sciences, Cranfield University, Cranfield, MK43 0AL, UK
{a.tiwari,j.r.alcock}@cranfield.ac.uk

**Abstract.** Following an initial review and evaluation of current techniques for modelling delivery information flow in microsystems technology (MST) companies, this article analyses the 'information channel diagram' (ICD) approach – as a diagrammatical technique for modelling information flow through an empirical study that compares the ICD with existing information flow models used by MST companies.

**Keywords:** information flow, process modelling, conceptual design, function-orientation, delivery phase, microsystems technology.

## 1 Introduction

In an online survey of 100 MST companies, Durugbo *et al.* [1] identified two main formal techniques applied by analysts and managers to model information flow within the MST domain: data flow diagrams (DFDs) and design structure matrices (DSMs).

However, for effective use of diagrammatic models, it has been suggested that existing diagrammatical techniques be assessed based on their ability to aid perceptual (for thorough grasp of meaning) and conceptual (for hypotheses development) cognitive processes [2]. This assessment aids designers and researchers in systematically identifying modelling requirements of intended technique users that may then be applied in: selecting techniques that meet user requirements, combining techniques to create a hybrid version for use in modelling organisation characteristics, modifying techniques to meet user requirements, or developing new techniques to fill existing gaps or fulfil user requirements.

The aim of this paper is to analyse the 'information channel diagram' (ICD) approach (introduced in [3]) – as a diagrammatic information flow modelling technique. In order to accomplish this, models of delivery information flow created using DFDs, DSMs and ICDs were compared in case studies of 3 MST firms. These firms are all based in the United Kingdom with a targeted global market, and deliver MST based products and services as business-to-business solutions for customers that are mainly original equipment manufacturers or an academic institution.

P. Cox, B. Plimmer, and P. Rodgers (Eds.): Diagrams 2012, LNAI 7352, pp. 315–317, 2012.

## 2 Case Studies

In all cases (the companies where the interviews were conducted - Company A, Company B and Company C), 9 questions were used to assess the techniques face-to-face with 18 company staff (6 from each company): (1) Can delivery personnel roles be identified? (2) Can the delivery information flow paths be identified? (3) Can multiple communication channels during delivery be identified? (4) Can delivery process timing be identified? (5) Can collaborative delivery processes be identified? (6) Can the synchronisation of communication channels during delivery be identified? (7) Can the internal and external delivery information flows be identified? (8) Can the context for delivery information be identified? (9) Can the sharing of delivery information be identified?

The scoring system used at each company, was calculated as a fraction of 54. This is based on responses to the questions by the six participants from each company i.e. 6 participants × 9 questions = 54. This number represents the total number of possible responses from each company for each compared technique.

At Company A, where designers and engineers are allowed to make use of intuitive and individual approaches, ICD scored 50/54, DFD scored 16/54, and DSM scored 4/54. For Company B where the DFD is used in software design, ICD was selected for each of the 9 questions posed to each of the 6 participants from Company B – giving the ICD a score of 54 out of a possible 54 times. DFD scored 18/54 whereas DSM scored 3/54. Company C had previously used DFD and DSM approaches and the responses of participants revealed that ICD scored 42/54, DFD scored 20/54, and DSM scored 4/54.

For the DFD, the main positive attribute highlighted by participants was serial representation that made the flow of information easy to follow. Negative attributes of the DFD noted by participants included the absence of process times for establishing the duration of tasks and the duplication of entities in produced diagrams.

In the case of the DSM, participants responded negatively towards the technique with major difficulties in establishing the path and context for information flow. An absence of entities or roles and insufficient level of detail was also a negative attribute of the DSM noted by participants. However, participants commented positively on the ease with which the flow captured by the DSM can be converted into a software code (i.e. programmability) and the compact representation of the DSM.

For the ICD, the main positive attribute emphasised by participants was the ability of the technique to clarify flow depiction through the use of colour and the distinction/demarcation of roles and jobs through the use of swim-lanes. Other positive attributes of the technique commented on by participants included the depiction of process times for capturing task durations and the depiction/description of media forms for capturing the types of information content. Participants also commented on the unsuitability of the ICD to model backend tasks and interactions such as manufacturing and assembly.

In terms of overall visualisation and potential use, the ICD scored highest in two companies (Company A and Company B), as shown in Fig. 1. In the third company (i.e. Company C), the ICD tied with the DFD approach.

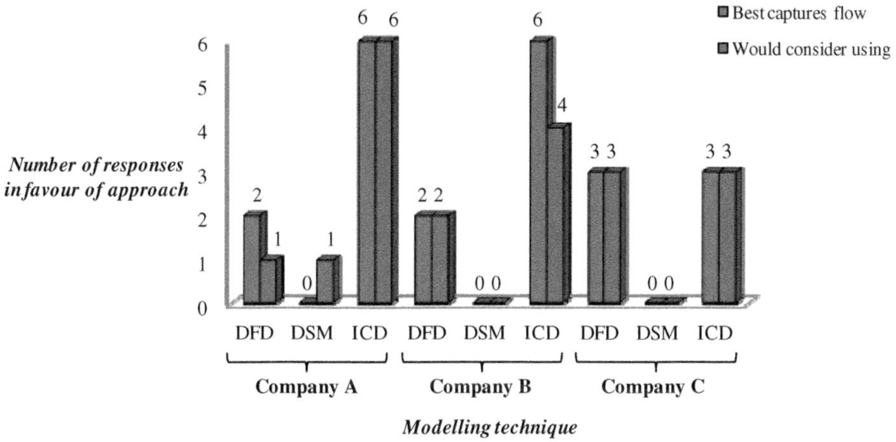


**Fig. 1.** Responses from case companies

## 3 Conclusions

In this paper, the 'information flow channel' (ICD) approach, a diagrammatical technique for modelling information flow, has been compared with data flow diagrams and design structure matrices created in an empirical study of delivery information flow in three United Kingdom based MST companies. Findings and discussions with 18 participants that took part in the empirical study revealed that the ICD scored highest in participant responses to questions involving roles of company personnel, possible paths for information flow, the presence of multiple channels, process times, collaborative processes, synchronised communication channels, harmonised internal and external flows, information sharing and contextualised information.

**Acknowledgments.** The authors would like to extend their sincere thanks to the Engineering and Physical Sciences Research Council (EPSRC), for its support via the Cranfield Innovative Manufacturing Research Centre (CIMRC), towards the work carried out in the preparation of this paper.

## References

1. Durugbo, C., Tiwari, A., Alcock, J.R.: Survey of Media Forms and Information Flow Models in Microsystems Companies. In: Camarinha-Matos, L.M., Pereira, P., Ribeiro, L. (eds.) DoCEIS 2010. IFIP AICT, vol. 314, pp. 62–69. Springer, Heidelberg (2010)
2. Hungerford, B.C., Hevner, A.R., Collins, R.W.: Reviewing software diagrams: a cognitive study. IEEE T. Software Eng. 30, 82–96 (2004)
3. Durugbo, C., Hutabarat, W., Tiwari, A., Alcock, J.R.: Information Channel Diagrams: An Approach for Modelling Information Flow. J. Intell. Manuf. (2012) doi:10.1007/s10845-011-0523-7

# Completeness Proofs for Diagrammatic Logics

Jim Burton, Gem Stapleton, and John Howse

Visual Modelling Group, University of Brighton, UK
{j.burton,g.e.stapleton,john.howse}@brighton.ac.uk

**Abstract.** We identify commonality in the completeness proof strategies for Euler-based logics and show how, as expressiveness increases, the strategy readily extends. We identify a fragment of concept diagrams, an expressive Euler-based notation, and demonstrate that the completeness proof strategy does not extend to this fragment.

## 1 Introduction

There have been a number of sound and complete logics based on Euler diagrams developed to date, such as [4,1]. All of the proofs of completeness have used constructive strategies, providing a proof that the theorem follows from the axioms. Moreover, they all adopt a similar framework, converting the diagrams involved into normal forms that are easily comparable. The completeness proof for each considered logic is an extension of the completeness proofs for its fragments. We show this by detailing the strategies used for Euler diagrams [1] and spider diagrams [2].

Euler diagrams are comprised of closed curves, each with a label. Examples can be seen in figure 1, where $d$ expresses that (the sets) $A$ and $C$ are disjoint, $B$ is a subset of $A$, and $D$ is a subset of $C$. The diagram $d'$ expresses that $D$ is a subset of $C$. To prove completeness of this logic, Hammer [1] proceeds by constructing a proof-writing algorithm: given an axiom $d$ and a theorem $d'$, carry out the following steps to prove $d'$ follows from $d$. First, add one curve labelled $L$ for each curve label, $L$, in $d'$ that is not in $d$, to give a diagram $d_c$. Next, erase all curves from $d_c$ that have labels not appearing in $d'$, to give a diagram $d_e$. Finally, add minimal regions to $d_e$ until it is the same as $d'$. The proof of completeness shows that it is possible to apply this algorithm whenever $d \vDash d'$ (i.e. $d$ semantically entails $d'$), thus establishing that $d \vdash d'$ (i.e. there is a proof

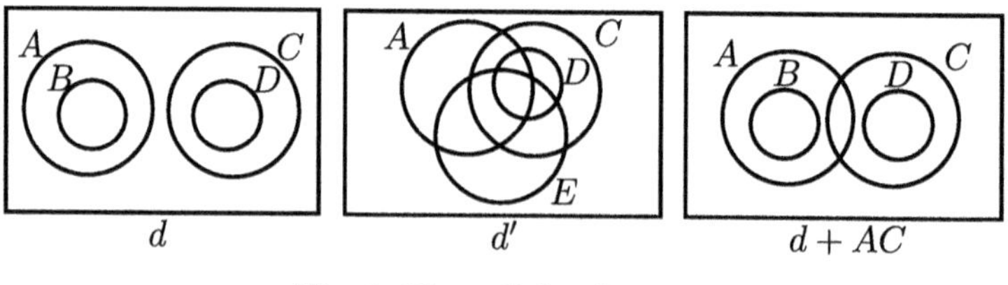


**Fig. 1.** Three Euler diagrams

P. Cox, B. Plimmer, and P. Rodgers (Eds.): Diagrams 2012, LNAI 7352, pp. 318–320, 2012.

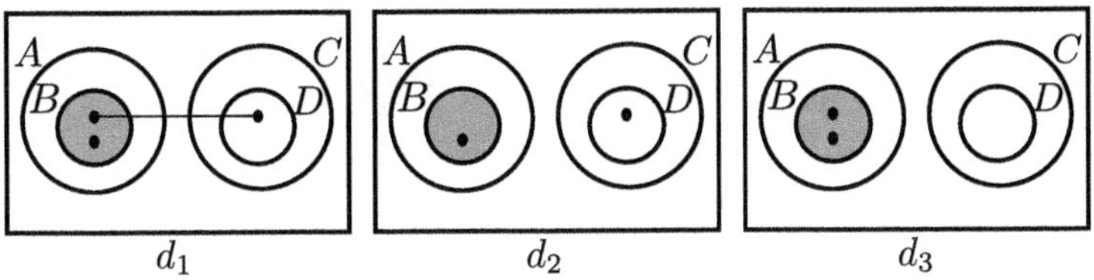


**Fig. 2.** Three spider diagrams

that $d'$ follows from $d$). We observe that we can interchange the last two steps in the completeness proof without significantly impacting its details. Thus, we *maximise* the syntax in the axiom diagram so that only inference steps that erase syntax are required in order to obtain $d'$.

Spider diagrams extend the Euler diagram logic by adding spiders, shading and logical connectives between diagrams. Examples of spider diagrams can be seen in figure 2 where $d_1$ expresses – using spiders – that there are at least two elements, one of which is in $B$ and the other of which is in $B \cup D$. Diagram $d_1$ also expresses – using shading – that no further elements are in $B$.

The completeness proof strategy for spider diagrams, from [2], starts with axiom $d$ and theorem $d'$, so $d \vDash d'$. In brief, the process starts off by converting $d$ to a normal form where the only logical connective used is $\vee$ and the spiders each comprise just a single node, giving diagrams we will denote by $d_{NF}$ and $d'_{NF}$. Furthermore, $d_{NF}$ and $d'_{NF}$ are the disjunction of sets of diagrams, $\{d_1, \ldots, d_i\}$ and $\{d'_1, \ldots, d'_i\}$ respectively where, for each $d_i$ and $d'_i$ in $d_{NF}$ and $d'_{NF}$ respectively, the sets of regions are the same. We then begin to add the spiders and shaded regions which are present in $d'_{NF}$ but not $d_{NF}$ until it can be established that each unitary diagram, $d_i$, in the axiom logically entails a unitary diagram, $d'_i$, in the theorem. Once this maximal form is achieved it is merely a matter of erasing syntax from $d_i$ to obtain $d'_i$. Then we have $d_i \vdash d'_{NF}$ and it can be trivially shown that $d_{NF} \vdash d'_{NF}$, as required. We refer to [2] for full details.

## 2 More Expressive Notations and the End of the Strategy

Concept diagrams [3] build on spider diagrams by adding further syntax: arrows, to place constraints on binary relations, and unlabelled curves. The diagrams in figure 3 are concept diagrams. Diagram $d$ expresses that there are two sets, $x$ and $y$, which are disjoint subsets of $A$, that the image of the relation $f$, when its domain is restricted to $A$, is $B$, and that there is an element in $x$ such that the image of $f$, when its domain is restricted to that element, is the empty set.

It is possible to extend the notion of maximality to concept diagrams by taking arrows into account, using the notion of injective mappings from the arrows of one diagram to another. The process of obtaining the maximal form of a diagram, then, includes the task of adding as many arrows as is possible without changing the meaning of the diagram. We call these arrows *potential* arrows.

In figure 3, $d \vDash d'$ but there is no mapping from the arrows of $d'$ to $d$, and so we maximise $d$ by adding potential arrows. The arrows $(f, x, B)$ in diagram $d_1$ and

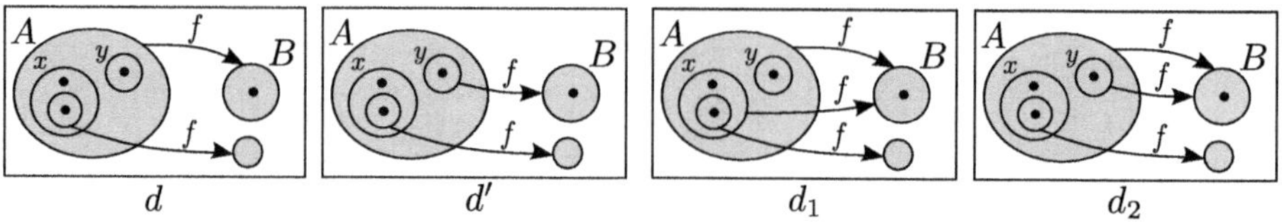


**Fig. 3.** Four concept diagrams

$(f, y, B)$ in diagram $d_2$ are potential arrows for $d$, since either arrow can be added to $d$ without changing its meaning. After adding either arrow, however, the other arrow ceases to be a potential arrow. If we are to extend the completeness proof strategy by adding arrows to the axiom, *all* of the diagrams obtained from $d$ using this process must have an arrow set that can be injectively mapped to by the arrows of $d'$ in the appropriate way; this is because the diagrams obtained are semantically equivalent to $d$ and, therefore, semantically entail $d'$. We can see that we can remove syntax from $d_2$ to obtain $d'$, but there is no general strategy that can be used to transform $d_1$ into $d'$, even though $d_1 \vDash d'$.

## 3 Conclusion

We have identified commonality in the completeness proof strategies of various logics based on Euler diagrams. As expressiveness increases the strategy readily extends in some cases, but breaks down for concept diagrams, which are syntactically richer and more expressive than earlier logics based on Euler diagrams. Thus, we have established that the existing completeness proof strategies are limited. The non-unique ways of adding syntax to concept diagrams results from the syntactic richness of the notation and from their expressiveness power. We believe that the same phenomena will arise in equally expressive logics.

## References

1. Hammer, E.: Logic and Visual Information. CSLI Publications (1995)
2. Howse, J., Stapleton, G., Taylor, J.: Spider diagrams. LMS Journal of Computation and Mathematics 8, 145–194 (2005)
3. Howse, J., Stapleton, G., Taylor, K., Chapman, P.: Visualizing Ontologies: A Case Study. In: Aroyo, L., Welty, C., Alani, H., Taylor, J., Bernstein, A., Kagal, L., Noy, N., Blomqvist, E. (eds.) ISWC 2011, Part I. LNCS, vol. 7031, pp. 257–272. Springer, Heidelberg (2011)
4. Shin, S.-J.: The Logical Status of Diagrams. Cambridge University Press (1994)

# Modelling Information Flow: Improving Diagrammatic Visualisations

Christopher Durugbo

Centre for Concurrent Enterprise, University of Nottingham, Nottingham, NG8 1BB, UK
christopher.durugbo@nottingham.ac.uk

**Abstract.** In this paper, the needs and modelling considerations of diagrammatic representations of information flow are assessed. It presents the main recommendations of 18 engineers, scientists, and managers during case studies involving 3 semiconductor companies. The study shows that effective modelling processes and primitives require the effective use of colour coding within diagrams, identification of dichotomies for information classification, simplification of information content and communication, and the case-by-case use of tool during modelling. The paper concludes by discussing the implications of the findings for research and practice.

**Keywords:** information flow, flow analysis, conceptual modelling, process modelling, diagrammatic reasoning, communication.

## 1 Introduction

The use of diagrams to model information flow makes it easier for organisational personnel to relate to and understand organisational requirements [1]. This offers a unique opportunity to gain insights into activities and interactions that impact on operations, management and support processes. A combination of strategies for re-engineering information and information flows can then be leveraged to improve organisational performance.

The process of modelling information flows for organisations is important for three main reasons. Firstly, it aids organisations to analyse their current state of information flow. Secondly, it enables organisations to identify and eliminate redundant and ineffective information flows as well as minimising the duplication of information. Thirdly, it helps an organisation to make assessments and recommendations for improving future internal/external communication and overall organisational performance. This activity is useful for implementing organisational strategies such as resource allocation and job description.

Driven by insights from a comparative analysis of diagrammatic information flow models, this paper seeks to offer fresh insights into the visualisation and analysis needs of information flow diagrams.

P. Cox, B. Plimmer, and P. Rodgers (Eds.): Diagrams 2012, LNAI 7352, pp. 321–323, 2012.

## 2 Information Flow Diagrammatic Modelling Needs

In an empirical study of 3 companies within the semiconductor domain [2], the needs of diagrammatic representations was analysed with 18 participants (engineers, scientists, and managers) through face-to-face semi-structured interviews. To initiate discussions on the modelling needs of information flow diagrams, participants were presented with 3 modelling tools (data flow diagrams (DFDs), design structure matrices (DSMs) and information channel diagrams (ICDs)), as described in [2]. The main points from these discussions were:

- Colour coding effectiveness,
- Simplification of information,
- Information classification dichotomies, and
- Case-by-case tool use.

The use of colour in representations was generally favoured by participants of the study. Within the ICD, use of colours reinforces the different roles of personnel. Participants however cautioned on the use of colour because individual colours could symbolise different properties and may have different roles in other tools used by engineers and scientists. For instance, the colour red, as noted by a scientist within the study, could be perceived as important roles or associated with critical tasks such as in the project evaluation and review technique.

The study found simplification as an important factor in the presentation of information content and means for communication. The suggestion was that models must be free of clutter and simplified as much as possible if they are to be useful. However, this conflicts with the findings of the comparison of [2] where the DSM, a tool that was conceived to minimise clutter (see for instance [4]), scored lowest among the compared tools. Comments by participants offered clues to this contradictory finding. Firstly, participants noted that although the compactness of the DSM makes it simple, the tool lacks enough primitives to characterise 'what was going on'. Secondly, the DSM according to some participants was difficult to understand and follow.

Two managers noted that interactions involving information flow with customers and staff can be modelled according to front-end flows for administrative, accounting, distribution and sales functions and back-end flows for design, manufacturing and technical service functions. Similarly, the technological/business distinction of data content aided the case companies in maintaining their day-to-day operations. Business content largely relate to data from front-end interactions whereas technological content is mainly associated with back-end interactions.

Participants noted that in day-to-day operations, the choice and use of diagrams must be based on a case-by-case basis depending on the level of complexity of concepts and system implementation. This is because the compared tools (DFDs, DSMs and ICDs) all depict the flow of information in different ways. Furthermore, in practise, groups (such as manufacturers) or users (such as customers) are typically only concerned with some aspects of the information model. This supports the idea that an all-encompassing information model is unnecessary and impractical for designers [3].

## 3 Implications for Researchers and Practitioners

In this paper, an attempt has been made to assess the needs and modelling considerations of diagrammatic representations of information flow. It concludes with the following implications for researchers and practitioners:

- *Information flow for firms is non-monolithic and dependent on companies' strategies for maintaining firm competitiveness.* During the discussions with participants, different starting points for information flows were identified. These points varied for the individual companies and were intended to establish industry scopes according to focus on customer requests, service contracts and work product releases. However, the general purpose of each flow was to maintain the competitiveness of the company with a view to maintaining sustainable operations.
- *A demarcation of roles is vital to modelling information flow.* This is because a wide range of information flows to and from companies during day-to-day operations. These flows are managed by roles and systems that coordinate interactions between information sources and destinations. Consequently, depictions to analyse information flow for organisations must include the information source, destination and management roles.
- *The use of colour enriches representations.* Colours offer opportunities for characterising the properties of concepts such as processes, objects and materials. Participants particularly favoured the use of colour in the ICD because it improved visual perception and reinforced the role of personnel.
- *Simplified communications and organisational models are necessary for effective operations.* Modern day businesses, in an attempt to remain competitive, undertake processes and projects that may be complex and/or large in scale. Communication and models of information flow if complicated in these cases creates additional tasks, wastes company time and reduces overall productivity.

**Acknowledgments.** The authors would like to extend their sincere thanks to Prof. Ashutosh Tiwari, Dr Jeffrey Alcock and the Engineering and Physical Sciences Research Council (EPSRC), for their support via the Cranfield Innovative Manufacturing Research Centre (CIMRC), towards the work carried out for this research.

## References

1. Sen, T.: Diagrammatic knowledge representation. IEEE T. Syst. Man Cyb. 22, 826–830 (1992)
2. Durugbo, C., Tiwari, A., Alcock, J.R.: Modelling Delivery Information Flow: a Comparative Analysis of DSMs, DFDs and ICDs. In: Cox, P., Rodgers, P., Plimmer, B. (eds.) Diagrams 2012. LNCS (LNAI), vol. 7352, pp. 315–317. Springer, Heidelberg (2012)
3. Durugbo, C., Tiwari, A., Alcock, J.R.: A review of information flow diagrammatic models for product-service systems. Int. J. Adv. Manuf. Technol. 52, 1193–1208 (2011)
4. Steward, D.V.: The design structure system: a method for managing the design of complex systems. IEEE T. Eng. Manage. 28, 71–74 (1981)

# A Graph Calculus for Proving Intuitionistic Relation Algebraic Equations*

Renata de Freitas and Petrucio Viana

Institute of Mathematics and Statistics,
UFF: Universidade Federal Fluminense, Niterói, Brazil

**Abstract.** In this work, we present a diagrammatic system in which diagrams based on graphs represent binary relations and reasoning on binary relations is performed by transformations on diagrams. We proved that if a diagram $D_1$ can be transformed into a diagram $D_2$ using the rules of our system, under a set $\Sigma$ of hypotheses, then it is intuitionistically true that the relation defined by diagram $D_1$ is a sub-relation of the one defined by diagram $D_2$, under the hypotheses in $\Sigma$.

**Keywords:** Proofs with graphs, Relation algebra, Intuitionistic logic.

**Introduction.** *Boolean reasoning*, i.e. reasoning involving plain sets and the Boolean operations of union, intersection and complement, may be performed through the algebraic language of Boolean algebras [1]. *Relational reasoning*, i.e. reasoning involving relations, the Boolean operations, and the Peircean operations of composition and conversion, may be performed through the more elaborated algebraic language of De Morgan-Peirce-Schröder-Tarski relation algebras [5]. A main difference between these environments arises from the fact that, although there is a number of algorithms to decide validity or perform inferences in the calculus with sets [1], the analogous tasks for the relational language are highly undecidable [6]. Hence, the problem of *building mechanisms which may help in the design of relational algebraic proofs from scratch* arises (cf. [4,2]).

A mechanism based on diagrams to perform relational reasoning is proposed in [3]. In that work, we handle inclusions rather than equalities, and we use diagrams in the left and right hand sides of an inclusion, instead of relation algebraic terms. Starting with the diagram in the left hand side of the target inclusion, by successive applications of our transformation rules, mediated by the inclusions taken as hypotheses, either we end up with the diagram in the right hand side of the target inclusion, when the inclusion is a consequence of the hypotheses or, otherwise, build a possibly very large non constructive counter model (cf. [3] for details). In this work, we modify that system to deal with intuitionistic relational algebraic inferences.

**Example.** A diagrammatic proof that $r^{-1} \circ t^{\mathsf{C}} \subseteq s^{\mathsf{C}}$ is an intuitionistic consequence of $r \circ s \subseteq t$ is displayed in Figure 1. The tag $*$ is introduced in the

* Research partially sponsored by CNPq and FAPERJ.

P. Cox, B. Plimmer, and P. Rodgers (Eds.): Diagrams 2012, LNAI 7352, pp. 324–326, 2012.

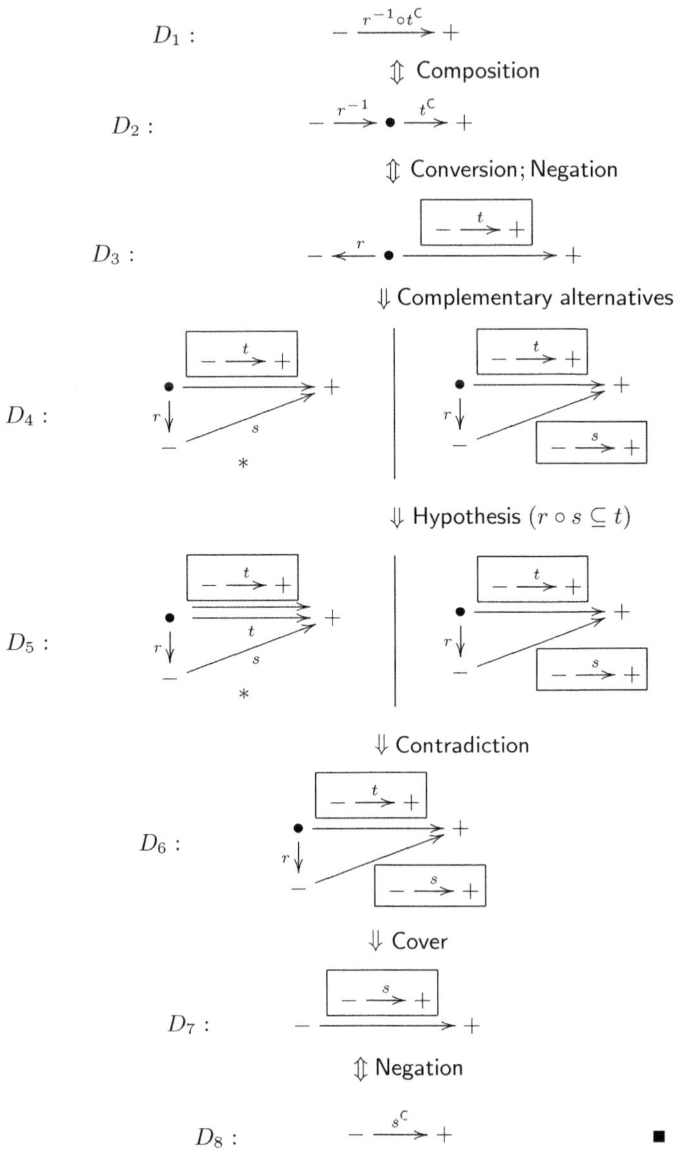


**Fig. 1.** Diagrammatic proof of $r \circ s \subseteq t \ \vdash_i \ r^{-1} \circ t^{\mathsf{C}} \subseteq s^{\mathsf{C}}$

proof when we apply the rule that allows transforming a diagram by breaking it in two parts, each one containing alternative complementary information: one positive and the other negative. This passage occurs in the example when we transform diagram $D_3$ into diagram $D_4$. The tag labels the subdiagram which does not have the occurrence of the negative information, expressed in the form of a boxed term or diagram. The tag is erased by the application of the rule that allows erasing contradictory subdiagrams. This passage occurs in the example when we transformed diagram $D_5$ into diagram $D_6$.

The part of the diagrammatic proof which consists of the tagged subdiagrams resembles the intuitionistic reductio ad absurdum. The appearance of the arc labeled by $s$ in the tagged subdiagram of $D_4$ corresponds to "assume $(-,+) \in s$ for a contraction". When a contradiction arises, at the tagged subdiagram of $D_5$, we derive the complementary alternative, $(-,+) \notin s$, represented by the arc labeled by $\boxed{- \xrightarrow{s} +}$ in $D_6$.

**Soundness.** We have defined a translation of diagrams $D$ into first-order formulas $tD$ and proved soundness of this calculus by showing that, given a diagrammatic proof of $D'$ from $D$ based on a set of hypotheses $\Gamma$, i.e. a sequence of diagrams $(D_1, \ldots, D_n)$ s.t. each diagram $D_{i+1}$ is obtained from $D_i$ by the application of some rule of the calculus, one can transform the sequence $(tD_1, \ldots, tD_n)$ of first-order formulas into an intuitionistic proof that the relation represented by $D'$ contains the relation represented by $D$. (For details, cf. `www.uff.br/grupodelogica/FV12a.pdf` )

**Summary.** The main characteristic of the system presented in [3], besides its accordance with reasoning from hypotheses, is an explicit diagrammatic representation of complement. One of the main transformation rules of that system allows us to change any given diagram into another one, by reproducing it in two parts, representing two complementary alternatives: some relation or its complement holds between some pair of points in the domain. In this work, we introduce tags in the diagrams occurring in proofs to distinguish intuitionistic from non intuitionistic applications of that rule.

## References


1. Brown, F.M.: Boolean reasoning: the logic of Boolean equations. Dover (2003)
2. Foster, S., Struth, G., Weber, T.: Automated Engineering of Relational and Algebraic Methods in Isabelle/HOL. In: de Swart, H. (ed.) RAMICS 2011. LNCS, vol. 6663, pp. 52–67. Springer, Heidelberg (2011)
3. de Freitas, R., Veloso, P.A.S., Veloso, S.R.M., Viana, P.: A Calculus for Graphs with Complement. In: Goel, A.K., Jamnik, M., Narayanan, N.H. (eds.) Diagrams 2010. LNCS(LNAI), vol. 6170, pp. 84–98. Springer, Heidelberg (2010)
4. von Oheimb, D., Gritzner, T.F.: RALL: Machine-Supported Proofs for Relation Algebra. In: McCune, W. (ed.) CADE 1997. LNCS, vol. 1249, pp. 380–394. Springer, Heidelberg (1997)
5. Schmidt, G., Ströhlein, T.: Relation and Graphs: discrete mathematics for computer scientists. Springer, Berlin (1993)
6. Tarski, A., Givant, S.: A Formalization of Set Theory Without Variables. American Mathematical Society (1987)

# Genetic Algorithm for Line Labeling of Diagrams Having Drawing Cues

Alexandra Bonnici and Kenneth Camilleri

Department of Systems and Control Engineering, University of Malta, Malta
{alexandra.bonnici,kenneth.camilleri}@um.edu.mt

**Abstract.** Drawings are an integral part of the design process, helping designers communicate abstract concepts to others. In this paper we propose a genetic algorithm that successfully exploits cues present in drawings in a line labeling algorithm for sketches.

**Keywords:** genetic algorithms, line labeling, drawing cues.

## 1 Introduction

Cues help artists portray intent and hence aid the drawing interpretation. Cues may include *line phrasing* where designers adjust the stroke width according to the depth of the object edge[3], *table lines* which indicate the spatial relation of the object with respect to its background and tone or illumination changes. In pen-and-paper sketches, tone changes are created by hatching techniques [4] and serve to give the impression of depth as well as to emphasize the shape and form of the object and its spatial relationship with other objects in the sketch. Here we propose a method to exploit such cues in an off-line line labeling algorithm suitable for hand-drawn sketches.

## 2 Genetic Algorithm Approach for Line Labeling

Line labeling algorithms are used to describe each edge in the drawing in relation to its neighbouring edges. Huffman [5] and Clowes [1] created a junction dictionary $\mathbf{\Gamma}$ by which these edge can be labeled. Waltz [8] and Cooper [2] enhanced the labeling by introducing hard constraints that determine the appropriate edge label in scenes containing shadows and contrast failures. However, hard constraints are inappropriate for use with concept sketches where cues may be geometrically incorrect. Hancock and Myers [7] propose the use of a genetic algorithm (GA) to determine the edge labels, representing edges as an arbitrarily ordered sequence of genes forming a fixed length chromosome. Each gene may take a value $\lambda_i \in \mathbf{\Lambda}$ where $\mathbf{\Lambda}$ is the list of all possible edge labels such that a chromosome is defined by $E = \{\lambda_1, \lambda_2, \cdots \lambda_N\}$, where $N$ is the number of edges in the drawing. The drawing can be described as a list of junctions $J_k, k = 1 \cdots K$, where $K$ is the number of junctions. The edge labels $E(J_k) \in E$ therefore list the edge labels of edges forming junction $J_k$ in chromosome $E$. The fitness of the chromosome is then defined as the Hamming distance between the possible labels defined in junction dictionary $\mathbf{\Gamma}$ and $E(J_k)$ in chromosome $E$ [7].

P. Cox, B. Plimmer, and P. Rodgers (Eds.): Diagrams 2012, LNAI 7352, pp. 327–329, 2012.

## 3 Introducing Drawing Cues to Enhance Line Labeling

The GA of [7] may converge to a legal labeling which does not match the design intent portrayed by the cues. We enhance this GA such that drawing cues may guide it towards an intended solution. We focus here on three cues, namely cast and attached shadows and table lines which are predominantly used in drawings [6].

Cues constrain the relevant edges to assume a subset of allowed labels. These constraints have been used to compile a cue constraint filter (CCF) to restrict the allowed edge labels of the corresponding edges. This CCF is used to limit the possible edge labels of each gene $g_n$ in the chromosomes that form the initial population, thus obtaining an initial population which is close to the expected solution. Since the GA is allowed to change the chromosome through cross-over and mutation operations, the initial information prompted by the cues may be lost through the evolutionary process. For this reason, besides applying the CCF to the initial population, it is also used to obtain a subset of labels $\Lambda(n)$ from $\Lambda$ that may be assigned to a gene $g_n$ given the set of cues $C(n)$ that bear upon the edge represented by $g_n$. This is represented as $CCF(\Lambda|g_n, C(n)) = \Lambda(n)$. We then define a penalty function as $P_n = \{\frac{1}{N}\text{if}\Lambda(n) \neq \varnothing\lambda_i \notin \Lambda(n); \text{otherwise}0\}$ which acts as a soft constraint on the edge label. The fitness function is then defined as

$$F(E) = \alpha\left(\frac{1}{2N}\sum_{k=1}^{K}\min_{l=1,\cdots,|\mathbf{\Gamma}|} H(E(J_k), \mathbf{\Gamma})\right) - (1-\alpha)\left(\sum_{n=1}^{N} P_n\right) \quad (1)$$

$H$ denotes the Hamming distance and $\alpha$ is a weight factor that determines the confidence in the cues. The value of $\alpha$ was arbitrarily set to 0.6.

## 4 Results and Discussion

The cue-based GA was evaluated on diagrams such as those shown in Fig. 1 depicting drawings which, although having the same geometric shape, have different cues and require different interpretations. The GA was implemented with proportionate fitness selection, a 1-point crossover with a rate of 0.9 and a mutation rate of 0.03. The cue-based GA was performed over 50 trials of 500 generations each and performance was compared to a cue-less GA implementation with the same parameters.

In the cue-less GA, all the trials converged to a mean maximum fitness of 1, achieving geometrically correct solutions in all trials. However, only 18% achieved the intended solution portrayed by the cues. In contrast, the cue-based GA achieved a mean maximum fitness of 0.9889 with an average of 76% converging to the intended solution with the remaining 27% converging to a solution which, although geometrically correct, was contradictory to the cues. In the case of Fig. 1(c), which is an example of an ambiguous drawing, 46% of the trials disregarded the evidence of cue (2) which is in conflict with the other cues while

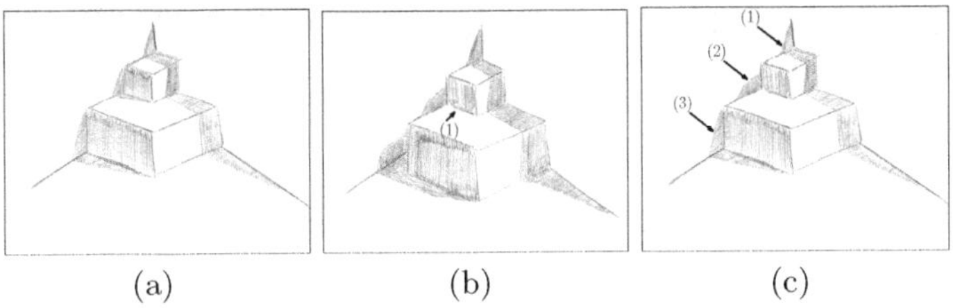


**Fig. 1.** A sample of diagrams on which the cue-based GA has been tested. Diagram (b) has a missing cast shadow at (1) and Diagram (c) has conflicting cast shadows at (1), (2) and (3)

the remaining 54% match all the cues but in so doing create mismatches with the junction dictionary. This experiment shows that the proposed GA may handle ambiguous drawings gracefully.

## 5 Conclusion

The results obtained encourage the use of cues in the interpretation of the drawings and give scope for future work to this effect. One possible improvement to this approach is to make use of a mechanism that would allow the cue-based interpretation of the edge to co-evolve with the junction interpretation of the drawing. This would allow stronger cooperation between the two aspects of the population fitness, hence enhancing the chances of identifying solutions that match the interpretations implied by the cues in the drawing.

## References

[1] Clowes, M.B.: On seeing things. Artificial Intelligence 2(1), 76–116 (1971)
[2] Cooper, M.: The interpretation of line drawings with contrast failure and shadows. International Jouranl on Computer Vision 43(2), 75–97 (2001)
[3] Costa Sousa, M., Prusinkiewicz, P.: A few good lines: Suggestive drawing of 3d models. Computer Graphics Forum 22(3), 381–390 (2003)
[4] Guptill, A.L.: Rendering in Pen and Ink. Watson-Guptill (1997)
[5] Huffman, D.A.: Impossible objects as nonsense sentences. Machine Intelligence 6, 295–323 (1971)
[6] Mamassian, P., Knill, D.C., Kersten, D.: The perception of cast shadows. Trends in Cognitive Sciences 2(8), 288–295 (1998)
[7] Myers, R., Hancock, E.R.: Genetic algorithms for ambiguous labelling problems. Pattern Recognition 33(4), 685–704 (2000)
[8] Waltz, D.: Understanding line drawings of scenes with shadows. In: The Psychology of Computer Vision, 2nd edn., pp. 19–91. McGraw-Hill (1975)

# A Logical Investigation on Global Reading of Diagrams

Ryo Takemura[1], Atsushi Shimojima[2], and Yasuhiro Katagiri[3]

[1] College of Commerce, Nihon University, Japan
[2] Faculty of Culture and Information Science, Doshisha University, Japan
[3] Department of Complex and Intelligent Systems, Future University Hakodate

**Abstract.** We call the extraction of higher-level information from diagrams "global reading," and investigate it from the viewpoint of logic.

**Introduction.** By "global objects," we mean those patterns or structures in diagrams (e.g., multiple dots in a scatter plot; multiple columns in a vertical bar graph; rows of cells in a table; sequences of edges in a directed graph, etc.) that allow the extraction of higher-level information about the represented domain. The extraction has been variously called "macro reading" [8], "pattern perception" [2], "direct translation" [5], and "cognitive integration" [6], and contrasted to the extraction of more concrete information from local objects (such as individual dots, bars, cells, edges, etc.). Although both designers and researchers agree that the former largely accounts for the inferential advantages of information graphics [1,8,5,9,3,2,6], few attempts have been made to flesh out what exact computational advantages it provides.

We call the extraction of higher-level information from diagrams "global reading." One of the difficulties in studying global reading is to define which collection of components of a diagram is regarded in general as a meaningful unit, i.e., a global object. Given a diagram, we may find a variety of global objects that could be interpreted and used to help reasoning. As a first step toward understanding the rich potential of global reading, we confine ourselves to an investigation of one of the most abstract global objects—not all those that are visually meaningful, but those that are invariant under any representation in the given diagrammatic system. We call these invariant global objects "units," and define units for systems of tables, directed graphs, and Euler diagrams, as they are broadly conceived. Then, we give a mathematical characterization of units of our diagrams, and study relationships between these units.

The idea is to compare, from a mathematical point of view, different diagrammatic systems for (1) their potentials of global reading and (2) the computational advantages resulting from them. For (1), we define a common set of basic data and, for representations thereof, we introduce abstract syntax for respective diagrammatic systems. The common basic data and abstract syntax based on them reveal abstract structures of units in different systems, and make it possible to study relationship between them. As for (2), we apply the complexity analysis in computer science. Investigations on algorithms and data structures

P. Cox, B. Plimmer, and P. Rodgers (Eds.): Diagrams 2012, LNAI 7352, pp. 330–333, 2012.

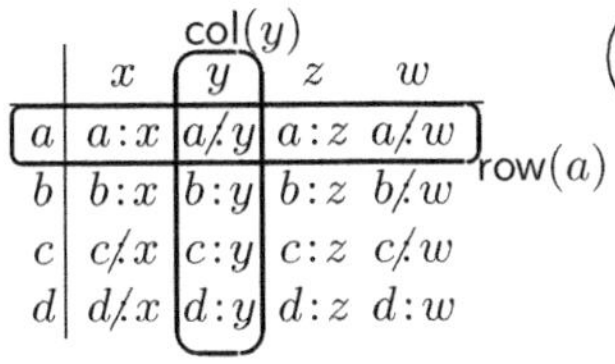


**Fig. 1.**

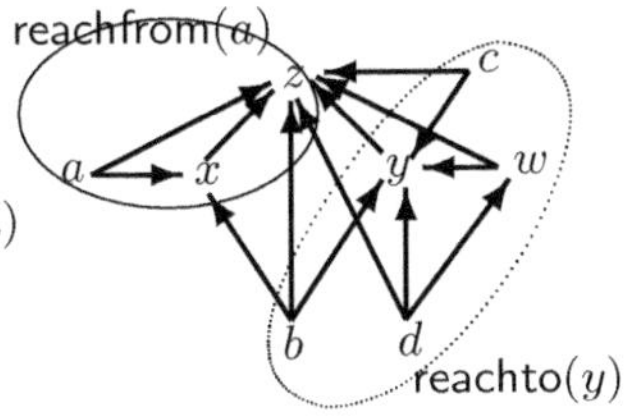


**Fig. 2.**

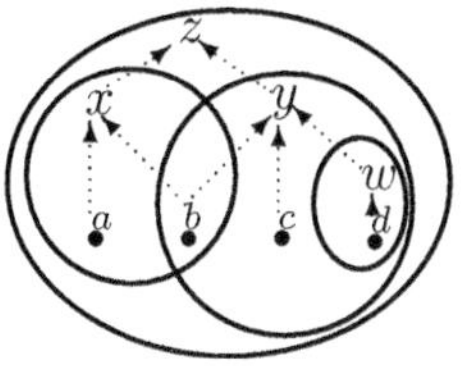


**Fig. 3.**

(such as arrays, lists, and trees etc.) are well-developed in the literature on computer science. Our work is unique in applying this aspect of computer science to the analysis on actual human reasoning that exploit units in diagrams. Our general finding is that units in a diagram considerably reduce the number of steps required in searching objects by putting some objects together, which helps reasoning tasks associated with that diagram. (See [7] for the detail of our study.)

**Basic Data.** Our basic data are $a{:}x$ (meaning "$a$ is $x$") and $a{\not:}x$ ("$a$ is not $x$") with $a \in A$ (the set of "objects") and $x \in X$ (the set of "properties"). A basic data set is represented by a sequence of data as follows:
$\{a{:}x,\ c{:}z,\ b{:}z,\ a{\not:}w,\ a{:}z,\ d{:}y,\ c{\not:}x,\ d{:}w,\ a{\not:}y,\ c{\not:}w,\ d{\not:}x,\ b{\not:}w,\ d{:}z,\ c{:}y,\ b{:}y,\ b{:}x\}$

Our reasoning tasks are enumerations of some data from a given sequence of basic data or a given diagram: Enumeration of all properties that an object $a$ has ($a$ is $\vec{x}$); Enumeration of all objects that satisfy a property $x$ (All of $\vec{a}$ are $x$); Enumeration of all properties that are implied by $x$ ($x$ is $\vec{x}$); Enumeration of all properties each of which implies $x$ (All of $\vec{x}$ are $x$).

The most basic and intuitive algorithm to enumerate target data from a given sequence of data is the so-called sequential search. Given a sequence of $l \times k$ data (with $l = |A|$ and $k = |X|$), we look through the sequence from the left to right. Hence, the **search space**, i.e., the number of data to be searched in a single search, or equivalently the number of steps required in a single search, of a basic data set is $l \times k$ in general. One of the remarkable features of our basic data set is the difficulty to read off relationships between properties.

**Tables.** Our table is defined as an $A \times X$-matrix over a basic data set, that is, a rectangular arrangement of basic data in which rows and columns are indexed by sets $A$ and $X$, respectively. We regard each collection row($a$) of data in a row and col($x$) of data in a column as a unit in a table as illustrated in Fig. 1.

In general, for reasoning tasks involving a table, we first focus on a row (resp. a column) by looking through all $l$ rows (resp. $k$ columns), and then enumerate target data from all $k$ data of the row (resp. $l$ data of the column). Hence, the search space of a table is generally $l + k$. As with a basic data set, it is not easy to read off relationships between properties by using a table.

**Directed Graphs.** Our directed graphs, called P-graphs (cf. Fig. 2), are essentially so-called Hasse diagrams for posets. Among various candidates for units, we define a unit in a P-graph as the set $\mathsf{reachfrom}(x)$ of nodes that are reachable from a node $x$ by traversing $\rightarrow$-edges. In view of the theory of ordered sets, the set $\mathsf{reachfrom}(x)$ is exactly the principal filter (or upset) generated by $x$.

In general, for reasoning tasks involving a P-graph, we first look through all $l+k$ nodes, i.e., units, and focus on the target. Then, we enumerate target data by searching the unit consisting at most of $k$ nodes. Hence, the typical search space of a P-graph is $(l+k)+k$. In contrast to data sets and tables, it is relatively easy in P-graphs to derive relationships among properties.

**Euler Diagrams.** We define Euler diagrams following [4], in which a diagram is specified by the set of inclusion and exclusion relations holding between circles and points in the diagram. We define a unit $\mathsf{circle}(x)$ of an Euler diagram as the set of circles and points that are inside a circle $x$. It is shown that the unit corresponds, in view of the theory of ordered sets, to the principal ideal generated by $x$, which is the dual notion of the principal filter, and it corresponds, in view of a P-graph, to the set $\mathsf{reachto}(x)$ of nodes reachable to $x$. (Cf. Fig. 3.)

In general, in a reasoning task involving an Euler diagram, we first focus on a target circle by looking through all $k$ circles, i.e., units. Then, by looking through a maximum of $k$ circles or $l$ points in the unit, i.e., inside the circle, we enumerate target data. Hence, the typical search space is $k+k$ or $k+l$. Since our units in Euler diagrams are the dual of units in P-graphs, they work effectively in tasks that have dual forms with respect to the successful tasks with P-graphs.

**Conclusion.** The relationship among our units in respective diagrams as well as their well-established mathematical counterparts are summarized as follows:

| Tables | | P-graphs | | Euler diagrams | Mathematical counter part |
|---|---|---|---|---|---|
| $\mathsf{row}(a)$ | $\simeq$ | $\mathsf{reachfrom}(a)$ | | | principal filter |
| | | $\updownarrow$ dual | | | $\updownarrow$ dual |
| $\mathsf{col}(x)$ | $\simeq$ | $\mathsf{reachto}(x)$ | $\simeq$ | $\mathsf{circle}(x)$ | principal ideal |

Our computational complexity analysis are summarized as follows.

| | $a$ is $\vec{x}$ | All of $\vec{a}$ are $x$ | $x$ is $\vec{x}$ | All of $\vec{x}$ are $x$ |
|---|---|---|---|---|
| Set | $l \times k$ | $l \times k$ | $(((l \times k) \times 2) \times l) \times k$ | $(((l \times k) \times 2) \times l) \times k$ |
| Table | $\boxed{l+k}$ | $\boxed{l+k}$ | $k + (((l+k) \times l) \times k)$ | $k + (((l+k) \times l) \times k)$ |
| P-graph | $l+k+k$ | $(l+k+k) \times l$ | $\boxed{l+k+k}$ | $(l+k+k) \times k$ |
| Euler | $(k+l) \times k$ | $\boxed{k+l}$ | $(k+k) \times k$ | $\boxed{k+k}$ |

Research on visual processing have mostly focused on perceptual mechanisms underlying visual object identification and recognition, and interaction between higher level cognitive processes and lower level perceptual processes have not studied extensively. Logical reasoning with diagrams provides us with a good domain to investigate the interaction by integrating logical, computational and psychological approaches.

## References


1. Bertin, J.: Graphics and Graphic Information. Walter de Gruyter, Berlin (1981) (originally published in France in 1977)
2. Cleveland, W.S.: The Elements of Graphing Data. Hobart Press, Summit (1994)
3. Guthrie, J., Weber, S., Kimmerly, N.: Searching Documents: Cognitive Processes and Deficits in Understanding Graphs, Tables, and Illustrations. Contemporary Educational Psychology 18, 186–221 (1993)
4. Mineshima, K., Okada, M., Takemura, R.: A Diagrammatic Inference System with Euler Circles. Journal of Logic, Language and Information, doi:10.1007/s10849-012-9160-6
5. Pinker, S.: A Theory of Graph Comprehension. In: Freedle, R. (ed.) Artificial Intelligence and the Future of Testing, pp. 73–126. L. Erlbaum Associates (1990)
6. Ratwani, R.M., Trafton, J.G., Boehm-Davis, D.A.: Thinking Graphically: Connecting Vision and Cognition During Graph Comprehension. Journal of Experimental Psychology: Applied 14(1), 36–49 (2008)
7. Takemura, R., Shimojima, A., Katagiri, Y.: A logical investigation on global reading of diagrams, technical note (2012), http://abelard.flet.keio.ac.jp/person/takemura/index.html
8. Tufte, E.R.: Envisioning Information. Graphics Press, Cheshire (1990)
9. Wainer, H.: Understanding Graphs and Tables. Educational Researcher 21, 14–23 (1992)

# Pictures Are Visually Processed; Symbols Are also Recognized

Peter W. Coppin

Faculty of Information, University of Toronto
Toronto Ontario, Canada
peter.coppin@utoronto.ca
www.petercoppin.org/academic

**Abstract.** *What makes a representation pictorial?* I respond to this question as a small step toward a perceptual-cognitive understanding of graphic representation properties that play important roles in the usability of information systems. Here, I focus to capabilities that play a role in whether material objects are visually processed or recognized as pictorial or symbolized representations. I distinguish pictorial and symbolized information in terms of how each makes use of "less-learned" *perceptual emulation* capabilities that evolved to enable reaction to real-time environmental changes, and *more-learned* capabilities to *recognize* features in order to *predict and plan ("simulate")* future changes from *memory traces* of past percepts. *Pictorial information* makes use of these capabilities to cause perceptual emulation of environmental surfaces that are not part of the marked surface and are referred to here as *"pictured." Symbolized (visual) information* is conceived here as visual information from a visual representation, that, through learning and recognition, causes retrieval of memory traces that serve as resources for the construction of mental simulations beyond (or other than) what is pictured. By locating information and representation at the intersection of perceiver and environment, a preliminary model to address the perplexing problem of distinguishing pictorial from symbolized representations is introduced.

## 1 Introduction

You are reading a "sentence." Previously you might have been viewing "pictures," or clicking an "icon." These graphic representations play important roles in the usability of information systems. Sentences, relative to pictures, seem "symbolic," "language-like," and "more-learned." But to what degree are symbols unlike pictures? Are icons pictures, or are they symbols? I ask: *What properties distinguish pictorial graphic representations from symbolized ones*?

Understanding what properties distinguish symbolized and pictorial representations, and the role that learned, or less-learned, possibly biologically grounded, capabilities play in their effectiveness would be an important step toward developing a scientific framework to inform the design and evaluation of information displays. In visual information design, the majority of the research, training, and

P. Cox, B. Plimmer, and P. Rodgers (Eds.): Diagrams 2012, LNAI 7352, pp. 334–336, 2012.

practice is focused on the techniques for *creating* graphic representations, such as how to draw them, arrange them, or create computer programs to generate them. Typically, very little effort is put into understanding *how* and *why* people perceive, cognitively process, react to, and socially interact through, graphic representations [8]. In the absence of a scientific understanding of graphic representation properties, and how those properties engender perceptual-cognitive reactions in audiences, rules of thumb are used instead (see [2,6,9] for examples).

As graphic representations play crucial roles in critical information systems, through human-computer interfaces, visual modeling languages, and information visualizations, there is a growing demand for a scientific understanding that could inform their design, and serve as a principled approach to evaluate their usability, or effectiveness [11,7]. It is within this gap, between applied visual information design and the science of perception and cognition, that seeks to understand why and how humans and other organisms interact with the world, where my response to this question resides.

**Thesis.** Here, I argue that the properties of graphic representations, and their affordances, emerge at the intersection of our capabilities to visually process and recognize physical properties, such as marks on a surface, intentionally configured, to cause an intended percept. In particular, the framework distinguishes graphic representation types in terms of two interrelated perceptual-cognitive capabilities: less-learned perceptual emulation and more-learned predictive mental simulation. I conceptualize these as capabilities that enable reactions to environmental changes in dynamic environments with two distinct, but interrelated, properties: current change, that impinges on an organism in real-time, and that the organism must react to in real-time; and future possible changes, that the organism must predict, and plan reactions to, using mental simulation.

- I conceive of *pictorial information*, as visual information, that makes use of less-learned visual processing capabilities, to isomorphically perceptually emulate a current change in order to react to a current change.
- I conceive of *symbolized information*, as visual information, that makes use of more-learned recognition capabilities, to non-isomorphically mentally simulate ("predict") a possible change.

**Outline.** First, I use literatures from the applied graphic arts (e.g., [6]) to identify issues to be addressed by using literatures from the science of perception and cognition as basic ingredients (e.g., [3,4,5,1]). To discuss representations as artifacts that appeal to biologically grounded perceptual, recognition-oriented, and predictive capabilities, I talk of perception by assuming a roughly homeostatic conception of *organisms* in their *environments* (e.g., [10]). I will view organisms as *perceiving information* from their environments in order to react to environmental *changes* so that they can maintain their internal conditions required for their survival (adapted from [4]).

Next, I will discuss "less-learned" capabilities to *perceive* "real-time" environmental changes, and *learned* capabilities that produce *predictions* from *memory* traces of past percepts to respond to dynamic changes that are unlike past changes.

Having conceived of the perception and prediction of information in general, I discuss *visual information* in particular, conceived here as a way to perceive and predict environmental changes. I will talk of *representations* that are intentionally created to make use of these "less-learned" perceptual and learned predictive capabilities.

From here, the stage will be set for a discussion of *visual representation*, taken here as a particular subset of representations that produce visual information for perception. I will focus on *pictorial* graphic information, conceived here as a type of information that appeals to "less-learned" perceptual abilities, and *symbolized* graphic information, conceived here is a type of information that appeals to "more-learned" learned capabilities that evolved to support recognition and prediction, but that are made use of by authors of graphic representations. These explanations and distinctions are then used to describe affordances of pictorial and symbolized graphic representations, such as why and how each type requires more or less mental work, relative to a context and purpose. Finally, comparisons to semiotics, and to the distinction between icons and symbols (and signs), as well as the distinction between data structures and knowledge representations, will be made.

## References


1. Barsalou, L.W.: Simulation, situated conceptualization, and prediction. Philosophical Transactions of the Royal Society B: Biological Sciences 364(1521), 1281 (2009)
2. Bertin, J.: Semiology of graphics: diagrams, networks, maps. University of Wisconsin Press (1983)
3. Gibson, J.J.: The ecological approach to the visual perception of pictures. Leonardo 11(3), 227–235 (1978)
4. Goodale, M.A., Króliczak, G., Westwood, D.A.: Dual routes to action: contributions of the dorsal and ventral streams to adaptive behavior. Progress in Brain Research 149, 269–283 (2005)
5. Kosslyn, S.M., Thompson, W.L., Ganis, G.: The case for mental imagery, vol. 39. Oxford University Press, USA (2006)
6. McCloud, S.: Understanding comics: The invisible art. HarperPerennial (1993)
7. Moody, D.: Theory development in visual language research: Beyond the cognitive dimensions of notations. In: Proceedings of the 2009 IEEE Symposium on Visual Languages and Human-Centric Computing (VL/HCC), pp. 151–154. IEEE Computer Society (2009)
8. Ramadas, J.: Visual and spatial modes in science learning. International Journal of Science Education 31(3), 301–318 (2009)
9. Tufte, E.R., Robins, D.: Visual explanations, vol. 25. Graphics Press, New York (1997)
10. Varela, F.J., Thompson, E., Rosch, E.: The embodied mind: Cognitive science and human experience. MIT press (1999)
11. Zuk, T., Schlesier, L., Neumann, P., Hancock, M.S., Carpendale, S.: Heuristics for information visualization evaluation. In: Proceedings of the 2006 AVI Workshop on BEyond Time and Errors: Novel Evaluation Methods for Information Visualization, pp. 1–6. ACM (2006)

# How Do Viewers Spontaneously Segment Animated Diagrams of Mechanical and Biological Subject Matter?

Jean-Michel Boucheix[1] and Richard Lowe[2]

[1] University of Burgundy, France
Jean-Michel.Boucheix@u-bourgogne.fr
[2] Curtin University, Australia
r.k.lowe@curtin.edu.au

**Abstract.** A challenges for learning from animated diagrams is to first parse the continuous flow of information into discrete event units. Inadequacies in this parsing process can prejudice the quality of the mental model constructed from the depiction. One approach that has been proposed for ameliorating such problems is for the designer to pre-segment the animation. However, the pre-segmentation techniques used tend to be either intuitive or based on an expert's understanding of the subject matter. Neither of these approaches takes proper account of the psychological processing that must occur for an external animation to be properly internalized. This poster reports a study of the processes that learners spontaneously use when asked to segment whole animations into events. It compared segmentation of two contrasting diagram types, one representing a mechanical system and the other a biological system. The number of events identified was low relative to the number that were actually present. There were deficiencies in participants' placement of event boundaries and in their characterization of inter-event relationships. Identification of events in the mechanical system proceeded from micro to macro, this order was reversed with the biological system.

**Keywords:** Animation, parsing, segmentation, events units, mechanical system, biological system, boundaries, macro-events, micro-events.

## 1 Introduction

Animated diagrams represent dynamics explicitly. However, the benefits anticipated from such animations too often fail to materialize [1]. In an attempt to fulfill their educational potential, techniques such as cueing and segmentation have been applied to animations [1] but with limited success. Fundamental problems remain with how learners process animations. According to the Animation Processing Model [2], a key step in learning from animation is its initial parsing into events units (i.e., entities plus their associated behavior). The transient nature of such depictions means that segmentation of animations is challenging for learner. One approach for addressing these issues is to pre-segment animations before they are presented to learners [3]. It is quite possible that the segmentation so applied could conflict with the unaided parsing that occurs when learners try to understand the animation themselves. The

P. Cox, B. Plimmer, and P. Rodgers (Eds.): Diagrams 2012, LNAI 7352, pp. 337–339, 2012.

present research therefore investigated how event units are identified in an animation and how well the relations between them are characterized.

# 2 Materiel and Method

Two types of animation were used, a mechanical and a biological system: Newton's Cradle and Kangaroo hopping (with normalized times, see Figure 1). 34 undergraduate French students ($M$ = 20.5 years) participated in this study. Each participant (with no domain-specific prior knowledge) segmented both animations, with the presentation order being counterbalanced across all participants. A T120 eye tracker was used to log eye movements and participants' verbal descriptions of their segmentation activities were recorded. Cards were supplied upon which participants wrote the names of events identified, and then sorted them into an event hierarchy.

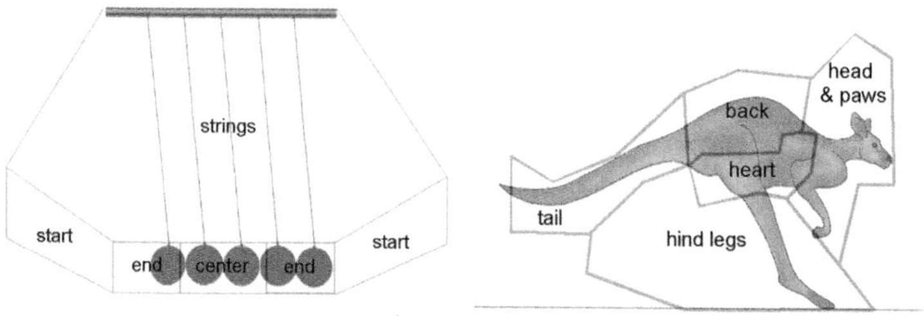


**Fig. 1.** Frames from Newton's Cradle and Kangaroo animations (also showing areas of interest used for eye tracking)

## 2.1 Procedure, Analysis and Scoring

The procedure had 4 stages. *(i)* In the *training stage,* before commencing the main segmentation tasks, participants were trained in the required procedure using videos of a ball race device (mechanical) and a high-jumping athlete (biological). *(ii) In the main segmentation of animations task into events*, participants were instructed to keep their eyes on the screen the whole time while breaking the animation into as many component events as possible and naming each of the events found. They undertook four sets of trials within which each animation was presented three times (including a confirmation of event names). *(iii)* In the *nomination of event boundaries* stage, participants used a computer program to scroll freely through the animation frames in order to determine the starting and ending boundaries if each previously identified event. These boundary frames were recorded on cards. *(iv)* In the *grouping of events* stage, participants used the record cards to classify the events identified and sort them (with no access to the animation) across time in order to generate 'families' of events.

# 3 Results

*(i) Number and scale of events identified.* The number of events identified by participants was low relative to the number of events present in each animation: for Newton's Cradle, $M$ = 6.05/13 (46.54%) and for Kangaroo, $M$ = 4.78/9 (25.16%).

Percents were higher for Newton's cradle than for Kangaroo ($t(33) = 2.98$, $p < .01$). *(ii) Scale of the events.* Figure 1 shows the sequence in which macro and micro events were identified. For the biological system, macro events were identified before micro-events, (Cochran $q(1) = 8$, $p < .05$ and $q(1) = 7.63$, $p < .05$). This order was reversed for the mechanical system ($q(1) = 6.23$, $p < .05$ and $q(1) = 10$, $p < .05$).

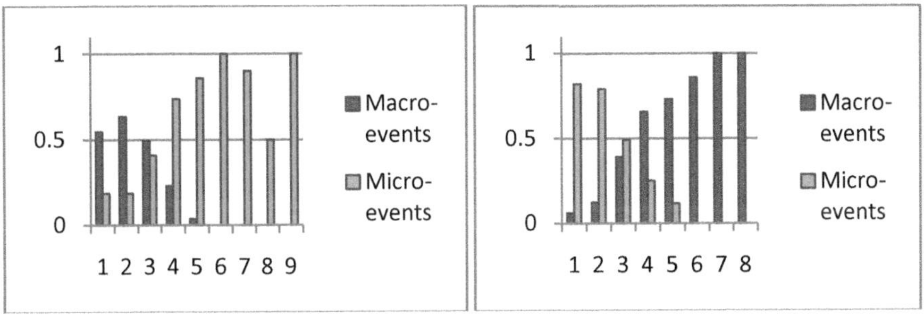


Kangaroo animation Newton's cradle animation

**Fig. 2.** Mean number of macro and micro events identified as a function of their identification order during the event segmentation task

*(iii) Events boundaries and relation between events.* The number of right boundaries was higher than the number of wrong boundaries, for the kangaroo, $M$ correct = 2.79, $M$ no-correct = 1.88, ($t(33) = 4{,}56$, $p < 0{,}01$); and for the Newton's cradle, $M$ correct = 5.31, $M$ no-correct = 1.55, ($t(33) = 3{,}74$, $p < 0{,}01$). However, participants did not always identify event boundaries correctly. They identified a mean of 5.94 relations (2.85 correct) for the Kangaroo and a mean of 7.47 relations (4.40 correct) for the Newton's cradle animation. Participants' difficulties also extended to characterizing how those events are related.

## 4 Conclusion

Breaking down an animation into its component event units is a non-trivial task. Both the correct identification of event boundaries and the establishment of relationships between the depicted events are challenging activities. Further, the scale at which viewers begin their segmentation of an animation was content-dependent.

## References

1. Bétrancourt, M., Tversky, B.: Effect of computer animation on users' performance: A review. Le Travail Humain 63(4), 311–329 (2000)
2. Lowe, R., Boucheix, J.-M.: Learning from Animated Diagrams: How Are Mental Models Built? In: Stapleton, G., Howse, J., Lee, J. (eds.) Diagrams 2008. LNCS (LNAI), vol. 5223, pp. 266–281. Springer, Heidelberg (2008)
3. Spanjers, I.A.E., van Gog, T., van Merrienboer, J.J.G.: A theoretical analysis of how segmentation of dynamic visualizations optimizes students' learning. Educational Psychology Review (2010)

# Which Diagrams and When?
## Health Workers' Choice and Usage of Different Diagram Types for Service Improvement

Gyuchan Thomas Jun[1], Cecily Morrison[2], Christopher O'Loughlin[3], and P. John Clarkson[2]

[1] Loughborough Design School, Loughborough University, Loughborough, UK
g.jun@lboro.ac.uk
[2] Engineering Design Centre, University of Cambridge, Cambridge, UK
{cpm38,pjc10}@cam.ac.uk
[3] Cambridge Intake and Treatment Team,Cambridge and Peterborough NHS trust, UK
Christopher.O'Loughlin@cpft.nhs.uk

**Abstract.** Diagrammatic representations, such as process mapping and care pathways, have been often used for service evaluation and improvement in healthcare. While a broad range of diagrammatic representations exist, their application in healthcare has been very limited. There is a lack of understanding about how and which diagrams could be usable and useful to health workers. In this study, ten mental health workers were asked to discuss positive and negative issues around their service delivery using one or two diagrams of their choice out of seven different diagrams representing their service: care pathway diagram; organisation diagram; communication diagram; service blueprint; patient state transition diagram; free form diagram; geographic map. Their interactions with diagrams were video-taped for analysis. The patient state transition diagram was the most popular choice in spite of relatively low previous familiarity. The overall findings provided insight into a better use of diagrams in healthcare.

**Keywords:** diagrams, problem identification, healthcare, service design.

## 1 Introduction

In a domain like engineering design, there has been a long tradition of using diagrams to understand, communicate and design complex systems [2]. On the other hand, the usage of diagramming for complex healthcare service design has been very limited to flowchart-based representation. Given the complex arrangement of technology, information and people in health service delivery, the need has been raised for better application of diagrammatic representations to the design of healthcare service delivery [1]. However, diagrammatic representations, if they were to be used for the design of healthcare service, need to be usable and useful to clinicians and healthcare managers. Jun et al. [3] made some previous effort to evaluate the usability and utility of a broader range of diagram types in healthcare, but it was based on the perception

P. Cox, B. Plimmer, and P. Rodgers (Eds.): Diagrams 2012, LNAI 7352, pp. 340–342, 2012.

of healthcare professionals rather than real application. Therefore, this study aims to investigate which diagrams(s) healthcare professionals actually choose and how they use diagrams for service improvement.

## 2 Methods

A case study was carried out with a team which provides adult mental health service for intake and treatment. Three major stages of the case study were service delivery understanding for diagramming, diagramming and diagram evaluation. First, two semi-structured interviews with a psychiatric consultant were carried out to collect information on how the service delivery works. Based on this initial information, seven different diagrams were generated or prepared including organisation diagram, communication diagram, service blueprint, state transition diagram, geographic map (added on the request by the psychiatric consultant) and care pathway (pre-existing in the team); Table 1 shows the composition of each diagram. Through diagram-based one-to-on interviews, ten healthcare professionals of the team with various experiences in the NHS and diagrams were introduced to the seven diagrams and asked to discuss service-related issues (either negative or positive) using a diagram or two of their choice. The way they use the diagrams and their comments were videotaped for analysis.

## 3 Results and Discussion

Table 1 shows the participants' previous familiarity to each diagram type, their choice and positive (+) and negative (–) comments. The patient state transition diagram was again found the most popular choice for health service evaluation, which coincides with the previous perception-based diagram evaluation study [3]. Two major factors which influenced the participants' diagram choice include how much diagrams are consistent with participants' mental model diagrams and how relevant diagrams are to the discussion topic. One of the participants, clinical psychologist, clearly indicated that her choice of the state transition diagram was based on its consistency with her existing mental model of feedback loops in care processes. On another occasion, a participant indicated that he chose the organisation diagram to discuss good collaboration among various team members because it was considered to best represent various specialties of the team members.

From the analysis of the recorded video clips, it was found that the participants commonly used diagrams as structured headlines for a big picture understanding rather than accurate descriptions of the reality. What is missing from the diagrams was filled in by questions and detailed verbal explanation. Given the same organization diagram was positively (simple and easy to follow) and negatively (too simplistic) commented for its simplicity by different participants, the participants had different preference for what headlines should look like. How different factors (diagrams, participants, reality) influence the diagram choice and usage by different individuals remain to be further researched.

**Table 1.** Participants' familiarity, choice and comments (+: positive. –: negative)

| Diagram type | Diagram composition | # of participants familiar | # of participants choose | Participants' comments |
|---|---|---|---|---|
| 1. Care Pathway | Timeline, action, outcome and facilitator | | | + Very familiar<br>– Not detail enough<br>– No feedback loops |
| 2. Organisation Diagram | Hierarchical structure of teams and staff members | | | + Straight and simple<br>– Too simplistic |
| 3. Communicatio n Diagram | Information flows between people, teams and organisations | | | + Easy to follow<br>+ Relevant to issues<br>– Too simplistic |
| 4. Service Blueprint | Interfaces between actions of service user and providers | | | – Inconsistent with mental models |
| 5. State Transition Diagram | Patient's state changes and transition conditions and actions | | | + Consistent with mental models<br>– Confusing feedback loops |
| 6. Free Form Diagram | Random combination of information, actions, decisions and teams | | | + Explicit description<br>– Potentially confusing to new staff |
| 7. Geographic Map | Geographic boundaries and locations of service providers | | | – Irrelevant to issues |

**Acknowledgement.** 'The research (was funded by and) took place at the National Institute for Health Research (NIHR) Collaboration for Leadership in Applied Health Research and Care based at Cambridgeshire and Peterborough. The views expressed are those of the author(s) and not necessarily those of the NHS, the NIHR or the Department of Health.'

## Reference

1. Clarkson, P.J., Buckle, P., Coleman, R., et al.: Design for Patient Safety: A Review of the Effectiveness of Design in the UK Health Service. J. Eng. Des. 15, 123–140 (2004)
2. Friedenthal, S., Moore, A., Steiner, R.: A practical guide to SysML: The systems modeling language. Morgan Kaufmann (2008)
3. Jun, G.T., Ward, J., Clarkson, P.J.: Systems Modelling Approaches to the Design of Safe Healthcare Delivery: Ease of use and Usefulness Perceived by Healthcare Workers. Ergonomics 53, 829–847 (2010)

# Eye Movement Patterns in Solving Scientific Graph Problems

Miao-Hsuan Yen, Chieh-Ning Lee, and Yu-Chun Yang

Graduate Institute of Science Education, National Taiwan Normal University, Taipei, Taiwan
myen@ntnu.edu.tw

**Abstract.** Eye movement patterns of science- and non-science students in solving scientific graph problems were compared. Experts (science-students) tended to spend more time, compared to novices, to comprehend the questions during the first run / inspection. Concerning the main graph region, both the True and False subregions (corresponding to correct and wrong answer choices, respectively) were inspected carefully during the first run. Significant differences were observed in the second run, in which the False region was fixated longer when participants made wrong responses.

**Keywords:** graph comprehension, expertise, eye movements.

## 1 Introduction

Interpreting scientific graphs (evidence) to evaluate the validity of claims (theory) is an important process skill in science education. In this preliminary study, how science-major college students solve scientific graph problems was compared with that by non-science-major students. Their eye movements were recorded to reveal the relevant areas inspected for graph comprehension as suggested by the findings in the literature that experts attend more to relevant areas (Gegenfurtner, Lehtinen & Säljö, 2011; Jarodzka, Scheiter, Gerjets, & van Gog, 2010).

## 2 Method

The independent variable was expertise (experts vs. novices) in scientific descriptions and graphs. Twenty college or graduate students participated in this experiment. Half of them majored in science or engineering and were considered as experts, while the other half were considered as novices in the present study.

All participants viewed the same 9 graphs with their eye movements being recorded. As is shown in figures 1 (a)-(c), a graph was presented on the monitor with a three-line multiple choice question. The first line mentioned the variables depicted in the graph. The second line was a question about the graph and the third line was composed of 3 choices. Three types of graphs / questions were created focusing on (a) the rate of change in the x-y relationship, (b) the rate of change in a contour map or isothermal chart, and (c) inference from two lines depicted (e.g., inferring population growth rate from birth and death rates). There were 3 trials for each type.

P. Cox, B. Plimmer, and P. Rodgers (Eds.): Diagrams 2012, LNAI 7352, pp. 343–345, 2012.

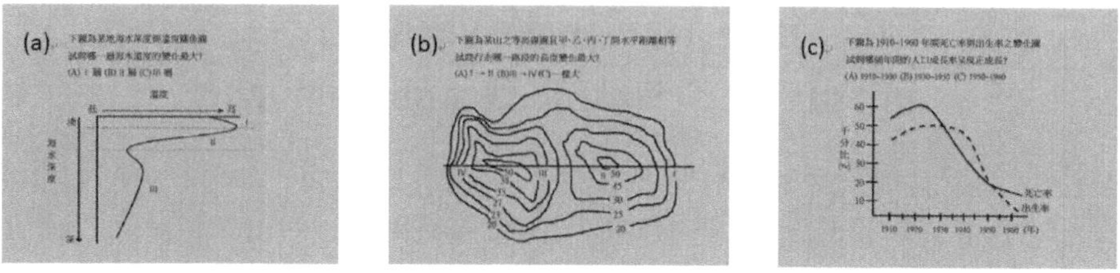


**Fig. 1.** Three types of graphs used in the experiment

Eye movements were recorded with an EyeLink 1000 desktop mount eye tracker. The sampling rate was 1000 Hz. The graphs were displayed on a 19-inch monitor at a resolution of $1024 \times 768$. The viewing distance was about 70 cm.

Participants were tested individually and instructed to comprehend the graphs carefully at their own pace to answer the questions. After camera setting, a 9-point calibration was conducted, followed by a validation routine and a practice trial. Each trial began with a fixation dot at the first character in the 3-line question. Then, the experimenter initiated the presentation of the material. The participants pressed one of the three answer buttons to end the trial. The experiment lasted about 20 minutes.

# 3 Results and Discussion

## 3.1 Interest Areas

Four interest areas were identified; namely, questions (the first two lines), answer choices (the third line), axes and main graph regions. Because reaction time differed among trials and participants, viewing time and numbers of fixation in each interest area were normalized by considering RT and all fixations in the trial. Fixations in each interest area can be further grouped as first-run and second-run fixations, which were separated by leaving and re-entering (if any) the interest area. Trials (28.5%) in which participants selected the wrong answer were excluded from this analysis.

Independent-sample T tests were conducted. As is shown in Table 1, experts spent significantly more time than novices in the question regions during the first run and the entire trial ($ps < 0.05$). In addition, they spent slightly less time in the answer regions than novices ($p > 0.7$). Other comparisons were not significant. The results suggest that experts spent time to interpret the questions thoroughly, especially during the first run. However, there was no difference in time spent in the graph region.

## 3.2 Responses and Inspection Regions in the Main Graph

In the main graph, Ture and False subregions were defined according to the correct and wrong answer choices, respectively. Normalized viewing time and numbers of fixations were calculated seperately for each subregion and according to the participants‘ response (correct or wrong) in each trial. Three way ANOVAs (responses $\times$ subregions $\times$ expertise) were conducted.

Concerning viewing time, significant interaction between response and subregion was found in total time and second-run viewing time ($Fs > 5$, $ps < 0.05$). All other comparisons were not significant. A similar pattern of results was observed for numbers of fixations. When participants made wrong responses, there were more and longer fixations in the False regions. Note that there was no difference during the first run, presumably because both subregions had to be inspected carefully to determine the correct answer. Differences emerged during the subsequent viewing.

**Table 1.** Mean proportion of viewing time and proportion of fixations in 4 interest areas

| | Interest area | Question | | Answer | | Axes | | Main graph | |
|---|---|---|---|---|---|---|---|---|---|
| | Expertise | Exp | Nov | Exp | Nov | Exp | Nov | Exp | Nov |
| Viewing time (%) | First run | 15.5 | 9.8 | 1.5 | 2.6 | 5.0 | 4.5 | 4.2 | 5.3 |
| | Second run | 6.7 | 5.7 | 1.8 | 2.2 | 3.2 | 2.7 | 5.2 | 4.4 |
| | All | 30.6 | 22.3 | 7.3 | 8.1 | 12.4 | 11.7 | 25.8 | 24.6 |
| Numbers of fixation (%) | First run | 14.2 | 10.4 | 1.8 | 2.6 | 5.4 | 4.9 | 4.3 | 5.3 |
| | Second run | 6.3 | 5.9 | 2.1 | 2.1 | 3.5 | 2.8 | 5.0 | 4.2 |
| | All | 28.8 | 23.1 | 7.9 | 7.3 | 13.4 | 12.3 | 25.3 | 23.5 |

**Table 2.** Mean proportion of viewing time and proportion of fixations in True and False subretions as a function of participants' responses

| | Response | Wrong | | | | Correct | | | |
|---|---|---|---|---|---|---|---|---|---|
| | Subregion | False | | True | | False | | True | |
| | Expertise | Exp | Nov | Exp | Nov | Exp | Nov | Exp | Nov |
| Viewing time (%) | First run | 4.2 | 4.0 | 4.4 | 3.4 | 2.9 | 3.5 | 2.6 | 3.9 |
| | Second run | 3.6 | 5.1 | 2.4 | 3.4 | 2.8 | 3.1 | 3.8 | 3.2 |
| | All | 18.7 | 24.4 | 16.6 | 14.7 | 14.3 | 14.7 | 17.6 | 19.0 |
| Numbers of fixation (%) | First run | 3.7 | 3.8 | 4.3 | 3.7 | 3.1 | 3.6 | 2.5 | 3.8 |
| | Second run | 3.6 | 4.6 | 2.7 | 2.8 | 2.8 | 3.2 | 3.6 | 2.8 |
| | All | 18.3 | 22.3 | 16.4 | 13.8 | 14.2 | 14.3 | 17.3 | 18.2 |

## Reference


1. Gegenfurtner, A., Lehtinen, E., Säljö, R.: Expertise differences in the comprehension of visualizations: A meta-analysis of eye-tracking research in professional domains. Educational Psychology Review 23(4), 523–552 (2011)
2. Jarodzka, H., Scheiter, K., Gerjets, P., van Gog, T.: In the eyes of the beholder: How experts and novices interpret dynamic stimuli. Learning and Instruction 20(2), 146–154 (2010)

# Formalising Simple Codecharts

Jon Nicholson and Aidan Delaney

Visual Modelling Group, University of Brighton
{j.nicholson,a.j.delaney}@brighton.ac.uk

**Abstract.** Codecharts are a formal diagrammatic language for specifying the structure of object-oriented design patterns, frameworks, and programs. Codecharts are attractive for applications in both forward (e.g. design verification) and reverse engineering (e.g. program visualization). Although the definition of Codecharts has been adequate for these applications, there is a need to develop the language further in more precise terms. This paper outlines our work in refining the definition of Codecharts. We informally describe the concrete syntax and semantics of Codecharts, and provide a new formal abstract syntax. We conclude with a brief discussion on future work.

Codecharts are a formal diagrammatic language for use in tasks such as software design and re-engineering of legacy source-code [1]. In particular, Codecharts can be used to represent software design patterns [2]. Given a Codechart representation of a pattern we can reason where implementations exist in legacy codebases. We provide an abstract syntax for Codecharts following the approach discussed in [3], in which the concrete and abstract syntax for a diagrammatic logic are considered separately.

Figure 1a is an example Codechart. Informally, this Codechart tells us that both the classes `WhiteRhino` and `BlackRhino` inherit from `Animal`. Furthermore, the classes `WhiteRhino` and `BlackRhino` have the property that they are `Endangered`. Codecharts use rectangles to represent single classes and rectangles with an offset to represent sets of classes. The set of classes that appear in the diagram is denoted $\mathcal{R}$. The set consisting of `WhiteRhino` and `BlackRhino` is an element of $\mathcal{OR}$. Relationships between classes and sets of classes are represented by labelled arrows. In this case, the arrow is an element of the set of single-headed arrows, denoted $\mathcal{A}_{\blacktriangleright}$. Unary predicates, such as `Endangered`, represented by inverted triangles are elements of the set denoted $\mathcal{T}_{\triangledown}$.

The example in figure 1b introduces triangles, ellipses and double-headed arrows. The triangle labelled `(Animal, {WhiteRhino,BlackRhino})` states that `WhiteRhino` and `BlackRhino` are classes within the inheritance class hierarchy (*hierarchy*) rooted at `Animal`. Such a hierarchy is an element of the set denoted by $\mathcal{T}_{\vartriangle}$. The ellipse labelled `eat` states that all classes in the respective hierarchy contain a method with the signature *eat*. The relationship between such method signatures and the classes they are attached to is denoted by $\mathcal{M}$, and $\mathcal{E}$ is the set of all method signatures that can appear in the diagram. The double-headed arrow labelled *givesBirth* represents a specific relation which can be restricted to

P. Cox, B. Plimmer, and P. Rodgers (Eds.): Diagrams 2012, LNAI 7352, pp. 346–348, 2012.

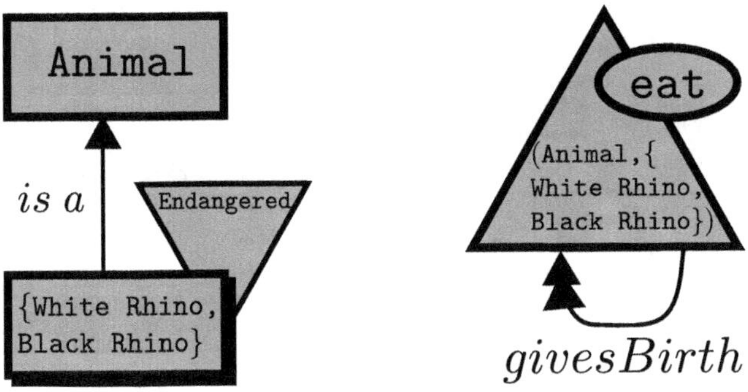


(a) (b)

**Fig. 1.** Two example Codecharts

a bijective function. In this case, it provides the information that a `WhiteRhino`, and similarly `BlackRhino`, can give birth to something of the same class. However, it also allows modelling of the case where the rhinos interbreed. That is to say, each type of animal can give birth to one type, however there can be cases where one type of animal can give birth to another. Considering a different example, a donkey can give birth to a donkey, or a hinny (female donkey crossed with a male horse). The set of such relations are represented by double-headed arrows is denoted by $\mathcal{A}_{\blacktriangleright\blacktriangleright}$.

Concrete Codecharts are formed from syntactic elements including labelled rectangles, triangles and ellipses: each of which can be drawn with an *offset*. Two types of arrows, single headed and double headed, and inverted triangles complete the set of syntactic elements as seen in figure 2a. As already seen, rectangles represent object-oriented classes, ellipses represent method signatures and triangles represent hierarchies. Offsets represent a set, thus an offset rectangle represents a set of classes. Offset triangles represent sets of hierarchies and offset ellipses represent sets of method signatures. Furthermore, we allow arrows to be sourced and targeted at any syntactic element other than arrows themselves or inverted triangles. Ellipses, offset ellipses and inverted triangles must overlap a single rectangle, triangle or offset of such. Other syntactic elements may not overlap. The examples in figure 2b demonstrate non well-formed diagrams.

The abstract syntax of a Codechart can be defined by the tuple

$$(\mathcal{R}, \mathcal{T}_{\vartriangle}, \mathcal{E}, \mathcal{OR}, \mathcal{OT}_{\vartriangle}, \mathcal{OE}, \mathcal{M}, \mathcal{T}_{\triangledown}, \mathcal{A}_{\blacktriangleright}, \mathcal{A}_{\blacktriangleright\blacktriangleright}).$$

As we have seen in the example, where $\mathcal{R}$ is a set of classes, the set $\mathcal{OR}$ is a set of sets of classes. The same holds for $\mathcal{T}_{\vartriangle}$, the set of class name hierarchies, and $\mathcal{OT}_{\vartriangle}$ the set of sets of class name hierarchies. Again, similarly, $\mathcal{E}$ is the set of method signatures, and $\mathcal{OE}$ the set of sets of method signatures. We now present the abstract syntax of the first of our example diagrams.

Let $D_1$ be the abstract Codechart in figure 1a. Then the abstract syntax of $D_1$ includes:

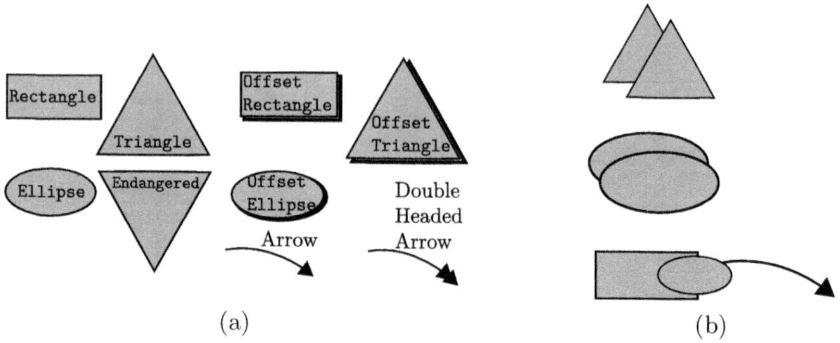


**Fig. 2.** Syntactic elements of codecharts and examples of non well-formed Codecharts

- $\mathcal{R} = \{WhiteRhino, BlackRhino, Animal\}$ the finite set of class names that appear in the diagram.
- $\mathcal{OR} = \{\{WhiteRhino, BlackRhino\}\}$ is a subset of the powerset of class names, labelling the offset rectangle.
- $\mathcal{T}_{\triangledown} = \{(Endangered, \{WhiteRhino, BlackRhino\})\}$ is an association of the unary predicate *Endangered* with the set of classes represented by the offset rectangle.
- $\mathcal{A}_{\blacktriangleright} = \{(\{WhiteRhino, BlackRhino\}, \text{“}is\ a\text{”}, Animal)\}$ is a single headed arrow sourced on the set of class names $\{WhiteRhinos, BlackRhinos\}$ in the concrete syntax, and targeted on *Animal*.

In this example, the sets $\mathcal{T}_{\vartriangle}, \mathcal{E}, \mathcal{OT}_{\vartriangle}, \mathcal{OE}, \mathcal{M}$ and $\mathcal{A}_{\blacktriangleright\blacktriangleright}$ are empty.

In this paper we have presented a formal syntax for abstract Codecharts. In future work we will present a formal concrete syntax and a formal model-based semantics.

## References

1. Eden, A.H., Nicholson, J.: Codecharts: Roadmaps and Blueprints for Object-Oriented Programs. Wiley-Blackwell (April 2011)
2. Gamma, E., Helm, R., Johnson, R., Vlissides, J.: Design Patterns: Elements of Reusable Object-Oriented Software, 1st edn. Addison-Wesley Professional (November 1994)
3. Howse, J., Molina, F., Taylor, J., Shin, S.-J.: Type-syntax and token-syntax in diagrammatic systems. In: Proceedings of the International Conference on Formal Ontology in Information Systems, FOIS 2001, vol. 2001, pp. 174–185. ACM, New York (2001)

# Notes about the London Underground Map as an Iconic Artifact

Breno Bitarello[1], Pedro Atã[2], and João Queiroz[2,*]

[1] State University of Rio de Janeiro, Superior School of Industrial Design, Rio de Janeiro, Brazil
[2] Federal University of Juiz de Fora, Institute of Arts and Design, Juiz de Fora, Brazil
queirozj@pq.cnpq.br

**Abstract.** The icon is defined as a sign whose manipulation reveals, by direct observation of its intrinsic property, some information on its object. The London Underground Map is an example of an artifact used to represent part-part/part-whole relations of the largest underground systems of the world. It provides a powerful semiotic niche built for extraction and manipulation of relations. This paper explores the design of the London Underground Map through the notion of iconic artifact.

**Keywords**: diagram, information design, cognitive artifacts, cognitive niche.

## 1 Iconic Operacionality of London Underground Map

The icon is operationally defined as a sign whose manipulation reveals, by direct observation of its intrinsic property, some information on its object [4]. This operational definition of the icon focuses solely on the capability of a sign to enclose information about its object, and represents a detrivialization of the concept of an icon as a similar entity. However, a second, stricter notion of icon have also been identified which allow considerations such as immediacy of the information presented and economy of elements. This stricter notion have been termed "optimal" iconicity. [4] As soon as an icon can be considered as consisting of interrelated parts, and since these relations are subject to experimental manipulation governed by laws, we are working with diagrams [2-3]. The London Underground Map provides a powerful semiotic niche built for extraction and manipulation of relations: "The Underground Diagram is more than a simplification of Underground railway routes. For most Londoners, it is an essential simplification of the city itself" [1: p.5]. Semiotically speaking, the London Underground Map is deeply dependent on the formal and material properties from which it was made: first of all, it's material features bear isomorphism to the environment (the same number of stations and the same possibilities of connections between the stations), making it capable of providing relevant information for users. Additionally, it has visual features destined to facilitate the process of

---

* Corresponding author.

P. Cox, B. Plimmer, and P. Rodgers (Eds.): Diagrams 2012, LNAI 7352, pp. 349–351, 2012.

uncovering the required information, i.e. it doesn't only regard that an isomorphism exist between the map and Underground system, but *how* both are connected.

## 2 London Underground Map: A Cognitive Tool for its Users

Using a diagram such as the London Underground Diagram commonly involves to depart from a set of information provided, to follow some general rules and -- hopefully – to arrive at some information desired. This process is dependent of inferential properties of diagrams and icons [4], namely, the inference to make explicit information that is implicit in the representation. According to Stjernfelt [4: pp. 397-398]: "in order to discover these initially unknown pieces of information about the object involved in the icon, some deductive experiment on the icon must be performed". In this process, the information provided includes where the user is at the moment. He or she is aided to discover it by the colors of lines (by knowing the color of the line being traveled one is easily able to stay up to date with his/her current position by following the sequence of stations), as well as by the fact that in spite of its simplification, the diagram retains some of the spatial relationships of the physical world. Colors are also an example of how the diagram itself provides some of the rules that guides the inference from the information provided to the information desired: they help to figure out the different steps of the journey by breaking general goals (for example, how to get to the Picadilly Circus station) into minor strategic goals (how to get to the blue line, where the Picadilly Circus station is).

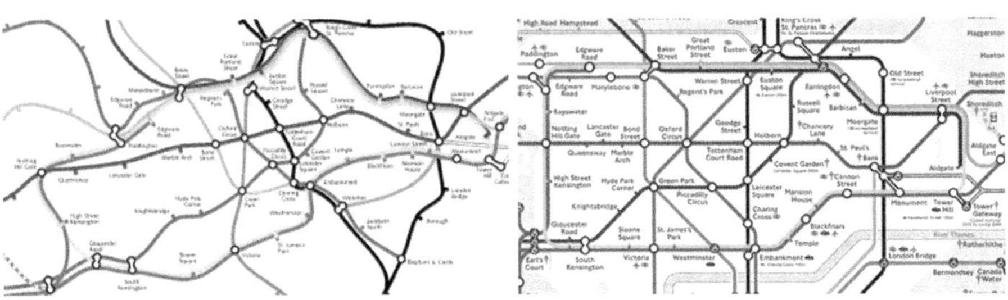


**Fig. 1.** Relations between the central area in a geographically accurate underground map of London (left) and in a standard Underground Map (right). Image adapted from: http://diagrams.org/fig-pages/f00022.html.

A distinctive trait of the modern version of the Underground route guide, however, concern not the color of the lines nor some spatial relationship correspondence with the physical world. The former was already present in the previous version of the guide and the latter was actually sacrificed by its redesign: before 1933, the guide was faithful to the actual physical location of stations and lines, as wells as the distances between then. The innovation of Henry Charles Beck was to favor clarity, by simplifying the lines only in verticals, horizontals and diagonals and expanding the central area [5]. In comparison to its predecessor, Beck's diagram has diminished the amount of implicit reachable information in the map, reducing the number of possible operations to be performed (to know about real distances, for example). Beck has added features that don't increase the amount of information, but rather decrease the difficulty of the search for the proper information, which influences in the whole process of problem-solving. That means to say that the behavior of the user as well as the task itself, are constrained and to a certain extent defined by the material iconic features of the representation. This redefinition of behaviors and tasks actually transforms the perception of city itself. According to Garland [1: p. 7], when the diagram was created, even the most experienced Londoners could not tell you where the center of London was.

## 3 Diagram as a Mind-Tool

Assuming that it is impossible for the user to keep and learn all the relations between all stations and lines of the Underground in an internal image or internal mind representation, we view the Underground Diagram as an external representation distributing the cognitive effort of the user. The London Underground Map is a powerful mind-tool changing radically how users move around the urban space. Diagrams can be viewed as a form of representation that provides new routes of problem-solving in specific contexts. The everyday users are influenced by the access to the transport system and the way he deals with it. In other words, acting as a navigation tool, the Underground Map makes the citizen more suited to the urban environment. We explored the London Underground Diagram in terms of 'semiotic artifacts' through the notions of 'cognitive external artifacts' which allow easy navigation through the object represented (London Railway System). This study also constitute an interesting example for the analysis of the interplay of different notions of iconicity: although Beck has reduced the 'similarity' between representation and object, he has increased the representation efficiency. As the possible uses of the guide were narrowed, it became more specialized, creating for the users an unprecedented understanding of the city itself. Future approaches will examine how these different notions of iconicity may be related to problem-solving efficiency of representations and the creation of new cognitive niches, through the exploration of other examples and concepts found in literature.

## References


1. Garland, K.: Mr. Beck's Underground Map. Capital Transport Publishing, London (1994)
2. Hookway, C.: Truth, Rationality, and Pragmatism – Themes from Peirce. Oxford University Press, Oxford (2002)
3. Stjernfelt, F.: Diagrammatology - An Investigation on the Borderlines of Phenomenology, Ontology, and Semiotics. Springer, Heidelberg (2007)
4. Stjernfelt, F.: On operational and optimal iconicity in Peirce's diagrammatology. Semiótica 186, 395–419 (2011)
5. Walker, J.: The London Underground Diagram: a semiotic analysis. Icographic 14/15 (1979)

# The Efficacy of Diagrams in Syllogistic Reasoning: A Case of Linear Diagrams

Yuri Sato and Koji Mineshima

Department of Philosophy, Keio University
{sato,minesima}@abelard.flet.keio.ac.jp

**Abstract.** We study the efficacy of external diagrams in syllogistic reasoning, focusing on the effectiveness of a linear variant of Euler diagrams. We tested subjects' performances in syllogistic reasoning tasks where linear diagrams were externally supplied. The results indicated that the linear diagrams work as effectively as Euler diagrams. It is argued that the relational information such as inclusion and exclusion is crucial for understanding the efficacy of diagrams in syllogistic reasoning.

## 1 Introduction

In psychology of deduction, it has long been known that solving categorical syllogisms is a difficult task for those who are untrained in logic. Certain external representations such as Euler and Venn diagrams are traditionally regarded as effective tools to support deductive reasoning. However, it is still open to discussion whether and how such diagrams could aid untrained people to conduct deductive reasoning in a successful way. For instance, Calvillo et al. [1] reported some negative effects of traditional Euler diagrams in syllogistic reasoning.

Sato et al. [4] examined the efficacy of Euler diagrams in solving syllogisms, in comparison to sentential reasoning and reasoning with Venn diagrams. In the experiments of [4], subjects were divided into three groups, the Euler group, the Venn group, and the Linguistic group. The Euler and Venn groups were presented with two sentential premises, together with the corresponding two diagrams, and asked to choose a valid conclusion. The Linguistic group was presented only with sentential premises and asked to choose an answer without any aid of diagrams. The results indicated that the performance of the Euler and Venn groups was significantly better than that of the Linguistic group, and that the performance of the Euler group was significantly better than that of the Venn group.

The differences in performance between the three groups can be explained on the basis of the dual roles played by diagrams in the overall process of reasoning, namely, interpretational and inferential roles [4,5]. More specifically, a categorical sentence in syllogisms is intended to be interpreted as denoting a *relation* between sets. Thus, a universal sentence of the form All $A$ are $B$ is to be interpreted as expressing $\mathbf{A} \subseteq \mathbf{B}$, i.e. the inclusion relation, and No $A$ are $B$ as $\mathbf{A} \cap \mathbf{B} = \emptyset$, i.e. the exclusion relation. Such relational semantic information

P. Cox, B. Plimmer, and P. Rodgers (Eds.): Diagrams 2012, LNAI 7352, pp. 352–355, 2012.

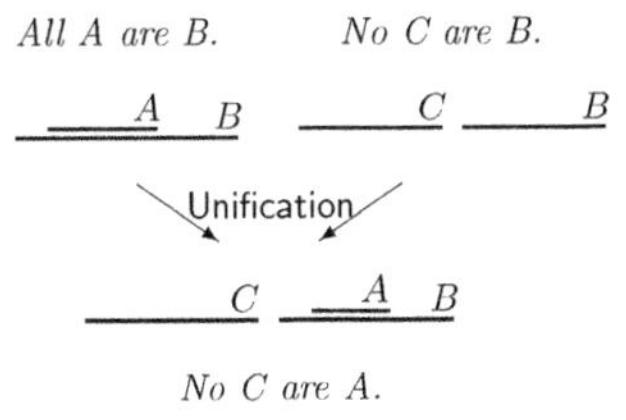


Fig.1. Solving a syllogism with linear diagrams

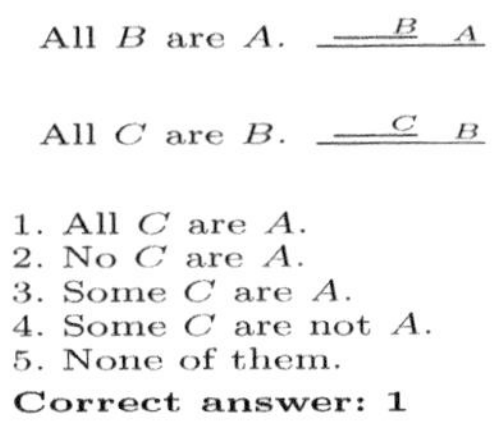


**Fig.2.** An example of a syllogistic reasoning task in the Linear group

is often not directly accessible to untrained reasoners [4]. Euler and Venn diagrams could then help the reasoners realize the semantic relationships implicit in quantificational sentences in terms of the spatial relationships between objects (circles or points), and thereby avoid reasoning errors due to misinterpretation. For the inferential side, note first that a deductive reasoning task in general requires the reasoner to assemble the information contained in the premises. In the case of the Euler group, then, such a task could naturally be replaced with the task of manipulating diagrams, specifically, of unifying premise diagrams and extracting information [5]. Moreover, the manipulations of diagrams are expected to be spontaneously triggered without much effort, if the spatial relations holding on external diagrams are governed by natural constraints—constrains that depend solely upon spatial properties of diagrams so that they are accessible even to untrained users. By contrast, Venn diagrams lack this kind of inferential efficacy, due largely to the fact that the manipulation of them to solve deductive reasoning tasks requires prior understanding of some conventions (cf. [4,5]).

The key assumption here is that the form of diagrams mirrors the semantic information required in a given reasoning task, specifically, the relational information in the case of syllogisms. If this is correct, then it is expected that any diagram which can make explicit the relational information encoded in a categorical sentence would be effective in supporting syllogistic reasoning. The aim of the present study is to investigate whether this holds good of a linear variant of Euler diagrams, where set-relationships are represented by one-dimensional lines, rather than by circles in a plane. Although it is known that linear diagrams have limited expressive power (cf. [2]), they are expressive enough to represent categorical syllogisms. We hypothesize that linear diagrams would be effective ways of representing relational structures and of reasoning about them. An example of linear diagrams and the process of solving a syllogism with them are indicated in Fig. 1 (cf. the case with Euler diagrams in [4]). Here by unifying the two linear diagrams in premises the reasoner could almost automatically obtain the desired information about the relationship between $A$ and $C$. If such linear diagrams would work as effectively as Euler diagrams, it could count as evidence that the effectiveness of external diagrams in syllogistic reasoning is not due to particular shapes such as circles of Euler diagrams.

## 2 Experiment and Result

The semantics of the linear diagrams used is essentially the same as that of Euler diagrams in [4,3]. The experiment was conducted in the same manner as that of [4]; the only difference is that in syllogistic reasoning tasks, Euler diagrams associated with premises are replaced by the corresponding linear diagrams. Note that we adopted a system of categorical syllogisms without the existential import, hence in the syllogism of Fig. 2 we do not count 3 as a correct answer.

**Method.** Thirty-three undergraduates (mean age $22.72 \pm 8.72$ SD) participated in the experiment, which we call the Linear group. Of them, we excluded five students who did not follow our instruction. Subjects were first provided with an instruction on the meaning of linear diagrams, and asked to take a pretest to check whether they understood the instruction correctly. Then the subjects were asked to solve syllogistic reasoning tasks supported by linear diagrams. An example is shown in Fig. 2. In this task, the subjects were presented with two sentential premises and asked to choose a correct answer. We presented 31 syllogisms in total. The test was a 20-minute test.

**Result.** In the following analysis, we exclude the seven subjects who failed the pretest. The average accuracy rate of the total 31 tasks in the Linear group was 80.7%. The data were compared with those of the Linguistic group, the Venn group, and the Euler groups reported in [4] by one-way Analysis of Variance. There was a significant main effect, $F\ (3, 140) = 37.734$, $p < .001$. Multiple comparison tests by Ryan's procedure yield the following results. (i) The accuracy rate of the Linear group was higher than that of the Linguistic group: 46.7% for the Linguistic group ($F\ (1, 64) = 7.112$, $p < .001$.). (ii) The accuracy rate of the Linear group was higher than that of the Venn group: 66.5% for the Venn group ($F\ (1, 49) = 2.741$, $p < .05$.). (iii) There was no significant difference between the accuracy rate of the Linear group and that of the Euler group: 85.2% for the Euler group. It should be noted that if we include those subjects who failed the pretest, we still obtain similar results in each comparison: for (i) and (ii), there were significant differences, $p < .001$; for (iii), there was no significant difference.

These results support our prediction that linear diagrams work as effectively as Euler diagrams in syllogistic reasoning. This in turn provides evidence that the efficacy of external diagrams in syllogistic reasoning depends upon the fact that the diagrams make explicit the semantic relations such as inclusion and exclusion relations in such a way that they are suitable for syntactic manipulation.

## References

1. Calvillo, D.P., DeLeeuw, K., Revlin, R.: Deduction with Euler Circles: Diagrams That Hurt. In: Barker-Plummer, D., Cox, R., Swoboda, N. (eds.) Diagrams 2006. LNCS (LNAI), vol. 4045, pp. 199–203. Springer, Heidelberg (2006)
2. Lemon, O., Pratt, I.: On the insufficiency of linear diagrams for syllogisms. Notre Dame Journal of Formal Logic 39(4), 573–580 (1998)

3. Mineshima, K., Okada, M., Takemura, R.: A diagrammatic reasoning system with Euler circles. Journal of Logic, Language and Information (to appear, in press)
4. Sato, Y., Mineshima, K., Takemura, R.: The Efficacy of Euler and Venn Diagrams in Deductive Reasoning: Empirical Findings. In: Goel, A.K., Jamnik, M., Narayanan, N.H. (eds.) Diagrams 2010. LNCS(LNAI), vol. 6170, pp. 6–22. Springer, Heidelberg (2010)
5. Sato, Y., Mineshima, K., Takemura, R.: Constructing internal diagrammatic proofs from external logic diagrams. In: Proceedings of 32nd Annual Conference of the Cognitive Science Society, pp. 2668–2673 (2010b)

# Author Index

GPSR Compliance

*The European Union's (EU) General Product Safety Regulation (GPSR) is a set of rules that requires consumer products to be safe and our obligations to ensure this.*

*If you have any concerns about our products, you can contact us on ProductSafety@springernature.com*

In case Publisher is established outside the EU, the EU authorized representative is:

Springer Nature Customer Service Center GmbH
Europaplatz 3
69115 Heidelberg, Germany

**Batch number: 10844619**

Printed by Printforce, the Netherlands